Plant Systematics and Evolution

Supplementum 6

P. F. Yeo

Secondary Pollen Presentation

Form, Function and Evolution

Springer-Verlag Wien GmbH

Dr. P. F. Yeo
University of Botanic Garden
Cambridge, United Kingdom

Originally published by Springer-Verlag Wien New York in 1993
Softcover reprint of the hardcover 1st edition 1993
Typesetting: Thomson Press (India) Ltd., New Delhi, 110001

With 55 Figures

ISSN 0172-6668
ISBN 978-3-7091-7375-6 ISBN 978-3-7091-6670-3 (eBook)
DOI 10.1007/978-3-7091-6670-3

Preface

Secondary pollen presentation is the presentation of pollen either on floral structures other than the anther-sacs (thecae) in which it is produced or by special mechanisms of expulsion involving contact of the pollen with other floral parts. This intriguing subject has until now remained largely unexplored and few botanists seem to be aware of it. This is in spite of the fact that secondary pollen presentation is virtually universal in the *Asteraceae* (*Compositae*), the largest family of flowering plants, and common in the *Leguminosae*, a family including many crops, for which successful pollination is of great economic importance. Secondary pollen presentation is universal in other major families such as the *Campanulaceae* and *Lobeliaceae*. It is also prevalent in the extraordinarily interesting southern hemisphere family *Proteaceae* and the very large tropical family *Rubiaceae*.

Information on secondary pollen presentation is scattered in the literature of floral biology; I have now brought it together for the first time and added some of my own. This has made it possible to consider the adaptive significance of the different forms of secondary pollen presentation and the patterns of its occurrence in relation to classification and evolution, affording a basis for comparative study.

We need to know a great deal more about secondary pollen presentation before we can fully understand it, and I give suggestions for further research (information is incomplete for almost all the plants dealt with in this book). Enquiries are needed at varied levels on topics ranging from structure to the mechanics of pollen delivery, and from pollinator behaviour to pollen dispensing, pollen transport, pollination success and pollen/ovule ratios. These are all fields of current interest and there are possibilities here for project work in schools and universities. I hope the book will also provide useful background for morphologists and taxonomists, encouraging more of them to follow the current trend of considering function when they are describing structures. Above all it is intended to serve as a source-book for floral biologists and pollination ecologists.

I began gathering information for this review in 1984. In 1987, when I had written about half of the first draft, I learned that Dr. CHRISTIAN WESTERKAMP had begun a similar project. As my work was considerably more advanced than his he generously put his ideas and a large bibliography at my disposal. He and all others who have helped in any way with the preparation of this work are gratefully thanked; many have gone to considerable trouble.

The following helped with general discussion and/or comments on the introduction and the discussion section: S. A. CORBET, A. DAFNI, A. J. LACK, CHR. WESTERKAMP.

Those who commented on and supplied information for particular families are D. BRIDSON (*Rubiaceae*), B. G. COLLINS (*Myrtaceae*, *Proteaceae*), H. A. FORD

(*Proteaceae*), P. E. GIBBS (*Papaveraceae, Vochysiaceae*), E. HOLM (*Myrtaceae*), Y. HESLOP-HARRISON (pollen of *Asteridae*), S. D. HOPPER (*Proteaceae*), W. J. KRESS (*Cannaceae, Marantaceae*), L. A. S. JOHNSON (*Proteaceae*), D. G. LLOYD (*Lobeliaceae*), D. MACFARLAND (*Proteaceae*), R. MAKINSON (*Proteaceae*), R. M. POLHILL (*Leguminosae*), T. J. ROSATTI (*Apocynaceae, Campanulaceae*), C. H. STIRTON (*Leguminosae*), S. STUMPF (*Goodeniaceae, Lobeliaceae*), E.-M. THIELE (*Asteraceae*), V. TURNER (*Proteaceae*), B. VERDCOURT (*Rubiaceae*), P. S. WESTON (*Proteaceae*), F. WHITE (*Meliaceae*), CHR. WESTERKAMP (*Liliaceae* and literature references for many families).

Permission to use illustrations has been received from the following (in many cases original drawings have been lent): for Fig. 1, Der Palmengarten, Stadt Frankfurt am Main, and Dr CHR. WESTERKAMP, for Fig. 9. *Eucnide urens*, The Journal of the Arnold Arboretum; for Fig. 10. *Acrotriche serrulata*, The Annals of Botany Company (Harcourt Brace Jovanovich) and Dr C. A. McCONCHIE; for Fig. 11. *Stylosanthes gracilis*, M. R. PEREIRA-NORONHA; for Fig. 14. *Bruguiera gymnorhiza* (from the Journal of South African Botany), The South African Association of Botanists and J. E. BARTON (née DAVEY); for Fig. 15. *Darwinia vestita*, Fig. 16. *Darwinia macrostegia* and *D. citriodora*, Fig. 17. *Verticordia huegelii* and *V. habrantha*, Fig. 18. *V. grandis*, Fig. 19. *Actinodium cunninghamii* and *Chamelaucium uncinatum*, Fig. 22. *Grevillea eriostachya*, Fig. 23. *Grevillea fasciculata*, Fig. 24. *Banksia coccinea* and Fig. 25. *B. prionotes* and *B. dryandroides*, E. HOLM; for Fig. 28. *Polygala vauthieri* and *P. monticola* var. *brizoides*, Springer-Verlag and N. B. M. BRANTJES; for Fig. 29. *Nerium oleander*, Prof. dr. L. J. G. VAN DER MAESEN (editorial board, Wageningen Agricultural University Papers) and Ir. FRANK PAGEN; for Fig. 30. *Nerium oleander*, The Linnean Society of London and J. HERRERA; for Fig. 31. *Stephanostema stenocarpum*, the Department of Plant Taxonomy, Agricultural University, Wageningen, by courtesy of Prof. dr. L. J. G. VAN DER MAESEN; for Fig. 36. *Wahlenbergia consimilis*, the Bentham-Moxon Trustees (Royal Botanic Gardens, Kew) and E. M. STONES; for Fig. 38. *Lobelia cardinalis*, S. STUMPF; for Fig. 39. *Isotoma petraea*, the editor and publisher of Acta Botanica Neerlandica and N. B. M. BRANTJES; for Fig. 40. *Isotoma axillaris*, Fig. 41. *Scaevola crassifolia* and Fig. 43. *Selliera radicans*, S. STUMPF; for Fig. 42. *Scaevola thesioides* var. *filifolia*, Fig. 44. *Dampiera cuneata* and Fig. 45. *Lechenaultia tubiflora*, E. HOLM; for Fig. 48. *Asteraceae*, J. Cramer (Gebrüder Borntraeger) and Dr E.-M. THIELE.

Cambridge, 1992 *P. F. Yeo*

Contents

Chapter 1

Introduction and explanations

1. Definition and scope

As used here the expression 'secondary pollen presentation' covers transfer of the pollen to some structure other than the thecae of the anthers before presentation to potential pollen vectors and the operation of mechanisms in which the style and/or the stamen-filaments act as a piston within a cylinder, so that these parts and those that form the cylinder are in contact with the pollen.

The phenomenon has not previously been extensively reviewed but it was discussed by CAROLIN (1960), YEO (1972) and BRANTJES (1982, 1983). I began gathering information for the present work in 1984. I do not deal with abiotic pollination systems, nor with flowers that disperse their pollen by a simple 'mess and soil' process, such as *Magnolia* and *Arum*, which drop a mass of pollen into the hollow base of the blossom (FAEGRI & VAN DER PIJL 1966) (see Table 1-1). *Cytisus*, in which a cloud of pollen is thrown over the insect, was later added to this category (FAEGRI & VAN DER PIJL 1971, 1979) but I do not accept this.

In many explosive flowers dry pollen is freed within the flower before the explosion is triggered. If floral parts other than the anthers are involved in retaining this freed pollen the flower is considered to show secondary pollen presentation. The families with explosive flowers in which this is apparently not the case are listed in Table 1-2.

Some *Apocynaceae* have secondary pollen-presentation but the related *Asclepiadaceae* have true pollinaria: there is a structure to which the pollen is secondarily attached that is removable by the pollen-vector. On the ground that this constitutes a distinct mechanism the *Asclepiadaceae* are omitted from this review. There exists a stepwise intergradation between such plants with a detachable pollen-bearing structure and those that apply an amorphous adhesive of non-thecal origin to the body of the pollen-vector at a point where it is likely to touch and carry away the pollen (see Chapter 2: '*Apocynaceae*'). The latter may or may not present their pollen on secondary structures. In *Orchidaceae-Orchidoideae* the pollinium stalk may be derived from material within the thecae and is then termed a caudicula, or be formed from part of the rostellum (median stigmatic lobe), in which case it is termed a stipes (cf. RASMUSSEN 1986). Orchids with a stipes form a strict parallel with the *Asclepiadaceae* and are excluded here on the same ground; those with caudiculae have no secondary pollen-bearing structure.

The probable functions of secondary pollen presentation appear to be (1) harmonization of sites of presentation and reception of pollen in the flower, (2) protection of pollen against robbery, (3) placement of the pollen on the vector so that the latter cannot misuse it, and (4) the issue of pollen in separate doses. As already stated, secondary pollen presentation also plays a part in some explosive

Table 1-1. Families with 'mess and soil' pollen presentation based mainly on information from Dr CHR. WESTERKAMP (not further considered here)

Annonaceae (GOTTSBERGER, 1988)
Araceae (examples in PROCTOR & YEO, 1973: 227–9, 306–8)
Aristolochiaceae (KNUTH, 1904)
Magnoliaceae (THIEN, 1974)
Rafflesiaceae (it seems likely that *Rafflesia* drops its pollen into the pollination chamber but I have not found a definite report that this is so)
Sarraceniaceae (KNUTH, 1904)

Table 1-2. Families considered as possibly having secondary pollen presentation (2PP) but for which available reports appear to rule it out

Lamiaceae: *Aeollanthus* SPRENGEL, *Eriope* BENTH. and *Hyptis* JACQ. have explosive flowers but there is no evidence of 2PP (BRANTJES & DE VOS, 1981, and references cited therein); indeed, in *Hyptis capitata* JACQ. it has been found that the anthers dehisce only on impact with the pollinator (KELLER & ARMBRUSTER, 1989).
Lecythidaceae: no 2PP reported by PRANCE (1976) or MORI, PRANCE & BOLTEN (1978) except for secondary presentation of a small amount of pollen in *Lecythis corrugata* POIT., of which the pollen has not yet been found on insect visitors (MORI, PRANCE & BOLTEN, 1978).
Loranthaceae: flowers are explosive but there is no evidence of 2PP (FEEHAN, 1985).
Lentibulariaceae: *Utricularia* L. (source of this suggestion lost) no evidence for 2PP in TAYLOR (1989)
Scrophulariaceae: *Pedicularis* L. no evidence for 2PP in many publications by L. W. MACIOR.
Stylidiaceae: ERICKSON (1958).

systems. For each family included in Chapter 2 a discussion is provided in which these and other possible roles are considered together with any evidence relating to Angiosperm classification and the evolution of pollination systems. Floral morphology and breeding system are also discussed and the number of nuclei in the pollen grain noted. The arrangement of the families (see 'Contents') and their taxonomic subdivision follows HEYWOOD (1978). Within each family information is given genus-by-genus on the structures that present the pollen, the method by which it is issued, the presentation sequence of pollen and stigmas (in the maturation sequence the androecium always takes precedence) and flower/pollinator interactions.

Figure 1 illustrates a useful and simple classification of methods of secondary pollen presentation due to WESTERKAMP (1989). In German these are (1) Pollenhaufen, (2) Nudelspritze and (3) Pseudo-Staubblatt, which may be rendered in English as (1) pollen-heap, (2) noodle-squeezer and (3) pseudostamen. The pollen-heap is any exposed deposit, including all of the here-excluded 'mess and soil' category. The noodle-squeezer is a piston operating in a cylinder and expelling sticky pollen [this

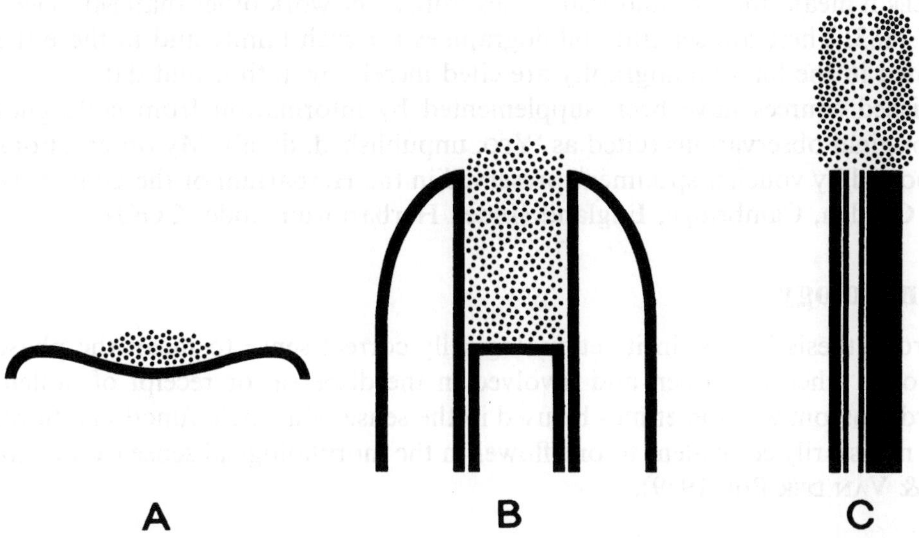

Fig. 1. WESTERKAMP's classification of secondary pollen presentation types. *A* Pollenhaufen (pollen-heap). *B* Nudelspritze (noodle-squeezer). *C* Pseudo-Staubblatt (pseudostamen). (C. GUTMANN, from WESTERKAMP 1989.)

is the 'forma a stantuffo', long ago recognized by DELPINO (1873–4)], and the pseudostamen is any structure that approximately mimics a stamen and which substitutes for a stamen in presenting the pollen (it is most usually the style). These categories are admittedly connected by transitional states. Where possible the method of presentation according to this classification will be indicated for each family.

2. Sources of information

When I began this work I found that apart from a few recent investigations (e.g. BRANTJES 1982, 1983; KENNEDY 1978; SCHICK 1980, 1982), studies of the floral mechanism of plants showing secondary pollen-presentation mostly dated from the second half of the nineteenth century. The only study covering several members of a major group that shared – at least in part – my present standpoint was that of JUEL (1908) on the *Asteraceae*. The spectacular revival of interest in pollination biology which began about the middle of our own century has mainly taken other directions. Consequently, I have had to seek information in studies of comparative morphology undertaken with classification and phylogeny primarily in mind, as well as by adding some of my own observations. However, the situation has already been improved significantly by the publications of ERBAR, KUNZE, LEINS, STUMPF, THIELE and others.

I have made much use of the classic reference works of MÜLLER (1883, a translation of MÜLLER 1873) and KNUTH (1908, 1909). Where the observations cited are not original I have said so but time has not always permitted me to go back to the source. These two reference works are translations from the German. Details of the original editions can be found in these translations and, in the case of KNUTH, in SCHMID & SCHMID (1970). When KNUTH is said to be citing the work of MÜLLER it means that the material comes from some work other than MÜLLER's book of 1883. There are separate bibliographies for each family and in these the works cited in the final bibliography are cited merely by author and date.

Published sources have been supplemented by information from colleagues and by original observations (cited as 'YEO, unpublished, date'). My observations are supported by voucher specimens preserved in the Herbarium of the University Botanic Garden, Cambridge, England (Index Herbariorum code: CGG).

3. Terminology

The word anthesis is used in its etymologically correct sense to mean the phase of the flower when it is open and involved in the dispersal or receipt of pollen. The word blossom will sometimes be used in the sense of a single functional floral unit not necessarily equivalent to one flower in the morphological sense (following FAEGRI & VAN DER PIJL 1979).

4. Standardized summaries of secondary pollen presentation

In the accounts of families, reports on secondary pollen presentation for species or species-groups are summarized on the plan shown below. Bold-type here shows the abbreviations used. Whatever expanded explanation is necessary then follows. As already mentioned the anthers can be involved mutually with other organs; they can also be involved when they have sterile appendages that take part in pollen presentation. Provision is also made for showing when part of the pollen is presented directly by the anthers, unassisted by other structures. Explosive release is in natural conditions always triggered by the pollinator. Flower-visitors are listed as given in the publications cited (or summarized); if they are not discussed further the presumption is that they are likely pollinators but not necessarily proved to be so.

STR – Structures involved in presenting pollen
p = perianth, **k** = calyx, **c** = corolla, **f** = filaments (including staminodes), **a** = anthers, **(a)** = anthers presenting pollen directly, **s** = style.
ISS – Method of issue
rel = non-explosive release of concealed pollen
rel(dose) = the same but in measured doses
the preceding may be suffixed as follows
– **poll** = powered by pollinator
– **trigg** = powered by internal tension and triggered by pollinator
– **irrit** = powered by plant's movement in response to touch (irritability)
growth = gradual expulsion by growth
expl cloud = release of pollen cloud explosively
expl depos = explosive deposition on vector
exposed = total pollen production in an accessible place or on an exposed site in the flower
SEQU = Presentation sequence of pollen and stigma
pln = pollen first
stg = stigma first

Table 1-3. Names of biotic pollination syndromes

cantharophily	beetles (Coleoptera)
chiropterophily	bats (Chiroptera)
entomophily	insects (Insecta)
lepidopterophily	Lepidoptera in general
melittophily	bees (Hymenoptera-Apoidea)
micromelittophily	small bees
myiophily	flies (Diptera)
ornithophily	birds (Aves)
phalaenophily	moths (Lepidoptera in part)
psychophily	day-flying Lepidoptera
sapromyiophily	saprozoic & coprozoic flies (Diptera)
sphingophily	hawkmoths (Lepidoptera-Sphingidae)
therophily	non-flying mammals

sim = simultaneous
APPL = Place of application to pollinator
general = unrestricted
a = anterior, **p** = posterior, **d** = dorsal, **v** = ventral, **l** = lateral (if more than one area is involved a comma is used, thus **a, d**; however, the combination **ad** means 'antero-dorsal')
RWD = Reward to pollinator
n = nectar
p = pollen
nil
other rewards to be stated
unknown: shown by a dash
SYN = Pollination syndrome
conventional terms are used, without the suffix '-phily' (Table 1-3)
VIS = Observed pollinators
NA or **na** = 'not applicable'
ignorance is indicated by ' – ' and uncertainty by '**?**'

Chapter 2

Descriptions of secondary pollen presentation

Myristicaceae

Systematic position: *Magnoliidae, Magnoliales*
Restriction of occurrence within the family: To genus *Myristica*

Floral features. All or nearly all species are dioecious. KNUTH (1904), however, cites a report of trees being male in their sex expression when young and female later. Perianth of one whorl of three largely united segments. Stamens two to thirty, partly or wholly united. Ovary superior, consisting of a single uniovulate carpel. Floral development has been described by ARMSTRONG & TUCKER (1986).
Type of pollen presentation. 'Pollenhaufen', using anther appendages and perianth.

Examples of Secondary Pollen Presentation in Myristicaceae

Myristica GRONOV.
M. fragrans Houtt. (ARMSTRONG & DRUMMOND 1986, illus.). **STR p, a, (a)?. ISS** exposed. **SEQU na. APPL general. RWD p** (in male flowers), **nil** (in female flowers). **SYN** cantharo. VIS the beetle *Formicomus braminus* (*Coleoptera-Anthicidae*).
The cream-coloured flowers of both sexes are small (6–10 mm long and 4–8 mm wide) and downwardly directed, with a fleshy urn-shaped perianth. The stamens (usually 9 in number) are completely fused, including the distal prolongations of the connective. Flowers open soon after dark and produce a scent. Male flowers are functional for only one night; they shed their sticky pollen before opening and it adheres to the united connective-appendages and to the throat region of the perianth. The beetles, which are 3–4 mm long, visit the flowers and pick up pollen grains on their body hairs. The gap between the throat of the female perianth and the stigma is less than that between the male perianth and the staminal column. It is presumed that the beetles eat some of the pollen but that their visits to the female flowers are unrewarded; the female flowers are thus practising mimicry of the males. It is consistent with this supposition that the female flowers are greatly outnumbered by the males, a situation which probably contributes to the fact that beetle visits to the female flowers were not observed. This study was carried out in South India, where *M. fragrans* (the nutmeg) is not native; there is, however, one native species of *Myristica* in the area.

Although pollen was the only pollinator-reward evident in this study, KNUTH (1904, citing WARBURG) stated that the smooth inside of the perianth appears to produce a sugary secretion. However, in a study of another species, *M. insipida* R. BR. in Queensland, ARMSTRONG & IRVINE (1989) found that liquid that occurred in female flowers was sugar-free, and was apparently rainwater. *M. insipida* was said by these authors to have exactly the same pollination arrangements as

M. fragrans but pollen presentation on the perianth is not actually mentioned. Small beetles of three families were seen visiting both male and female flowers and in both cases were sternotribic. Feeding on pollen was confirmed; no other floral material was eaten. Several of the features of the cantharophil syndrome are missing from *Myristica* and certain other genera pollinated by small beetles that do not practise general floral phytophagy; ARMSTRONG & IRVINE therefore propose the recognition of an additional syndrome: 'microcantharophily'.

Discussion of Myristicaceae

The function of secondary pollen presentation in *Myristica fragrans* appears to be to achieve a more general distribution of pollen over the insect's body than if all the pollen remained in the anthers. This could be advantageous if the beetles, when visiting female flowers, are liable to present either their dorsal or their ventral surfaces to the stigma. The function of secondary pollen presentation would then be the opposite of one of those that I regularly look for: exact positioning of pollen on the body of the vector. This could not apply to *M. insipida* where the beetles are consistently sternotribic. It has not been made clear whether all pollen is shed from the anthers of *M. fragrans* before transfer to the insect, but it seems highly probable that in fact some of it is presented directly, as is the case with some or all the pollen of *M. insipida*. Binucleate pollen is reported in *Myristica* (BREWBAKER 1967).

References for Myristicaceae

BREWBAKER, J. L., 1967.

ARMSTRONG, J. E., DRUMMOND, B. A., III, 1986: Floral biology of *Myristica fragrans* HOUTT. (*Myristicaceae*), the nutmeg of commerce. – Biotropica **18**: 32–38.

ARMSTRONG, J. E., IRVINE, A. K., 1989: Floral biology of *Myristica insipida* (*Myristicaceae*), a distinctive beetle pollination syndrome. – Amer. J. Bot. **76**: 86–94.

ARMSTRONG, J. E., TUCKER, S. C., 1986: Floral development in *Myristica* (*Myristicaceae*). – Amer. J. Bot. **73**: 1131–1143.

KNUTH, P., 1904.

Papaveraceae

Systematic position: *Magnoliidae, Papaverales*
Restriction of occurrence within the family: Subfamilies *Hypecooideae* and *Fumarioideae* (absent from the only other subfamily, *Papaveroideae*)

Distribution within the subfamilies. According to FEDDE (1909) subfamily *Hypecooideae* consists of the monotypic genus *Pteridophyllum*, which is without secondary pollen presentation, and *Hypecoum*, which comprises 15 species in all of which secondary pollen presentation is displayed. Subfamily *Fumarioideae* consists of several genera, of which all the more important are known to have secondary pollen presentation. [In the classification of THORNE (1983) *Pteridophyllum* is removed to a separate monotypic subfamily, while four tribes previously recognized in the *Papaveroideae* are treated as subfamilies].
Floral features. The floral formula is K2, C2+2, A4 or 6, G(2). Flowers disymmetric in *Hypecoum* and *Dicentra* but transversely zygomorphic in other *Fumarioideae*

Fig. 2. Floral diagrams of *Papaveraceae*. *A Hypecoum*; working outwards from ovary are seen four dithecous stamens (each theca 2-lobed), two inner petals with pouches laid open, two outer petals, two sepals (vertically hatched), two bracteoles and a bract. *B Dicentra*; perianth members and bracts as in 'A' except that pouches are absent and the androecium consists of two staminal complexes each of one dithecous stamen and two monothecous. *C Corydalis*; similar to 'B' but flower is transversely zygomorphic, with inner petals asymmetric, outer unequal and staminal complexes unequal. Nectaries in 'B' and 'C' horizontally hatched; asterisks indicate the median plane. (After Warming 1895.)

(Fig. 2). Stamens four (dithecous) in *Hypecoum*; six (two dithecous and four monothecous) in *Fumarioideae*, suggesting fission of two stamens. Style simple, bilobed in *Hypecooideae* and two- or more-lobed in *Fumarioideae*, where the lobed head is compressed in the plane on which the outer petals lie.

Type of pollen presentation. Pollen is presented either by the inner petals, acting as 'Nudelspritzen' (*Hypecoum*), or by the stylar head, acting as 'Pseudo-Staubblätter' (*Fumarioideae*), in which case the inner petals provide a displaceable cover for it.

Examples of Secondary Pollen Presentation in Papaveraceae

The structure of the flowers of these plants was beautifully described and illustrated by Hildebrand (1869) but he made no detailed observations of stigma receptivity, and he was in many cases unfortunate in not seeing insect visits.

Subfamily *Hypecooideae*

Hypecoum L. (Fig. 2).

H. procumbens L. (Hildebrand 1869, illus.; P. E. Gibbs, pers. comm.). **STR c. ISS rel(dose)-poll. SEQU stg. APPL? RWD? SYN** melitto. **VIS** –.

The anthers are extrorse and their pollen is placed on, and becomes enclosed by, the two inner (median) petals long before anthesis. The anthers that alternate with the inner petals deposit the pollen of each theca on the adjacent petal; thus each inner petal receives all the pollen of the stamen opposite it and half the pollen of each stamen alternating with it. The inner petals are peculiar in that each is formed by a pair of lobes and a stalked central part specially modified for its role. At the time the pollen is shed from the anthers the central part presents an innervated spathulate area set on a rectangular lamina; subsequently the spathulate area becomes pouched, forming an elongate receptacle for the pollen, while the rest of the lamina remains patent on either side of the groove which marks the closure of the pouch. Between the two specialized petals stands the gynaecium. HILDEBRAND found that the petaline pollen pouches would yield up some of their pollen if pressure was applied to them from above, so the release of pollen appears to be in doses issued in response to the activity of an insect in the flower. He found that the stigmas were protected from pollen during the development of the flower and that they appeared not to be receptive until it opened. He observed arrangements which seemed to provide for self-pollination at the end of anthesis. However, EAST (1940) attributed reports of self-incompatibility in *H. procumbens* and *H. grandiflorum* L. to HILDEBRAND and DARWIN. From HILDEBRAND'S description of *H.. procumbens* one might suppose that during anthesis pollen presentation precedes stigma receptivity, but Dr P. E. GIBBS tells me that the flowers are functionally female first.

H. leptocarpum HOOK F. & THOMS. (YEO, unpublished 1992). (Fig. 3).

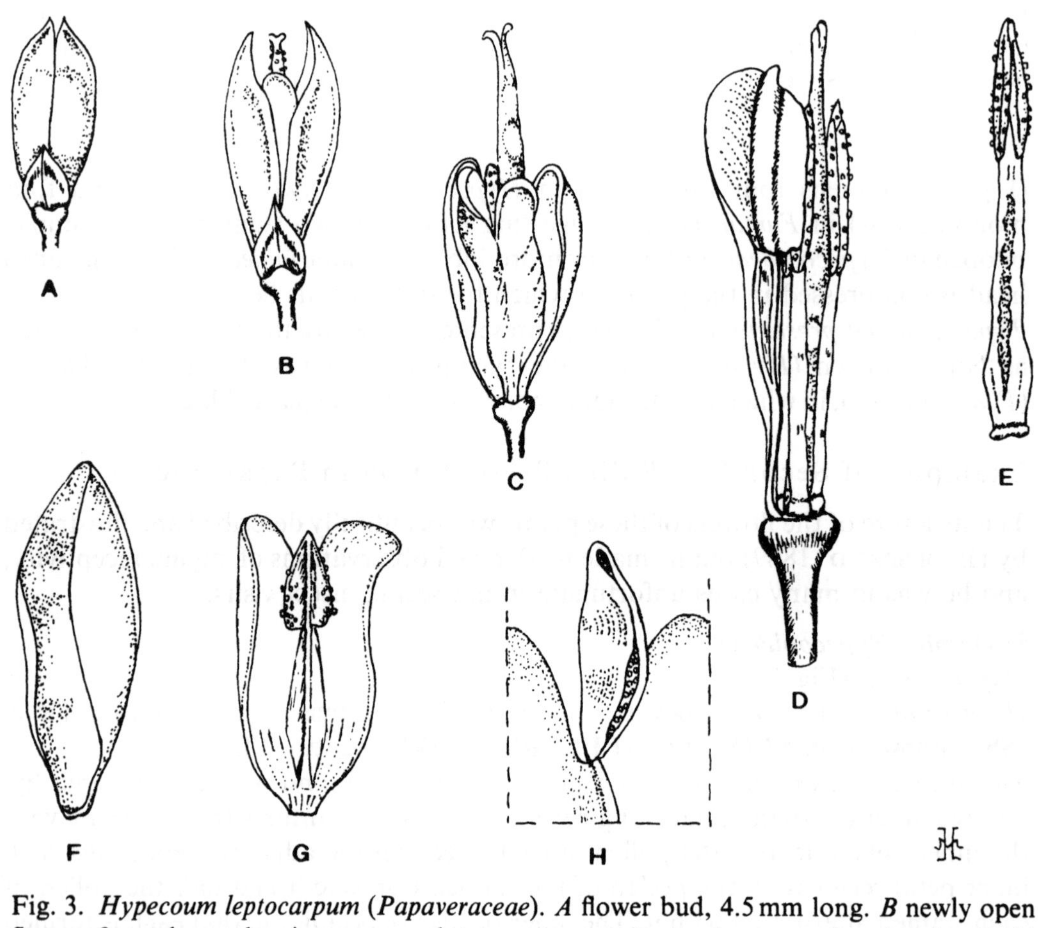

Fig. 3. *Hypecoum leptocarpum* (*Papaveraceae*). *A* flower bud, 4.5 mm long. *B* newly open flower, 9 mm long, showing one sepal, two outer petals, one inner petal and style. *C* old fertilized flower with developing fruit. *D* flower with both outer petals and one inner removed; in the centre is the ovary with one stamen either side of it, that on the left partly embraced by the pollen-pouch of the inner petal. *E* stamen from unopened flower. *F* outer petal. *G* inner petal from within, showing pouch with scattered pollen grains. *H* view of pollen-loaded pouch from an oblique angle, further enlarged. (M. HICKEY.)

This differs from *H. procumbens* in that lobes of the inner petal are relatively larger and the pollen pouch is sessile. At anthesis the anthers of the inner stamens are still in the mouth of the pollen pouch, suggesting that the system is not fully functional in this species.

Subfamily *Fumarioideae*

Dicentra BERNH. Fig. 2, 4 (*D. formosa*, mentioned under *D. eximia*).

D. spectabilis (L.) LEMAIRE. (HILDEBRAND 1869, as *Diclytra spectabilis*, illus.; KNUTH, for visitors). **STR s. ISS rel-poll. SEQU sim? APPL v. RWD n. SYN** melitto. **VIS** the bees *Bombus* and *Anthophora* (*Hymenoptera-Apoidea*).

The flowers are pendulous. The two outer petals are deeply concave and their edges overlap those of the inner petals; their bases are produced into a pouch-like spur on either side of the pedicel while their apices are tongue-like and reflexed. The two inner petals are much narrower and are divided into basal and apical parts of about equal length by a hinge which is created by conduplication. The basal half is flat and narrowly oblong, while the apical half, also narrow, is rather elaborate in form; its back is keeled, its edges are incurved and its tip is coherent with that of its partner. The two apical halves thus make a cowl around the distal parts of the androecium and gynaecium. The stamens form two groups, each consisting of a median dithecous and two lateral monothecous stamens. These also are differentiated into basal and apical parts. The basal is the longer and consists of an arched band formed by the flattened filaments and lying within the cavity formed by the outer petal to which it is opposite; this band is abaxially gutter-shaped and acts as a proboscis-guide to the nectary at its base. The apical part of each stamen is free and filamentous and these parts lie close together around the style in the cowl. The expanded head of the style is flattened and 4-lobed, with two lobes directed towards the apex and two towards the base. The surface bears protruberances (HALSTED 1889). Long before the flower opens the anthers dehisce and deposit their pollen on the inner walls of the cowl and on the stylar head. Deposition of pollen on the stylar head is assisted by the penetration at dehiscence of the superficial protruberances into the thecae (HALSTED 1889). These, together with the main lobes of the head, apparently scoop pollen out of the anthers as the distal parts of the filaments shrivel and draw the anthers back from the tip of the cowl (LINDLEY 1853 – referring to the entire family, *Fumariaceae*; HALSTED 1889).

When the flower is mature the cowl can be displaced about its hinge in either direction in the plane of the outer petals. Upon the removal of the displacing force it returns to its original position and again encloses the style. The style, on the other hand, is stiff and retains its position, and is thus exposed to contact with any insect that displaces the cowl. The keels of the cowl lie in the plane at right angles to the plane in which the cowl swings and therefore provide something on either side of the style on which the insect can press. The shape of the proximal parts of the stamen-filaments ensures that in order to reach the nectar the bee must cling to the cowl rather than to one of the outer petals. The stylar head will therefore touch it ventrally and can deposit and receive pollen. Although HILDEBRAND says that the flower deposits its own pollen partly on the insides of the cowl petals, any pollen that remains there cannot function in pollination and accordingly I

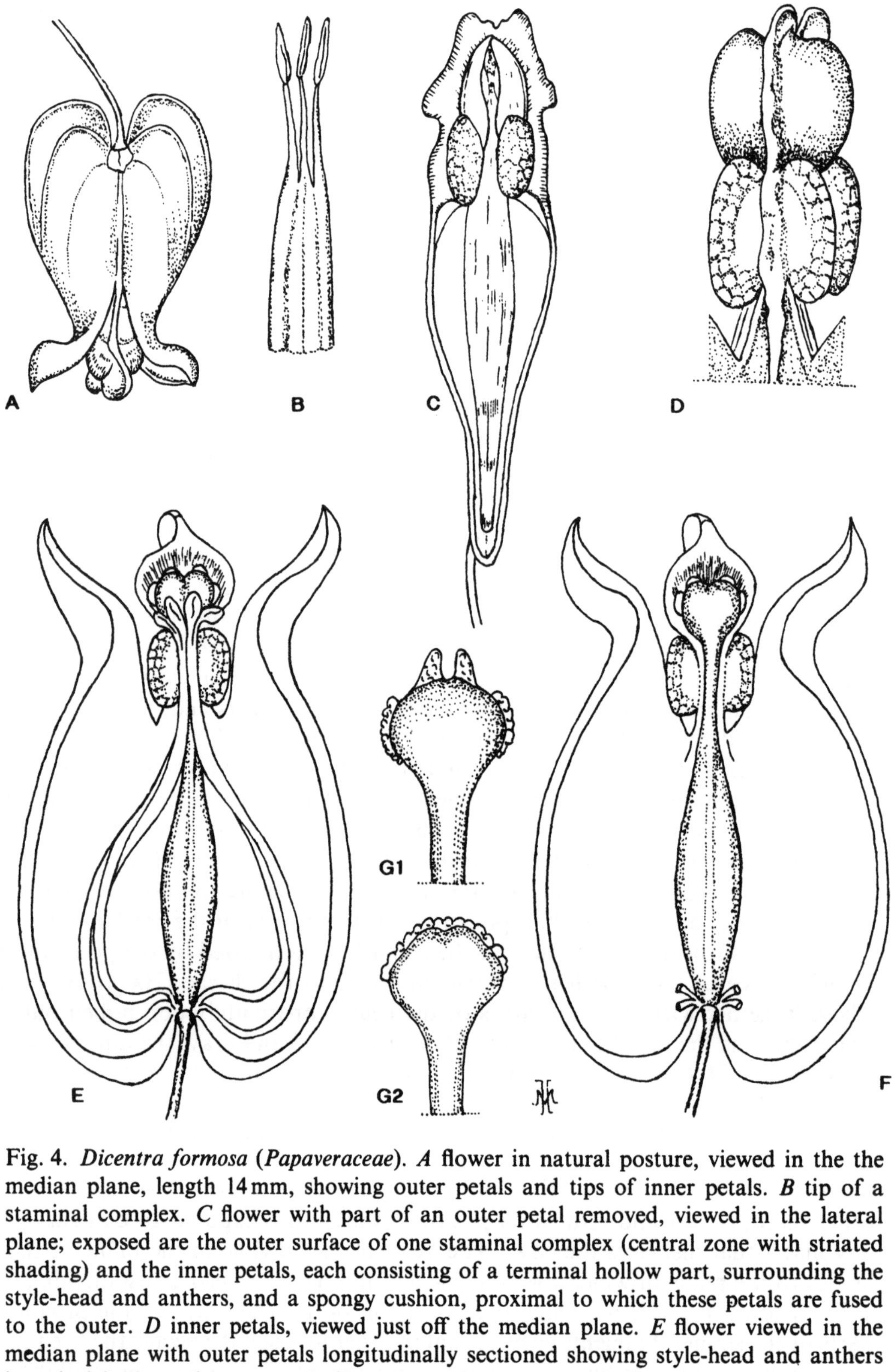

Fig. 4. *Dicentra formosa* (*Papaveraceae*). *A* flower in natural posture, viewed in the the median plane, length 14 mm, showing outer petals and tips of inner petals. *B* tip of a staminal complex. *C* flower with part of an outer petal removed, viewed in the lateral plane; exposed are the outer surface of one staminal complex (central zone with striated shading) and the inner petals, each consisting of a terminal hollow part, surrounding the style-head and anthers, and a spongy cushion, proximal to which these petals are fused to the outer. *D* inner petals, viewed just off the median plane. *E* flower viewed in the median plane with outer petals longitudinally sectioned showing style-head and anthers in apical hollow of inner petal; filaments and ovary also visible. *F* the same but with stamens removed. *G1, G2* two different forms of the style-head found in this material. (M. Hickey.)

have omitted 'corolla' from the coded summary for this species. It seems hardly possible that the receipt of foreign pollen and the removal of self-pollen can be separated in time but the matter is not explicitly considered by HILDEBRAND (see, however, *Corydalis bulbosa*). Pollinator visits were observed in Europe, outside the natural range of this East Asian species. No seeds whatever were set by the plants available to HILDEBRAND, even though they were obtained from various sources and were interpollinated as well as selfed, and exposed to insect visits or left undisturbed.

D. eximia (KER-GAWL.) Torr. (HILDEBRAND 1869, illus., again as *Diclytra*) and *D. formosa* (HAW.) WALP. (YEO, unpublished 1990). (Fig. 4).

The system of secondary pollen presentation in these two North American species is similar to that in *D. spectabilis* but all the petals are fused for about two thirds of their length and the cowl hardly projects from the gap between the two outer petals. Its scope for movement is accordingly greatly restricted. The stamen-filaments are free near the base and differ from one another in their degrees of arching. The construction of the hinge is simpler. However, *D. formosa* (Fig. 4) further differs in that each inner petal displays a massive spongy swelling just proximal to the firm hollowed tip in which the style-head lies. On the adaxial side this cushion fits over the tip of the staminal complex and, more importantly, over the very rigid zone in which the tip of the ovary tapers into the style. Its function is to resist to a degree the displacement of the inner petal tips and to restore them to their original position when pressure is relaxed. The approximately lens-shaped style-head carries a deposit of strongly coherent pollen round its edge, initially like the tyre of a wheel, though gradually becoming depleted. In fact part of the 'wheel' is recessed to accommodate the 'tyre'.

The function of the cushions of the inner petals is the same as in *Corydalis* after allowing for the zygomorphy of the latter (see later); their spongy external appearance is also the same. (L. MÜLLER found that the corresponding tissue in *Corydalis bulbosa* and *Pseudofumaria lutea* was full of air spaces). Also it is possible to see some resemblance in the form of the inner petals of *D. eximia* to those of *Hypecoum*.

D. eximia has been seen to receive visits from various long-tongued bees in its native habitats, and among them *Bombus* seems to be particularly important (CLEMENTS 1889; STERN 1961). These authors state that after probing one side of the flower, the bumblebees turn round while clinging to the petals, and probe the other side, thus displacing the inner petals in both directions in one flower-visit.

D. cucullaria (L.) BERNH. (HILDEBRAND 1869).

This species also apparently functions similarly to the preceding but it has free petals (the outer with long, widely divergent spurs) and straight, slender stamen-filaments lying close to the ovary; these are free except at the apex where they are loosely united into two bundles just under the four-lobed style-head. The flowers are visited by bumble-bees (MACIOR 1970).

Adlumia RAF.

A. fungosa (AIT.) GREENE. (HILDEBRAND 1869, illus., as *A. cirrhosa*). **STR s. ISS rel-poll. SEQU sim? APPL av. RWD n. SYN** melitto. **VIS** –.

This flower functions in essentially the same way as those of *Dicentra*, and of those described here it most resembles *D. eximia*. The outer petals have no spur but are still somewhat saccate at the base, and all four petals are firmly fused. The hingeing arrangements are very simple and only slight movement of the cowl is possible. The filaments of each staminal complex are flattened and fused to each other and to the corolla, thus forming a partition separating the passage to each nectary from the ovary.

Corydalis DC.

C. bulbosa (L.) DC. [Nomenclature according to MOWAT & CHATER in TUTIN & al. 1964]. (HILDEBRAND 1867, 1869; JOST 1907; KIRCHNER 1911; L. MÜLLER 1939; all illus., all as *C. cava*; KNUTH, for visitors; YEO, unpublished 1991). **STR s. ISS**

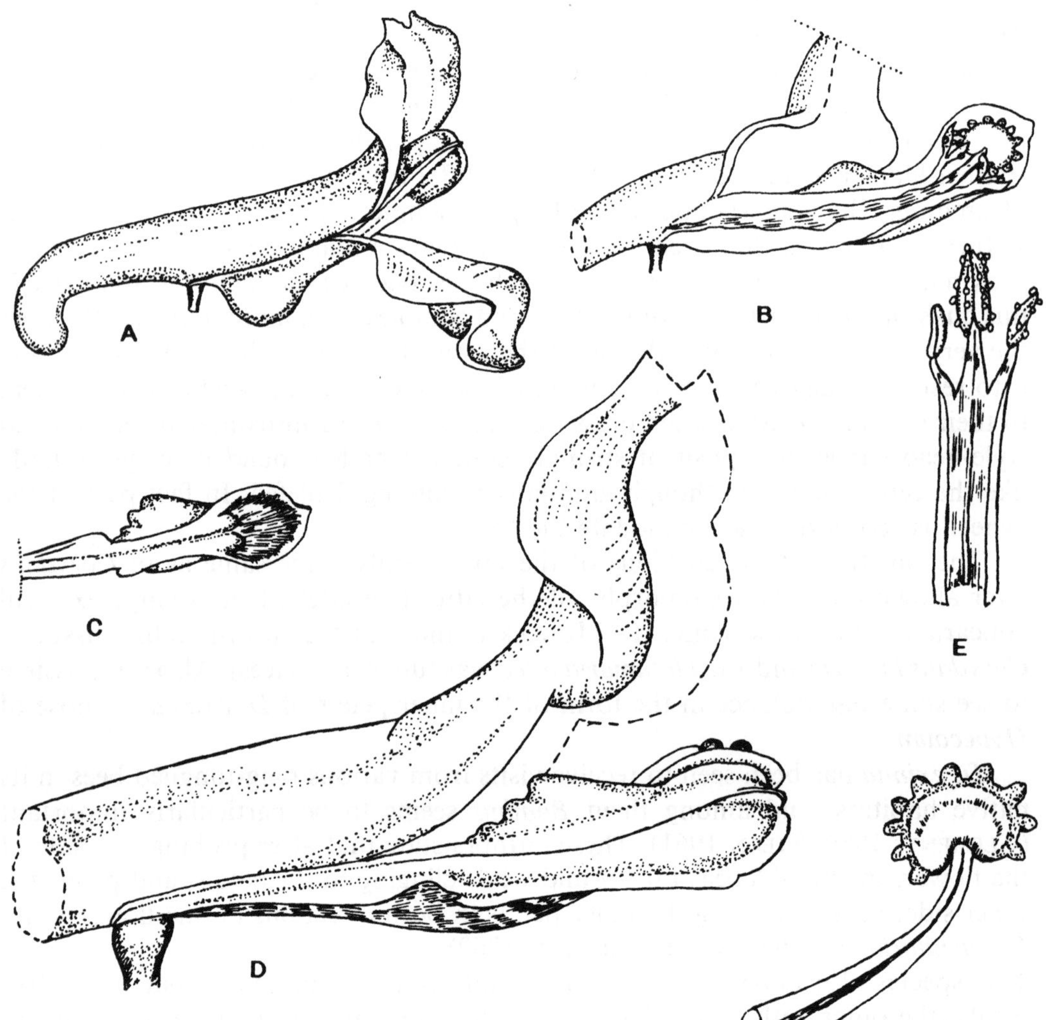

Fig. 5. *Corydalis bulbosa*-1 (*Papaveraceae*). *A* side view of flower. *B* the same with upper petal raised, lower petal and one inner removed, showing the other inner, the stamens and the style-head. *C* outside view of an inner petal, showing upper and lower spongy cushions; striated shading marks area of dark pigmentation. *D* like 'B' but the near inner petal present, showing longitudinal keel. *E* apex of one staminal complex. *F* style-head. (M. HICKEY.)

rel-poll. SEQU sim. APPL v. RWD n. SYN melitto. **VIS** the long-tongued bee *Anthophora pilipes* and the honeybee *Apis mellifera* (*Hymenoptera-Apoidea*). (Figs. 2, 5, 6).

HILDEBRAND (1869) gave a detailed account of the structure and functioning of the flower. His earlier (HILDEBRAND 1867) account of the effects of autogamous and xenogamous pollination was summarized in the later paper, as it was also by JOST (1907). L. MÜLLER (1939) analysed the interrelationships of the petals and reported on their histology. The flower is zygomorphic and horizontal in posture. The upper of the two outer petals has a long spur, whereas the lower is merely slightly pouched above the base, and each has a somewhat expanded limb. The inner petals are positioned transversely and cohere firmly at their tips forming a cowl over the style. They are identical with each other but individually asymmetric. About a third of their length from the base they have a peculiar folding of their edges which divides them into proximal and distal portions; this functions as a hinge, allowing the distal part to move down from the resting position in response to pressure but not up. In their proximal parts the inner petals are narrow and fused to the edges of the upper petal. Distal to the hinge the upper edge of each inner petal forms a slightly bulbous cushion over which the upper petal appears to be clamped. The tissue of this cushion is full of air spaces while the epidermis is large-celled (L. MÜLLER). By analogy with *Dicentra formosa* (above) these cushions should be responsible for offering resistance to the entry of a proboscis and for the return movement of the inner petals after they have been parted by a sufficiently strong insect visitor. Beyond the cushions the distal parts change their nature again to form the cowl. The filaments are free for a short distance near their tips but proximally are fused into an upper and a lower group and flattened so as to form two bands. The upper band is produced at its base into a long spur which is fused to the lower inside surface of the petaline spur, and secretes nectar at its tip. Above its attachment to the receptacle the upper band is fused to the edges of the upper petal to the same extent that the inner petals are. The stylar head has 8 or 10 lobes. In the bud the anthers deposit their pollen on it and then shrivel and shrink.

L. MÜLLER gave further details of the structure and movement. The upper petal has projections that fit over the cushions on the inner petals to make a tight closure of the flower entrance. When the cushions are depressed they twist outwards, so that the tips of the inner petals are parted slightly, making way for the emergence of the style-head. The upper edge of the inner petals at the hinge shows a deep emargination with a cushioned edge. Each inner petal carries a narrow flange on the outer surface that runs throughout its length. The flanges act as a guide, hindering lateral displacement of these petals, especially during the return to the resting position. If both flanges are cut in the hinge region the automatic return to the resting position only takes place if helped. Wounding only one wing reduces the effectiveness of the mechanism to a smaller degree. MÜLLER considers that while the articulation makes possible the tilting movement, the wings guarantee the return of the inner petals without dislocation in roughly the same way that the ligamentary apparatus in human and animal joints is important for their perfect functioning. The inner petal as a whole could be seen mechanically as an angle-lever with its fulcrum in the region of the hinge.

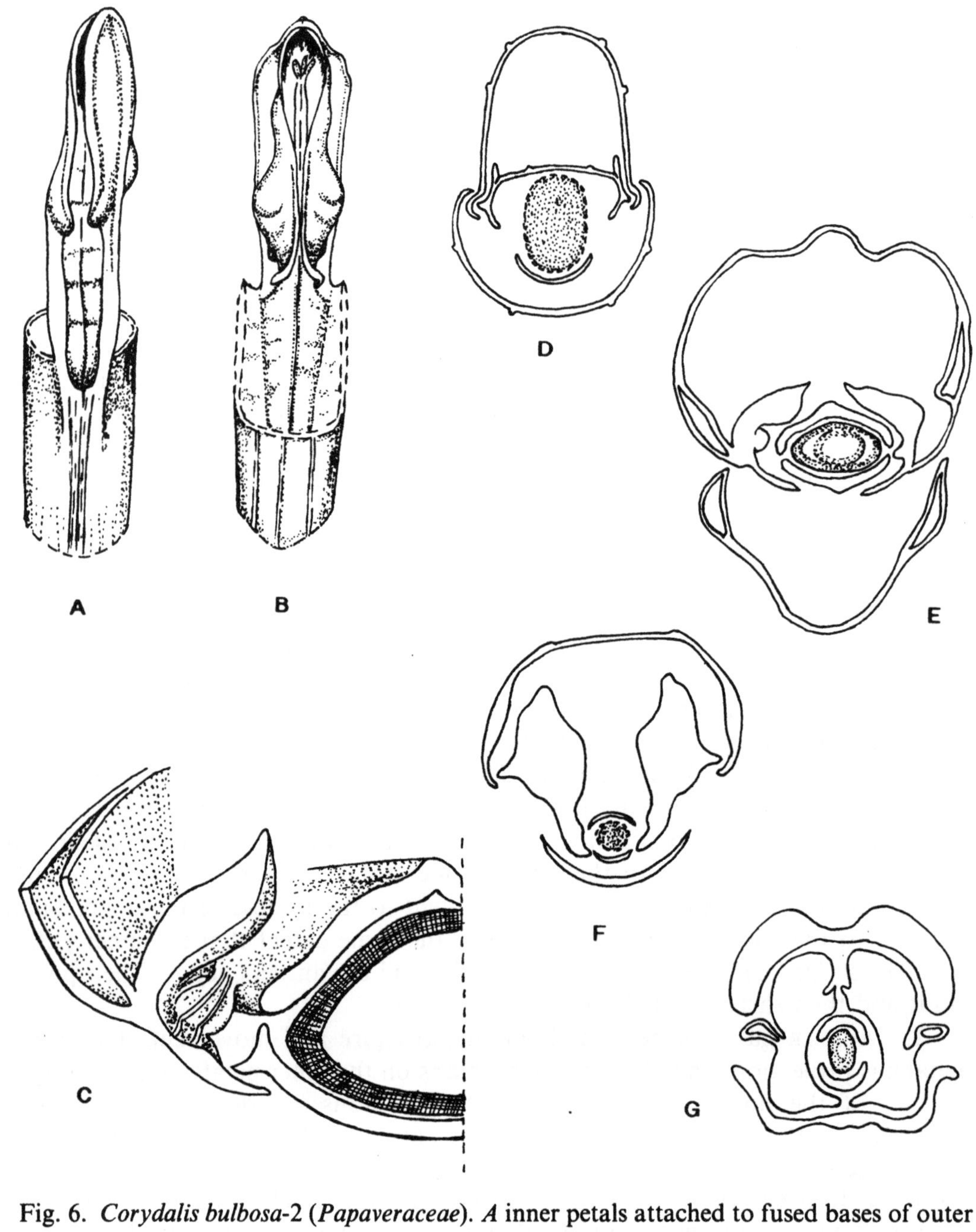

Fig. 6. *Corydalis bulbosa*-2 (*Papaveraceae*). *A* inner petals attached to fused bases of outer petals, from below; the space between them exposes the lower staminal complex. *B* the same from above; proximal to the spongy cushions are the folds that allow depression of the distal part of this pair of petals. *C* a dissection at the point where the inner and outer petals and the staminal complex cohere just proximal to the hinge, looking towards base of flower. *D* transection of flower, 1 mm above pedicel, coherence of outer petals not shown. *E* transection at same point as 'C'. *F* transection in the cushion region, showing keels of inner petals. *G* slightly more distal transection with upper and lower cushions about equal and keels larger. (M. HICKEY; *G* adapted from MÜLLER 1939.)

My own observations on the structure and mechanism of this flower were made when I had largely forgotten MÜLLER's account. Afterwards I re-read MÜLLER and, with help from my brother, J. J. YEO, clarified some of the detail of her paper, which in places is badly written. In broad agreement with MÜLLER, I found that when the flower is viewed from the side each inner petal appears to have two cushions of spongy tissue towards the base of the distal part, a massive one on the upper side, bearing some shallow transverse ridges on its proximal slope, and a smaller one on the lower side below the lateral keel. On the adaxial surface the upper cushion bulges inwards considerably and the lower less so; there is a furrow between them but it does not completely divide them. When the system is pushed downwards the inward bulges of the upper cushions are forced apart by the ovary and this separates the edges of the cowl so that the style-head can emerge. They must inevitably resist the downward movement of the cowl and they also help in its return but mainly at the end of the movement. The hinge is, as others have said, thin and flimsy. On the lower side it is narrowed by an emargination or incision and it seems to have a few faint folds at this point: this allows for compression. On its upper side the edge is looped into the shape of a U, as illustrated by HILDEBRAND (1869: plate 30, fig. 4) and by MÜLLER; this allows for stretching. The longitudinal flange, the importance of which was emphasized by MÜLLER, does not stand at right angles to the surface that bears it but is bent upwards like a turned-up hat-brim. When viewing the inner petals from the side one sees that the distal part is not in line with the proximal but is bent upwards through a small angle at the hinge. When the distal parts of the inner petals are jointly depressed the edges of the flanges must be stretched as the system is straightened; then, as it is bent downwards the 'hat-brim' flips to the 'turned-down' position, which might tend to hold the flower open. The very firm union of the tips of the petals enables the two lateral flanges to act together like a sling, tending maintain the upward – and the possible alternative downward – bend at the hinge. I too tried cutting the flanges at the level of the hinge and found that the first part of the return movement is suppressed or takes place relatively slowly; the later part of the movement, when the cushions begin to come above the ovary, is quicker. It is possible that the lower petal also plays a small part in the return movement. The function of the large adaxial upper cushions is not fully clear; the upper petal is shaped towards its edges to accommodate them but hardly, according to my observations, clamps them. Possibly they make contact with the ventral side of the pollinator and act as the point at which the downward pressure is received, but they do not meet on the midline and they do not block the entrance to the spur. The cowl has its edges produced to form an upper and a lower keel, but at this point the two petals are not fused, so that the pollen presenter can emerge from the upper keel. The pollen presenter is approximately lens-shaped in the vertical plane and the pollen-deposit on it is also lens-shaped and completely envelops the teeth. (Owing to the small amount of time available to me for studying this flower, my conclusions should be taken as tentative and in need of confirmation.)

When the flower has opened bees visit it and in probing for the deep-seated nectar depress the cowl, so that the stylar head touches the underside of the insect. HILDEBRAND found that fruit-set was very rare unless the stigma received pollen from another individual. In self-pollinated flowers the pollen tubes penetrated the stylar head and entered the style, but no fruit was formed. In nature, insect visits were necessary, and HILDEBRAND noted that the bees always went to the lowest flowers of a raceme and worked upwards. Artificial pollination with foreign pollen gave a fruit-set of about 90%. KIRCHNER reported that although the honeybee cannot reach the nectar it visits the flowers for pollen and in doing so causes pollination; he found that bumblebees pierced the flowers and stole the nectar and that honeybees also stole nectar through the bumblebees' perforations.

JOST (1907) repeated Hildebrand's pollination experiments and obtained similar results. He also studied the structure of the style and the growth of the pollen tubes. The epidermis of the stigma consists of T-shaped cells with spaces between them which connect with the cavities in an underlying spongy parenchyma. If the stigma is immersed in water it releases an oil which becomes emulsified but can be cleared with ether. Light pressure on the knobs on the stylar head destroys their epidermis and easily removes the parenchyma. Only when pollen comes in contact with the broken cells does it germinate. This arrangement seems to imply that foreign pollen arrives at the same moment that self-pollen is rubbed on to the pollinator.

JOST was able to stain the callose plug which is present at the tip of a pollen tube and by this means he established that foreign pollen could reach the ovary within 24 hours of pollination, whereas self-pollen grew only as far as the top of the style. JOST was unable to determine which substance in the stigma was responsible for stimulating pollen-germination, but he suggested that the function of the oil was to limit the rate of hydration of the pollen grains.

C. solida (L.) SWARTZ. [Nomenclature according to MOWAT & CHATER in TUTIN & al. 1964]. (HILDEBRAND 1869; KNUTH, for visitors).
This species is briefly mentioned by HILDEBRAND as being almost identical with *C. bulbosa* and as requiring cross-pollination for fruit-setting. This conclusion is broadly confirmed by RYBERG (1950: 271).
C. nobilis (L.) PERS. and *C. capnoides* (L.) PERS. (HILDEBRAND 1869, illus.).
HILDEBRAND also describes these two species as similar to the two preceding and he illustrates the second of them. Whereas *C. bulbosa* and *C. solida* are tuberous, spring-flowering species with simple stems, the present species have branched stems, *C. nobilis* being perennial and *C. capnoides* annual or biennial.
Pseudofumaria MEDIKUS. (Fig. 2).
This genus, recently reinstated by LIDÉN (1986), differs from *Corydalis* in delivering its pollen by explosive deposition.
P. alba (MILLER) LIDÉN (*Corydalis ochroleuca* KOCH). (HILDEBRAND 1869, illus.; MÜLLER 1939, illus.). **STR s. ISS expl depos. SEQU sim? APPL v. RWD n. SYN** melitto. **VIS** the honeybee *Apis mellifera* (*Hymenoptera-Apoidea*).
Compared with *Corydalis* the flower of this plant has the spur much shorter and the other petaline parts longer, the hinge of the inner petals differently formed, the filamental band of the upper staminal complex embracing the ovary and the stylar head differently formed. The upper staminal complex develops tension (see also *P. lutea*), such that it tends to lift the ovary and style, this force being opposed by the inner petals which clasp the staminal bands. Proximally the inner petals are clawed and their limb is auricled. When a bee probes for nectar it parts the inner petals (see also *P. lutea*), which are only loosely coherent at their tips, and this releases the upper filamental band which in turn carries the ovary and style upwards so that the stylar head hits the underside of the insect. On the departure of the insect the style is free to complete its movement and comes to rest with its head in a protective recess near the tip of the upper petal. The inner

petals, on their release, spring downwards, to become outwardly spread, with their inner surfaces facing downwards. None of the floral parts returns to its former position and no further pollen tranfer is possible. HILDEBRAND found that flowers protected from insect visits in a room set good seed, but he did not say whether the flowers were ultimately self-tripping. HILDEBRAND notes that the floral parts are not fused, and thinks this is to be expected in order to permit free movement when the fllower is tripped.

P. lutea (L.) BORCKH. (*Corydalis lutea* (L.) DC.). (HILDEBRAND 1869; KNUTH, for visitors; JOST 1907; L. MÜLLER 1939, illus.; YEO, unpublished 1990). (Fig. 7).

HILDEBRAND says that this differs from *P. alba* only in that the edges of the upper filamental complex cohere with those of the upper petal for a short distance near the base. L. MÜLLER's description agrees well with that of HILDEBRAND for *P. alba*. The style is thin and flexible at the base but is distally thickened and stiffened by lignification. The pollen is very sticky and most of it remains coherent during the explosion. I also found agreement in the main with HILDEBRAND's account of *P. alba*. The inner petals have a strongly developed cushion of aerenchymatous appearance (cf. *Corydalis bulbosa* and *Dicentra formosa*, above) on their upper edges distal to the auricles (Fig. 7C); in this region each filamental complex tapers away to a small thread while the cushions enlarge so that there is no empty space. The pair of cushions lies over the style and filaments, keeping them firmly on the lower side of the flower. The channelled upper petal clasps the inner. Slight swellings on the edges of these allow their tips to be locked into the apical hollows of the upper and lower petals. Apparently a bee has to free the upper petal in order to gain entry to the passage leading to the nectar-spur. The auricles of the inner petals stand up and obstruct this passage a little way above the ovary. When the flower is artificially probed the cushions offer resistance to the parting of the inner petals. However, if enough pressure is applied to the auricles the cushions are forced apart sufficiently to free the basal turgid regions of the filamental complexes. These then raise the style, the tip of which springs up abruptly. At the same time the inner petals bend downwards in the region of the auricle where there appears to be a rather thin pulvinus; unlike those of *P. alba* they do not separate at the tips. The filamental complexes are quite massive from the base to a point some way beyond the auricles and it is in this region that the upward tension develops. L. MÜLLER stated that both staminal complexes include a pulvinus. KNUTH lists various long-tongued bees as visitors. JOST found that abrasion of the stigma was necessary for pollen-germination and that the flowers were then self-fertile, like those of *P. alba*.

Fumaria L.

F. officinalis L. (HILDEBRAND 1869, illus.; KNUTH, for visitors). **STR s. ISS rel-poll. SEQU sim? APPL v. RWD n. SYN** melitto. **VIS** the honeybee *Apis mellifera* (*Hymenoptera-Apoidea*), rarely a few other insects.

This species is provided with a mechanism like that of *Corydalis* species in that the inner petals encase the pollen-bearing style and can be reversibly displaced from their protective position. It differs in the proportions of the parts and the more simply constructed hinge of the inner petals. Although it has an active nectary, HILDEBRAND never saw insect-visitors. Despite the existence of this floral mechanism the plants are self-fertile and, if left undisturbed, self-pollinating.

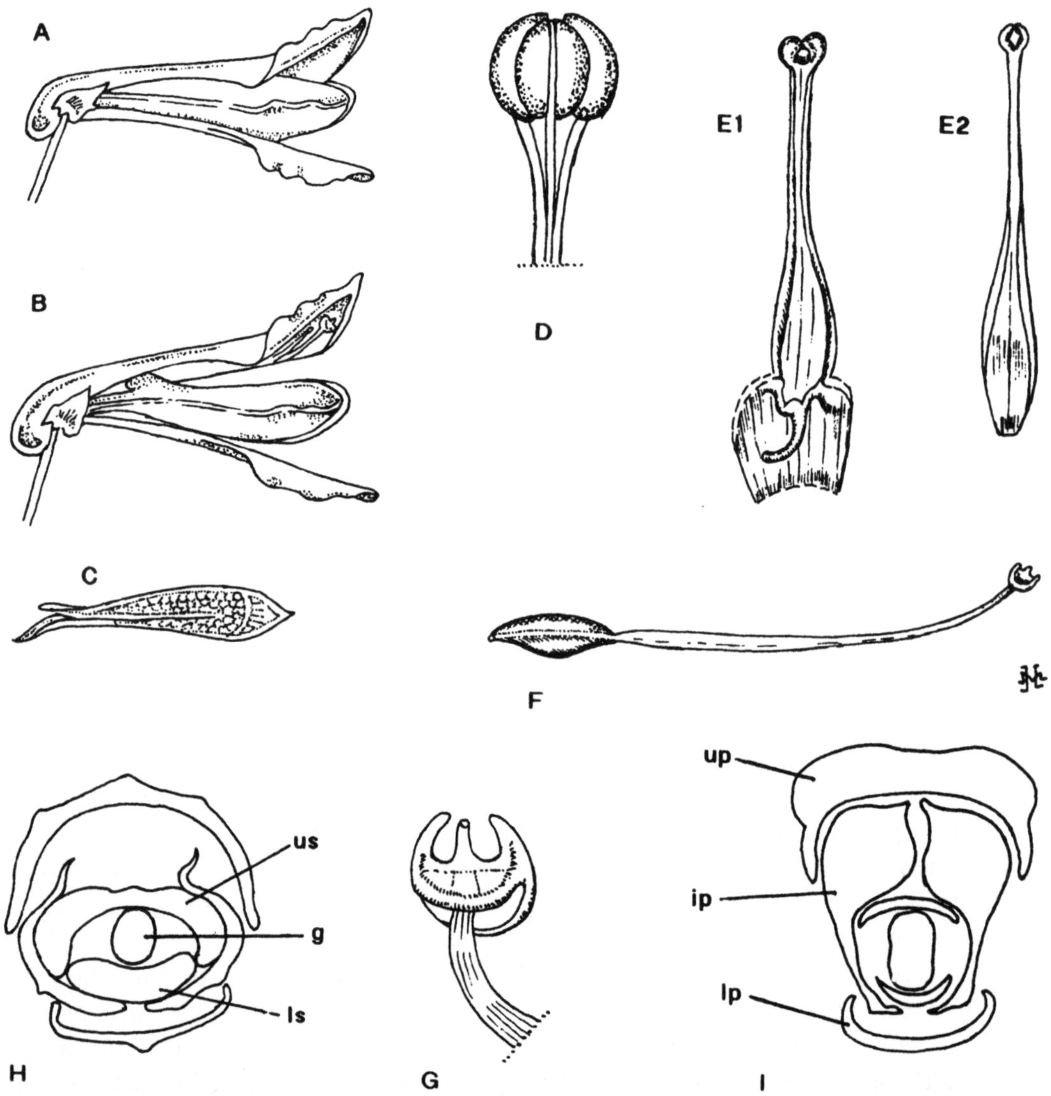

Fig. 7. *Pseudofumaria lutea* (*Papaveraceae*). *A* untripped flower; length 14.5 mm. *B* tripped flower; length 16 mm long. *C* view of inner petal from within showing auricle and spongy texture. *D* undehisced anthers of one staminal complex from bud. *E1* upper staminal complex with nectarial spur. *E2* lower staminal complex. *F* gynaecium. *G* style-head, shaped to support pollen mass. *H, I* transections just above ovary and half way along the flower. *up* = upper petal, *lp* = lower petal, *ip* = inner petal, *us* = upper staminal complex, *ls* = lower staminal complex, *g* = gynaecium. (M. HICKEY.)

F. capreolata L. and *F. parviflora* LAM. (HILDEBRAND 1869).
The flowers of these species are similarly constructed to those of *F. officinalis* but the elasticity of the inner petals is weak and after they have been depressed they return to their former position slowly or not at all, suggesting that the mechanism is losing its function.
Platycapnos (DC.) BERNH.
P. spicata (L.) BERNH. (HILDEBRAND 1869, illus., as *Fumaria spicata*). **STR s. ISS expl depos. SEQU sim? APPL v. RWD n. SYN** melitto. **VIS** –.

The floral mechanism of this plant is explosive, like that of *Pseudofumaria*. The plant is an annual and resembles a small-flowered *Fumaria* but the upper petal is only slightly pouched at the base. Upward pressure on the gynaecium is provided by tension in the filamental band of the upper staminal complex, while downward pressure resides in the inner petals that are wrapped round it; these have no hinge and when the flower is tripped they bend from the base. HILDEBRAND saw no insect visits but could trip the flowers with a needle. As in *Pseudofumaria*, when the probe was removed, the stylar head came to rest in a recess in the upper petal. Undisturbed flowers remained untripped but they set seed.

Discussion of Papaveraceae

The family consists almost exclusively of herbaceous plants. The contemporary view of the evolution of the family is that subfamily *Hypecooideae* represents the ancestral form and that this gave rise to subfamilies *Papaveroideae* and *Fumarioideae* as two separate lines, the first becoming actinomorphic nectarless pollen flowers with increased numbers of stamens and, sometimes, carpels, the second producing more extreme disymmetric flowers and transversely zygomorphic flowers (ROHWEDER & ENDRESS 1983: 169).

In comparing the floral diagrams of *Hypecoum* and *Dicentra* (*Fumarioideae*) (Fig. 2) it can be inferred that each median dithecous stamen of *Hypecoum* has given rise to two monothecous stamens in *Dicentra*. In the latter the fusion or coherence of the filaments of the lateral stamens with those of the adjacent monothecous median stamens gives rise to two staminal complexes. The thecae that participate in these are not the same as those which combine to allocate their pollen to the two inner petals in *Hypecoum*. There is insufficient information on how the pollen is actually arranged on the stylar head, but it could be that this corresponds with the arrangement in *Hypecoum*, leaving open the possibility of a direct evolution of the floral plan of *Dicentra* from that of *Hypecoum* despite the very unusual type of pollen presenter in the latter. The re-arrangement of the filaments in *Fumarioideae* is obviously conditioned by selection pressures associated with controlling the approach to the nectar. Within *Fumarioideae* the floral diagram of the zygomorphic *Corydalis* and *Pseudofumaria* (Fig. 2) differs little from that of the disymmetric *Dicentra*, and the one can easily be directly derived from the other. HILDEBRAND (1869) found a trace of a pouch on the lower outer petal of *P. alba* and a vestige of a nectary on the lower staminal complex; in *P. lutea* he found no inflation of the petal but a vestigial nectary, and in *C. bulbosa* some inflation of the lower petal but no trace of the lower nectary. As he said, it is easier to interpret these as vestiges persisting from an ancestor like *Dicentra* than to think of them as the initiation of a new development. Thus the disymmetric flower seems to have given rise to the transversely zygomorphic. In *Fumaria* these vestiges are entirely wanting. Other genera in the subfamily differ in aspects other than secondary pollen presentation. The fusion of parts in *Adlumia* and the formation of partitions within the flower suggest adaptation to hummingbird pollination. *Fumaria* and *Platycapnos* differ from *Corydalis* mainly by their fruits.

It must be mentioned here that the interpretation of the androecium given by EICHLER (HILDEBRAND 1869) is different from that given above. According to this the staminal complex in *Fumarioideae* consists of a tripartite stamen, the lateral portions of which represent stipular structures. In *Hypecoum* each of the two apparent median stamens then consists of an assembly of the two stipular structures from the neighbouring lateral stamens. Two pieces of evidence support this. The first is the tripartite nature of the inner and, less conspicuously, the outer petals of *Hypecoum* (interpreted by Eichler as modified leaves with stipules). The second is the fact that each median stamen of *Hypecoum* has two vascular bundles, whereas the lateral stamens each have one. The main fact against it is that nowadays stamens are not regarded simply as 'fertile leaves' (in which the stipules might have been retained and become fertile too). HILDEBRAND himself (1869), though quoting EICHLER's theory, does not dismiss the possibility that the two staminal complexes of *Fumarioideae* may correspond to the four stamens of *Hypecoum*. Another theory has been proposed by NORRIS (1941).

As to the function of secondary pollen presentation, in *Hypecooideae* it provides protection of the pollen in a magazine which requires a certain degree (and direction?) of pressure to open, and probably ensures its delivery in more or less measured doses.

In *Fumarioideae* secondary pollen presentation provides for exactly identical positioning of pollen and stigma, as noted for *Dicentra* by HILDEBRAND (1869), and this seems to be its main achievement, for the protection provided by the pivoted distal parts of the inner petals in this subfamily is independent of secondary pollen presentation. (Nevertheless, a unitary gynaecial presenter might be more effective than the two staminal complexes with their dissected tips.) However, protection may be rendered more necessary by the early release of the pollen from the anthers and its placement on the stigmas. A secondary achievement is that, in alliance with the protection provided by the inner petaline cowl, it has made possible explosive flowers that have to be tripped. (Explosive systems probably afford better protection to pollen overall, and may be slightly more specific in respect of pollinators, than non-explosive.) In the non-explosive members of the subfamily the capacity of the protective inner petals for repeated displacement makes it quite likely that pollen-release takes place in more or less measured doses, or at least that pollen-removal requires more than one pollinator-visit, but there is at present no information on this. The return of these petals after displacement could also be of value in protecting the stigma against the hazards of the environment. That it has such a role is suggested by the lodgement of the stigma in a recess of the upper petal after the flower is tripped in the explosive flowers of both *Pseudofumaria* and *Platycapnos*. Furthermore, in most of the zygomorphic genera there is a spot of blackish purple pigment inside the tip of the cowl invisible from outside (there may also be some pigment outside); LIDÉN (1986) wonders whether this might be for the protection of the stigma and the germinating pollen from radiation; it is absent from *Pseudofumaria* though not from *Platycapnos*.

All the structural and functional components associated with secondary pollen presentation in the *Fumarioideae* can also be found in *Leguminosae*, subfamily *Papilionoideae*. However, whereas in that very large group there is much parallelism, there is no reason to suppose that there has been parallel evolution in these components in *Fumarioideae* except in the case of the explosive system in *Platycapnos* and *Pseudofumaria* (see LIDÉN 1986: 16). In representatives of both families there are species in which the stigma has to be abraded before pollen tubes can penetrate it, such abrasion causing the release of lipidic material. Such species can be either self-compatible or self-incompatible. In the case of *Fumarioideae* the explosive flowers are self-compatible. They can therefore set seed in the event that no foreign pollen is received during their one and only pollinator-visit. If foreign pollen is prepotent any that reaches the stigma will cause fertilization before any of the self-pollen. The striking parallels between the *Papilionoideae* and the *Fumarioideae* appear to reflect the commitment of the two groups to melittophily and, to some extent, their basically separate-petalled structure.

Self-incompatibility is known in *Hypecoum procumbens*, in *Dicentra spectabilis* and certain individuals of *D. cucullaria* (East 1940), and in *Corydalis bulbosa*. In fact Lidén (1986: 108) states that nearly all members of tribe *Corydaleae* are strongly self-incompatible, though often interfertile with their nearest relatives, whereas members of the tribe *Fumarieae* (including *Fumaria*, *Pseudofumaria* and *Platycapnos*) are self-compatible, with a single exception, and show no taxonomic problems arising from hybridization. He considers that the difference is associated with life-form and habitat: the *Corydaleae* are perennials, some with strong vegetative reproduction; *Fumarieae*, on the other hand are annuals or short-lived chasmophytic perennials living in niches characterized by low interspecific competition and large fluctuations in population-size. This supports the view that, as in the *Leguminosae*, the occurrence of self-incompatibility is probably unrelated to the close proximity of pollen and stigma in the flower, self-pollination from this cause being prevented by other means. Its absence in explosive flowers is probably related to the single-visit system, as suggested above.

In view of the long distance from the mouth of the flower to the nectar, by which the flowers are often adapted to the visits of the larger and longer-tongued bees, it is not surprising that perforation of flowers by short-tongued species of *Bombus* is frequent in *Dicentra* (Müller; Stern 1961; Macior 1970) and *Corydalis* (Müller; Knuth). However, perforated flowers of *D. cucullaria* were found by Macior (1970) to have a good fruit-set. In Macior's (1970) study of *D. cucullaria* the single species of *Bombus* that was observed as a legitimate pollinator was sometimes found to be collecting its pollen. The honeybee (*Apis mellifera*, not native in North America where the study took place) regularly gathered pollen from this *Dicentra* (as recorded also by Clements 1889) and effectively pollinated it; it also took nectar through the perforations made by *Bombus*.

Conditions for stigmatic receptivity do not seem to have received attention since Jost's (1907) work and need to be investigated in a wider range of species. Binucleate pollen is recorded for *Dicentra* (Brewbaker 1967).

References for Papaveraceae

Brewbaker, J. L., 1967.

East, E. M., 1940.

Fedde, F., 1909: *Papaveraceae – Hypecoideae* et *Papaveraceae – Papaveroideae*. – In Engler, A., (Hrsg.): Das Pflanzenreich **IV.104** (**40**. Heft). – Leipzig: Engelmann.

Halsted, B. D., 1889: *Dicentra* stigmas and stamens. – Bot. Gaz. **14**: 129–130.

Hildebrand, F., 1867: Geschlechter-Verteilung bei den Pflanzen. – Leipzig: Engelmann.

Hildebrand, F., 1869: Über die Bestäubungsvorrichtungen bei den Fumariaceen. – Jahrb. Wiss. Bot. **7**: 423–471.

Jost, L., 1907: Über die Selbststerilität einiger Blüten. – Bot. Zeitung (Berlin) **65**: 77–117.
Kirchner, O. von, 1911: Blumen and Insekten. – Leipzig: Teubner.
Lidén, M., 1986: Synopsis of the *Fumarioideae* (*Papaveraceae*) with a monograph of the tribe *Fumarieae*. – Opera Bot. **88**: 1–133.
Lindley, J., 1853: The Vegetable Kingdom, 3rd. edn. – London: Bradbury and Evans.
Macior, L. W., 1970: The pollination ecology of *Dicentra cucullaria*. – Amer. J. Bot. **57**: 6–11.
Müller, L., 1939: Der Bewegungsmechanismus der *Corydalis*-Blüten und sein Feinbau. – Oesterr. Bot. Z. **88**: 1–23.
Norris, T., 1941: Torus anatomy and nectary characteristics as phylogenetic criteria in the *Rhoeadales*. – Amer. J. Bot. **28**: 101–113.
Rohweder, O., Endress, P. K., 1983: Samenpflanzen, Morphologie und Systematik der Angiospermen und Gymnospermen. – Stuttgart: Thieme.
Ryberg, M., 1950: Studies in the taxonomy and fertility of some Scandinavian *Corydalis* species of the sect. *Pes-Gallinaceus* Irmisch. – Acta Horti Berg. **15**: 207–284.
Stern, K. R., 1961: Revision of *Dicentra* (*Fumariaceae*). – Brittonia **13**: 1–57.
Thorne, R. F., 1983: Proposed new realignments in the angiosperms. – Nordic J. Bot. **3**: 85–117.
Tutin, T. G. & al., (eds.), 1964: Flora Europaea **1**. – Cambridge: Cambridge University Press.
Warming, E., 1895: A Handbook of Systematic Botany. Transl. & ed. M. C. Potter. – London: Swan Sonnenschein.

Sterculiaceae

Systematic position: *Dilleniidae, Malvales*
Restriction of occurrence within the family: Tribes *Dombeyeae* and *Byttnerieae* (out of 8 tribes recognized by Schumann 1890–1893)

Floral features. Flowers usually actinomorphic and pentamerous. Perianth may consist of sepals and petals or sepals only. Stamens said to form two whorls, but frequently their filaments are partly united to form a tube, and the petals may also be united to this at the base. Outer androecial whorl frequently staminodial. Ovary superior, usually of four or five united carpels, of which the styles are free or united to a varying extent.
Type of pollen presentation. In the large (c. 200 species) woody genus *Dombeya* the staminodes are more prominent than the stamens and are responsible for presenting a large proportion of the pollen ('Pseudo-Staubblatt'). In Tribe *Byttnerieae* the pollen is presented in concavities of the petals which are entered by small insects; part, at least, of the pollen is transferred to the petal surface ('Pollenhaufen') before being picked up by the insects.

Examples of Secondary Pollen Presentation in Sterculiaceae

Tribe *Dombeyeae*
Dombeya Cav.
D. mastersii Hooker. (*D. angulata* sensu Masters). (Masters 1867; Hooker 1868, illus.). **STR c, f, (a). ISS exposed. SEQU pln. APPL ? RWD n, p? SYN** melitto. **VIS** –.
The account here is by Masters; part of it was quoted by Hooker in the text accompanying the illustration. The nodding white fragrant saucer-shaped flowers

are 3 cm in diameter and are borne in umbel-like clusters. There are five linear-spathulate staminodes about as tall as the style, and 15 much shorter stamens. In the fully expanded flower the angular tip of each petal and the outer surface of the dilated part of each staminode are often dusted with pollen. In 'less fully developed flowers the barren stamens may be seen curving downwards and outwards so as to come in contact with the shorter fertile stamens, whose anthers open outwardly, and thus allow their contents to adhere to the barren stamens. These latter, provided with their freight of pollen, uncoil themselves, assume more or less of an erect position, and thus bring their points on a level with the stigma, whose curling lobes twist round them and receive the pollen from them.' Secondary pollen presentation is thus seen as an arrangement promoting self-pollination, as was frequently the case in the nineteenth century. The loading of the staminodes is different from that described for the next species.

D. burgessiae HARV. (YEO, unpublished, mainly 1985). **STR c, f, (a). ISS exposed. SEQU stg. APPL v. RWD n, p? SYN** melitto. **VIS** social wasps *Vespula* sp. (*Hymenoptera-Vespoidea*). (Fig. 8).

The flowers are saucer-shaped, about 3 or 4 cm in diameter and pendent. The reflexed sepals are pale green, the anthers cream and the petals and stigmas white. The 15 stamens are one quarter to one third as long as the petals and their filaments are united into a tube for just under half their length. The tube and the filaments

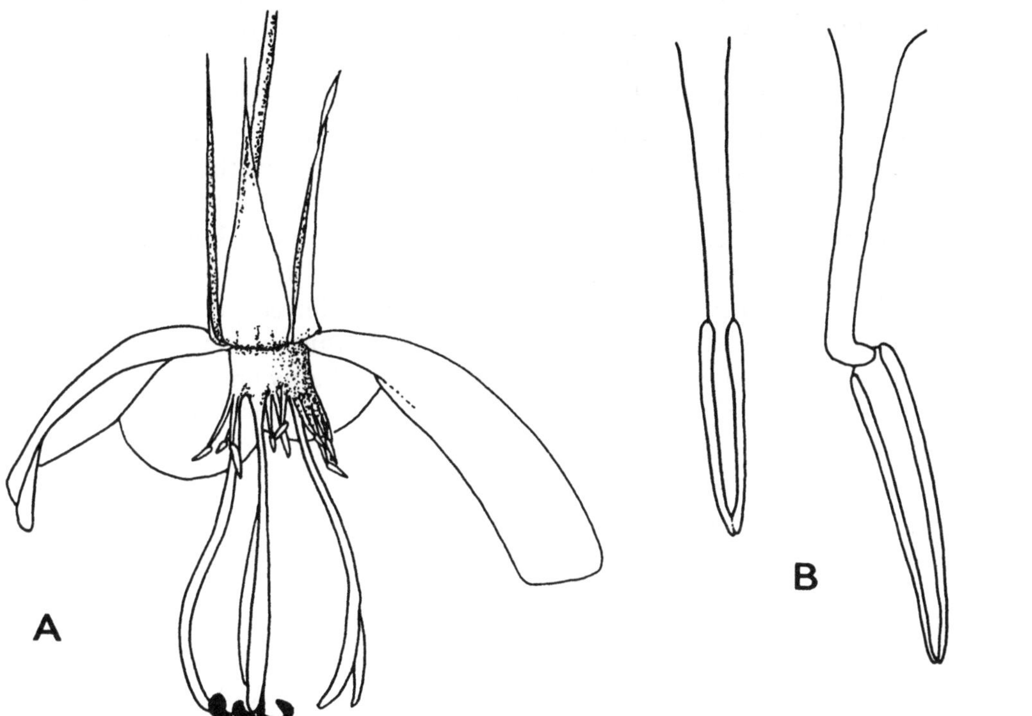

Fig. 8. *Dombeya burgessiae* (*Sterculiaceae*). *A* flower in natural posture with two petals removed. Pollen is presented by the anthers and by the greatly elongated staminodes; most of the style, the branches of which are shown in black, is hidden behind two staminodes. Staminodes are 12 mm long. *B* a simple and a geniculate stamen, viewed from the adaxial side; the thecae of the dehisced extrorse anthers are revolute. (P. F. YEO.)

are crimson, the latter fading to pale pink at their tips. The crimson colour also extends on to the base of the corolla and very slightly on to the sepals. The five staminodes lack all trace of anthers; they are narrowly spathulate and reach about $2\frac{1}{2}$ times as far from the base of the androecium as do the stamens and as far as the stigmas. They are crimson to the point where the stamen filaments cease to be so and above this are apparently white but in fact tinged faintly with pink. The styles are united for most of their length but split into five recurved stigmatic arms at the apex. The flower is fragrant and secretes five large well-exposed drops of nectar which rest at the junction of the corolla with the staminal tube.

The flower begins to open by throwing back its sepals; at this stage the corolla is inversely top-shaped and the style and stigmas emerge from the still closed apex. The petals then spread so that access to the nectar is possible. Although much pollen remains in the anthers, a very large amount is found now to have been deposited on the outer, convex, surface of the staminodes, and some is also present on the petals. That on the staminodes is loosely clumped and partly adherent to the fine hairs that clothe (rather sparsely) the abaxial surface. The petals are asymmetrically rhombic and pollen is deposited on a small area in one corner; in this region the margin, but not the surface, bears fine hairs similar to those on the staminodes, and some of the pollen clings to these. The petals are convolute in bud and the part that receives the pollen is the only part which is exposed to the anthers in the bud. In old flowers the hairs have disappeared from the staminodes, but a few papillae, from which a tiny point projects, can be seen, suggesting that the hairs are retracted as the flower ages.

The stamens are of two kinds, both having extrorse, dithecous, tetrasporangiate anthers. There are five, larger, inner ones which form part of the same whorl as the staminodes; these have the filament geniculate just under the anther, causing the anther to bend outwards slightly. The remaining ten stamens are slightly smaller, with perfectly straight filaments; they form an outer whorl in which they alternate with the inner stamens and staminodes. Pollen appears to be shed only when the bud is nearly full-sized, and to be transferred to the staminodes tangentially, a process facilitated by the strong recurvature of the thecal walls and the convexity of the staminode, the hairs of which probably enter the thecae. The inner anthers presumably load the distal part of the staminode and the outer the proximal part, though doubtless there is an overlap. The method of loading the staminodes is most certainly not as described above for *D. mastersii*; indeed, the staminodes bend inwards over the top of the ovary.

The above observations were made on a plant in the glasshouses of the University Botanic Garden, Cambridge, where it flowers in late summer and autumn. The wasps that visited the flowers cannot be taken as indicators of the likely natural pollinators, but their habit of alighting on the staminodes in a head-up position suggests how these might behave. The effect of secondary pollen presentation in *Dombeya burgessiae* is to bring some of the pollen near to the stigmas and greatly to extend the distribution of pollen on the long axis of the flower. Why this has not been achieved by modification of the stamens is to me unfathomable. Although a few hairs are present on the petals at the point where these bear pollen, I suspect that the pollen presented here is lost from the plant's life-cycle.

The second plant in which I found secondary pollen presentation is an unidentified specimen in the University Botanic Garden, Cambridge. It presents no significant differences in floral form or pollen presentation from *D. burgessiae* and may indeed be a hybrid of it. It differs mainly in habit, having larger leaves and slightly smaller flowers in much denser hemispherical clusters, and in flowering in mid-winter instead of autumn. My observations were made in January 1987.

Tribe *Byttnerieae*

For this tribe, SCHUMANN (1890–1893) made the general statements (1), that the anthers open in the bud, at which time the lower cap-shaped bases of the petals lie against the anthers and (2), when the bud opens the hoods of the petals bend outwards and carry the pollen, hidden in their depths, away from the empty anther cells.

Theobroma L.

T. cacao L. (SCHUMANN 1890–1893, illus.; YEO, unpublished 1990; YOUNG & al. 1984, illus.).

The report by YOUNG & al. indicates secondary pollen presentation but my observations contradict this.

The flowers are about 15 mm in diameter across the calyx which consists of five partly reflexed lanceolate white sepals strewn with tiny purple dots. The five lanceolate staminodes stand erect round the ovary and exceed the style in length; in fact their position ranges from connivent to somewhat divergent. They are largely purple and pilose but the subulate tips are white and glabrous. The petals are complex in shape and composed of two main parts. The basal part consists of a short patent claw, with two thick purplish red ridges on it, which is abruptly expanded and bent inwards to form an erect deeply pouched portion opening adaxially; the ridges extend into this, here being accompanied by a third median one. The distal part is composed of a narrow reflexed band that passes into a buff-coloured trullate lobe (YEO). The surface of the ridged interior of the basal pouch bears small conical trichomes and apparent stomatal nectaries suspected of supplying nutrients to the pollinators (YOUNG & al.). Pollen is deposited on this surface (YOUNG & al.) and is considered to be picked up from here by the pollinators, which alight first on the outside of the pouch and then enter it (YOUNG & al.). In order to pollinate the stigmas the insects must enter the flowers by a different route, namely between the staminodes and the gynaecium, perhaps attracted by glands on the ovary (YOUNG & al.).

My observations were made on a plant cultivated in the University Botanic Garden, Cambridge. SCHUMANN's above-quoted statements for the tribe are not true of this plant. In the most advanced buds available to me the anthers were undehisced (morning). It is true that by the development of the claw the petal pouch moves outwards from the androecium at a late stage of development but the stamen filaments lengthen and their distal parts bend outwards. Thus the apex of the stamen remains within the cavity of the petal, the anther being clasped by the spout-like apex of the pouch. As the anther is extrorse and has bent through nearly a right angle the pollen is exposed towards the base of the flower. However, it does not fall out of the anther because the flower is inverted! If, as stated by YOUNG & al., it is scattered inside the pouch this can only be as the result of insect visits. *T. cacao*, therefore, does not have secondary pollen presentation.

Herrania GOUDOT

H. purpurea (PITTIER) R. E. SCHULTES, *H. nitida* (POEPPIG) R. E. SCHULTES and *H. albiflora* GOUDOT. (YOUNG 1984, illus.). **STR c, (a?). ISS exposed. SEQU stg?**

APPL ad. RWD n. SYN myio, sapromyio. **VIS** the scuttle-flies (*Megaselia* sp. and *Dohrniphora* sp. (*Diptera-Phoridae*).
The first two species have dull purple flowers that give off a musty, aminoid, scent, while the third is white and scarcely scented. Flowers open at dawn and produce scent in the early morning and in the evening. On the day the flowers open the insides of the petal-pouches (see *Theobroma*, above) become coated with the creamy-white pollen. Only the two listed species of flies regularly visit the flowers and consistently behave in a manner conducive to pollination. Visits take place early in the morning and at dusk; the flies enter the petal-hoods and emerge well-covered with pollen on the head and dorsal thorax after an interval of 4–15 seconds. Then they fly to another flower in the same inflorescence or, more usually, leave the area. Pollen-laden flies were also seen moving through the central area of the flower and brushing against the style, after which they sometimes entered the petal-hoods. These flowers have long pendent petal-limbs, and approaching flies alighted either on these or on the staminodes. There are thus two routes of approach, as inferred for *Theobroma*. As with that genus the flies are thought to receive nourishment from stomatal nectaries. The fact that they carry pollen on a particular part of the body suggests that a definite orientation is imposed on them while they are in the pouch. YOUNG was not expressly concerned with secondary pollen presentation and did not make clear the relative importance of primary presentation, if any, and secondary. However, in order that the pollen is applied to a specific part of the body it should be presented from a single area within each petal-cavity and therefore perhaps all secondarily, as implied by SCHUMANN's statement for the tribe (above). Seed-setting behaviour suggests that there is self-incompatibility in this group, but verification of this is awaited.

Discussion of Sterculiaceae

As indicated above, it is not easy to suggest why there is a degree of secondary pollen presentation in *Dombeya*. Staminodes are commonly present in those members of the family that have bisexual flowers, and assuming they were ancestrally present, selection pressure to spread out the pollen within the flower could have led to a response on the part of the staminodes instead of the stamens. Spreading out the pollen along the floral axis might represent a crude version of the division of the pollen supply into a feeding allocation and a pollinating allocation, as found in some pollen flowers in *Leguminosae* subfamily *Caesalpinioideae* and *Melastomaceae* (VOGEL 1978). For a large, fragrant, nectariferous flower one cannot accept the notion of MASTERS that it is adapted for self-pollination. However, many plants adapted for outbreeding have provision for selfing if outcrossing fails, and it is just conceivable that secondary pollen presentation could be making provision for this in *Dombeya*. On the other hand, since the staminodes bring the pollen to the same level in the flower as the stigmas, their primary role could be in cross-pollination (the problem is, why is it not done by the stamens themselves?). In fact, I have never seen fruit development in *Dombeya* in the Cambridge Botanic Garden.

In the *Byttnerieae* we have the syndromes of myiophily and sapromyiophily. Here, the pollinators are typically not flower-adapted and their behaviour and movements have to be regulated by the flower, usually with an element of deceit.

The flower has developed the petal-pouch as a pollination chamber. It seems that the way this evolved led to the adoption of secondary pollen presentation in some cases.

There is a need for more studies of the floral biology of the *Sterculiaceae*. More attention should be given to the role of the staminodes and petals in this and, in general, to the transfers of function from one organ to another.

EAST (1940) stated that self-incompatibility was likely to be found in this family, incomplete indications of it having been observed in cultivated plants of the genus *Sterculia*. Here, we have noted indications of it also in *Dombeya* and *Herrania*. *Theobroma cacao* has long been known to have a self-incompatibility system of a distinctive type although this is not present in all stocks. The system is genetically complex and is unusual in that the male function fails only when the pollen tube has entered the embryo-sac (references in TARODA & GIBBS 1982, who report the same system in *Sterculia* – Tribe *Sterculieae*). Binucleate pollen is recorded from many genera (BREWBAKER 1967).

References for Sterculiaceae

BREWBAKER, J. L., 1967.
EAST, E. M., 1940.
HOOKER, J. D., 1868: *Dombeya mastersii*. – Bot. Mag. **93**: t. 5639.
MASTERS, M. T., 1867: *Dombeya angulata* CAV. – Gard. Chron. **1867**: 74.
SCHUMANN, K., 1890–1893: *Sterculiaceae*. – In ENGLER, A., (Hrsg.): Die Natürlichen Pflanzenfamilien **III.6**: 69–99.
TARODA, N., GIBBS, P. E., 1982: Floral biology and breeding system of *Sterculia chicha* ST. HIL. (*Sterculiaceae*). – New Phytol. **90**: 735–743.
VOGEL, S., 1978.
YOUNG, A., 1984: Mechanism of pollination by *Phoridae* (*Diptera*) in some *Herrania* species (*Sterculiaceae*) in Costa Rica. – Proc. Entomol. Soc. Wash. **86**: 505–518.
YOUNG, A. M., SCHALLER, M., STRAND, M., 1984: Floral nectaries and trichomes in relation to pollination in some species of *Theobroma* and *Herrania* (*Sterculiaceae*). – Amer. J. Bot. **71**: 466–480.

Loasaceae

Systematic position: *Dilleniidae, Violales*
Restriction of occurrence within the family: Subfamily *Mentzelioideae*

Floral features. Flowers actinomorphic, at least sometimes nectariferous. Sepals 5; petals 5, free, falsely sympetalous or sympetalous; stamens indefinite and not fascicled or (not the present subfamily) 2, 5 or many (then in 5 fascicles); filaments connate at the base into a short tube to which the petal-bases may be fused, or epipetalous; ovary inferior, syncarpous, with many ovules; style simple; fruit a capsule. (Plants are usually herbaceous and often bear stinging hairs).
Type of pollen presentation. Pollen is shed into the centre of the flower ('Pollenhaufen'), whence it is collected by bees.

Examples of Secondary Pollen Presentation in Loasaceae

The plants described here occur in the south-west United States and Mexico.
Eucnide ZUCC.

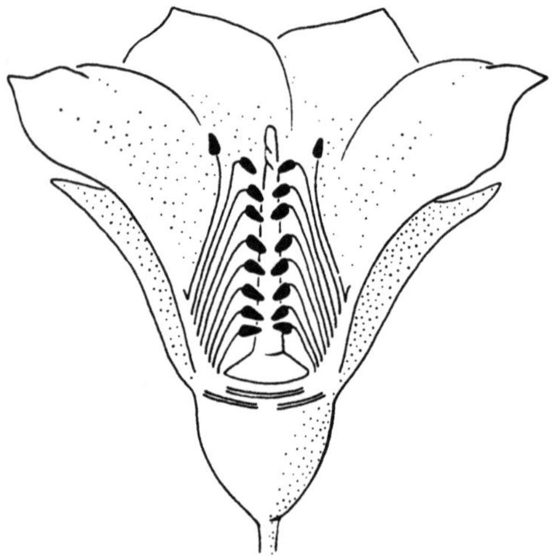

Fig. 9. *Eucnide urens* (*Loasaceae*). Pollen falls from the anthers to the centre of the flower. (THOMPSON & ERNST 1967.)

E. urens PARRY. (THOMPSON & ERNST 1967, illus.). **STR f**, top of ovary. **ISS exposed. SEQU sim. APPL general. RWD p, n? SYN** melitto. **VIS** a single species of bee *Hesperapis laticeps* (*Hymenoptera-Apoidea*: *Melittidae*). (Fig. 9).
The flowers are widely funnel-shaped and about 5 cm long, with cream-coloured petals. The numerous stamens are epipetalous; they are graded in length, those inserted higher up being the longer. The filaments are robust and inclined inwards from their peripheral points of insertion so as to reach the centre of the flower; the anthers (except the outermost) are nodding. The stamens thus form a cone rather more than half the length of the corolla below which is a chamber round the style. The stigma is level with the tips of the longest stamens. The flowers open in bright sunshine in the morning and remain open all day. As soon as they open the pollen is shed from the anthers into the central chamber, though some is caught between the filaments. The bees visit the flowers until early afternoon; they climb down the style, forcing their way between the stamens, and re-emerge backwards, and they touch the stigma on both entry and exit. The following information is not available: whether nectar is offered; if so, how it is separated from the pollen; when the stigma is receptive; what the bees do when in the chamber. Probably the stigma is receptive from the beginning of anthesis, nectar is absent, and the bees are females that actively gather pollen with their forelegs from the deposit; probably they also passively gather pollen held between the filaments and possibly that remaining in the anthers. All species of *Eucnide* are self-compatible but undisturbed flowers of *E. urens* do not set seed.
Mentzelia L.
M. tricuspis A. GRAY (ZAVORTINK 1972). **STR** top of ovary. **ISS exposed. SEQU sim? APPL a. RWD p. SYN** melitto. **VIS** a single species of bee *Megandrena mentzeliae* (*Hymenoptera-Apoidea: Andrenidae*).
The flowers are yellow and bowl-shaped with petals 1.5–2.5 cm long. The numerous stamens are connivent round the style and shed their pollen into the centre of the flower. The bees rake pollen from the spaces between the anthers with their forelegs. As with *Eucnide urens*, much desirable information is lacking.

Discussion of Loasaceae

Eucnide urens is the only species of its genus to be pollinated in the way described here, and the same is probably true of *Mentzelia tricuspis*. Secondary pollen presentation thus presumably arose independently in the two genera. Their flowers are similar in the way they deposit the pollen but different in the way the give access to it for the pollinators. They deposit pollen in the same way as flowers pollinated on the 'mess and soil' principle (FAEGRI & VAN DER PIJL 1966), but here the pollen is the floral reward (or forms part of it) and its collection by the pollinators is deliberate, not accidental. Secondary pollen presentation here could be part of a co-adaptation of each plant species with a particular species of bee and/or with a special method of pollen extraction. It may be doubted whether the monophily (FAEGRI & VAN DER PIJL 1966) of the plants is firmly established. In fact, the flower of *E. urens* shows and extraordinarily close resemblance in size and structure to that of *Ferocactus wislizenii* (ENGELM.) BRITTON & ROSE (Cactaceae; GRANT & GRANT 1979) which receives visits from *Megachilidae* and *Apidae* in central Arizona but has been reported as visited by *Halictidae* in southern New Mexico. Possibly the gathering of pollen by burrowing into a mass of stamens is widespread among the bees of the region.

Eucnide is shown above as self-compatible; binucleate pollen is reported in two genera not dealt with here (BREWBAKER 1967).

References for Loasaceae

BREWBAKER, J. L., 1967.

FAEGRI, K., VAN DER PIJL, L., 1966.

GRANT, V., GRANT, K. A., 1979: Pollination of *Echinocerus fasciculatus* and *Ferocactus wislizenii*. – Pl. Syst. Evol. **132**: 85–90.

THOMPSON, H. J., ERNST, W. R., 1967: Floral biology and systematics of *Eucnide* (*Loasaceae*). – J. Arnold Arbor. **48**: 56–88.

ZAVORTINK, T., 1972: A new subgenus and species *Megandrena* from Nevada, with notes on its foraging and mating behavior (*Hymenoptera: Andrenidae*). – Proc. Entomol. Soc. Wash. **74**: 61–71.

Epacridaceae

Systematic position: *Dilleniidae, Ericales*
Restriction of occurrence within the family: Known in only one genus

Floral features. Flowers actinomorphic, mainly pentamerous or tetramerous. Sepals free; petals united into a tube, often hairy within; stamens usually epipetalous, or attached to the receptacle; ovary superior, several- to many-seeded, often with a nectary at the base. Fruit a capsule.
Type of pollen presentation. Pollen is presented secondarily by hairs on the corolla-lobes ('Pseudo-Staubblatt', tending towards 'Pollenhaufen').

Examples of Secondary Pollen Presentation in Epacridaceae

Acrotriche R. BR.
A. serrulata R. BR. (MCCONCHIE & al. 1986, illus.). **STR c. ISS exposed. SEQU pln. APPL a. RWD n. SYN** thero. **VIS** – (Fig. 10).

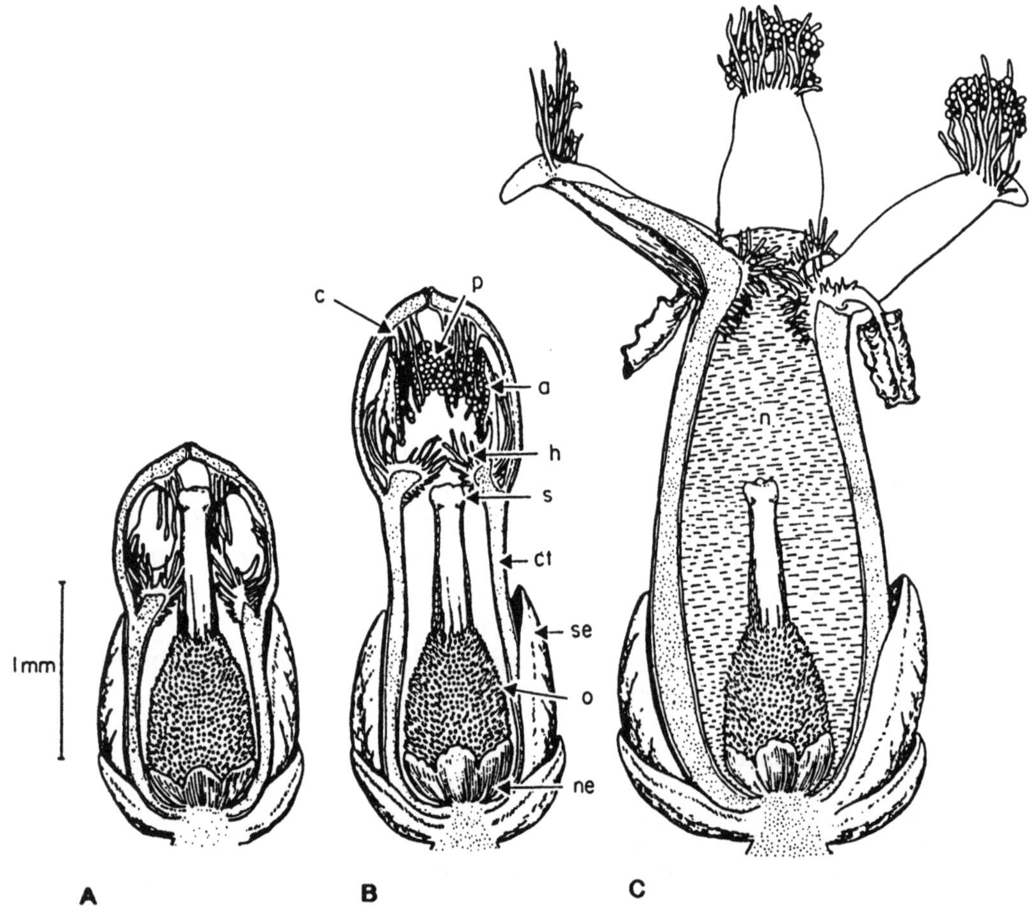

Fig. 10. *Acrotriche serrulata* (*Epacridaceae*). *A–C* sections through flower in two stages of bud and one of anthesis. *a* = anther, *c* = combs on corolla on which pollen is presented, *ct* = corolla tube, *h* = throat hairs, *n* = nectar, *ne* = nectary, *o* = ovary, *p* = pollen, *s* = stigma, *se* = sepal. (McConchie & al. 1986.)

The plant is a spreading shrub about 30 cm tall, growing in heathland in SE Australia. The small pentamerous flowers are borne in condensed spikes in the interior of the shrub; they open approximately simultaneously within the spike. The flowers are pale green, often with red stripes. The corolla tube is 4 mm long and urceolate, and the lobes are widely spreading, flat and papillose; each lobe bears subterminally a transverse row of erect hairs 0.6 mm long, forming a comb. The diameter across the flower from comb to comb is 4 mm. Additional hairs block the corolla-throat. The stamens are attached in the sinuses between the corolla-lobes. When the bud is 2.5 mm long the style reaches the tip of the bud, and its apex is covered with mucilage which contains pectic polymers; by this time it has virtually completed its growth. The hairs of the corolla-lobes are directed downwards between the style and the stamens. The hairs of the throat are mostly seated on coronal thickenings and the larger of them project upwards around the style and anthers. When the bud has reached a length of 4 mm the interior of the corolla has become more distinctly divided into two chambers by the coronal thickenings, and the tip of the style reaches only as far as these. The introrse anthers have opened and deposited their pollen (in tetrads), covered with a thick viscous coating, in the petaline hair-combs. The hairs of the throat help to press the anthers into contact with the combs. At the start of anthesis further growth of the corolla has occurred and the tube is now urceolate and the lobes spreading; the openings of all the flowers in the spike face upwards, a state of affairs achieved by differential growth of the corolla-tube as and where required. The corolla is filled with nectar to the throat or slightly beyond; this totally immerses the ovary. Despite this, the tip of the style, now only reaching half way up the corolla-tube, opens its five or six lobes slightly. Finally nectar-production ceases, the corolla drops off, apparently taking any unconsumed nectar with it, and the style is able to receive pollen. The plants are usually associated with anthills but ants have not been seen at the flowers and, despite the small size of the latter, the authors suggest that the flowers are pollinated by small marsupials. The features of the floral syndrome that support this suggestion are the inconspicuous green flowers held near the ground and concealed within the bush, a strong pungent floral odour and the copious nectar. The authors have unpublished evidence suggesting that plants are self-incompatible.

Discussion of Epacridaceae

None of the more prevalent functional explanations of secondary pollen presentation can apply here. In particular, the method and position of presenting the pollen is quite different from that of the stigmatic presentation. Assuming that the pollinators are indeed marsupials, one must suppose that flowers in the female stage receive pollen accidentally from the relatively large pollinators as they search for inflorescences that are in the male stage. Probably secondary pollen presentation in *Acrotriche serrulata* serves the function of keeping the pollen away from the nectar which accumulates in such large quantity that it more than fills the corolla-tube. This in turn results from the adoption of a very small flower-size in conjunction with a syndrome that involves offering large amounts of nectar. The above species is thought to be self-incompatible; binucleate pollen is reported from one genus of the family (Brewbaker 1967).

References for Epacridaceae

Brewbaker, J. L., 1967.

McConchie, C. A., Hough, T., Singh, M. B., Knox, R. B., 1986: Pollen presentation on petal combs in the geoflorous heath *Acrotriche serrulata* (*Epacridaceae*). – Ann. Bot. (Oxford) **57**: 155–164.

Leguminosae (alternative name Fabaceae)

Systematic position: *Rosidae, Fabales*
Restriction of occurrence within the family: Subfamily *Papilionoideae* (alternative name *Faboideae*) (synonym *Papilionaceae*, alternative name *Fabaceae* sensu stricto)

Distribution within subfamily. Papilionoideae have traditionally been divided into 10 or 11 tribes (Bentham 1865; Rendle 1925; Chant in Heywood 1978). The tribal

Table 2-1. Tribes of Leguminosae treated here (classification of POLHILL & RAVEN 1981)

Tribe 6,	Millettieae
Tribe 7,	Robinieae
Tribe 8,	Indigofereae
Tribe 9,	Desmodieae
Tribe 10,	Phaseoleae
Tribe 14,	Aeschynomeneae
Tribe 16,	Galegeae
Tribe 19,	Loteae
Tribe 20,	Coronilleae
Tribe 21,	Fabeae (Vicieae)
Tribe 23,	Trifolieae
Tribe 24,	Brongniartieae
Tribe 27,	Podalyrieae
Tribe 28,	Liparieae
Tribe 29,	Crotalarieae
Tribe 32,	Genisteae

classification adopted here is the one found in Advances in Legume Systematics (POLHILL & RAVEN 1981); with minor variations it is currently in general use (for example, MABBERLEY 1987). The number of tribes in the POLHILL & RAVEN version is 32, of which sixteen show secondary pollen presentation (Table 2-1).

Floral features. Zygomorphic; sepals (5) or (3)+(2), petals 5, stamens 10, either as (10) (=monadelphous) or (9)+1 (=diadelphous), ovary superior, monocarpellary, style therefore solitary. Nearly all members of the group have the characteristic 'papilionate' flower (Fig. 13): floral axis horizontal; petals clawed, one upper usually with spreading (i.e. erect), expanded limb ('vexillum', 'standard', 'banner', 'flag'), two laterals ('alae', 'wings') and two lower that are coherent along lower edge and usually part of upper edge to form a boat-like structure ('carina', 'keel'), to the basal part of which the wings are usually coherent; stamens lying in the keel, ascending at tips, free distally; style ascending at tip. Most monadelphous genera have no floral nectar and reward their pollinators with pollen; most diadelphous species produce nectar, to which access is given by an arching of the base of the free filament. This elsewhere abuts the edges of the united ones but its basal deformation creates a pair of 'nostrils'.

Type of pollen presentation. Most genera, including almost all those with secondary pollen presentation, are bee-pollinated. The insect visitor exerts itself so as to increase the separation between the standard and the wings, thereby depressing the latter, together with the keel, to which they are linked. This causes the tips of the stamens and style to move towards the apex of the keel and/or to emerge from it and touch the insect ventrally. The mechanism of primary pollen presentation in the *Papilionoideae*, in which the tips of the stamens and style emerge from the keel, is termed valvular (KNUTH 1908). There are five types of secondary pollen presentation: explosive discharge of dry pollen, explosive deposition, extrusion with style apex acting as a piston, extrusion with some part of the androecium acting as a piston and presentation on a stylar brush. In DELPINO's classification (DELPINO 1873–4) three terms were used: 'spring-mechanism' (for all explosive types, including those without secondary presentation), 'piston-mechanism' and 'trichostylous mechanism'. In WESTERKAMP's classification (see Chapter 1) the first two represent a form of the 'Pollenhaufen', the next two are 'Nudelspritzen' and the last is a 'Pseudo-Staubblatt'. The pollen-presenting brush should not be confused with a 'ciliate style' or a 'ciliate or penicillate stigma' (LAVIN & DELGADO 1990). In tribes 7. *Robinieae*, 10. *Phaseoleae* (in part) and 21. *Fabeae*, which have a pollen brush, there is a trend towards asymmetry of the flower and pleurotriby and nototriby.

Examples of Secondary Pollen Presentation in Subfamily Papilionoideae

Tribe 6, *Millettieae* (syn. *Tephrosieae*)

Secondary presentation of pollen by the style is briefly recorded for *Tephrosia virginiana* (L.) PERS. by KNUTH (1904, citing ROBERTSON); a stigmatic membrane that has to be abraded before it becomes receptive (compare *Lotus*, Tribe 19) occurs in the tribe (SHIVANNA & OWENS 1989: 174). LAVIN & DELGADO (1990) state that the pollen is presented on a brush and that this method is confined in *Tephrosia* to subgenus *Barbistyla*.

Tribe 7, *Robinieae*

The genera of this tribe that have a pollen-presenting brush are *Coursetia* DC., *Genistidium* I. M. JOHNST., *Olneya* A. GRAY, *Peteria* A. GRAY, *Robinia* L. and *Sphinctospermum* ROSE (LAVIN & DELGADO 1990). The flower is pleurotribic, the style (but not other parts) being asymmetric, and pollen is shed before anthesis into the keel and presented thence by the stylar brush (LAVIN & DELGADO 1990).

Robinia L.

R. pseudacacia L. (KNUTH 1908, illus., citing work of KIRCHNER). **STR c, s. ISS rel(dose)-poll. SEQU stg. APPL v. RWD n. SYN** melitto. **VIS** bumble-bees *Bombus* spp. and the honeybee *Apis mellifera* (*Hymenoptera-Apoidea*), outside the natural range of these species).

The style stands at right angles to the ovary and bears an asymmetric brush of hairs just below the apex on which pollen is presented to the visitor. There is a ring of bristles surrounding the capitate stigma (the peristigmatic fringe, see 'Discussion of Leguminosae'). Both pollen and stigma are presented during insect visits to the young flower but the stigma contacts the insect's body first. The stigma is said to remain receptive long after all pollen has been removed from the flower. For the participation of the corolla in pollen presentation see note on the tribe.

Coursetia DC.

The species differ in the direction of asymmetry of the style, brushing the insect on either the left-hand or the right-hand side (LAVIN & DELGADO 1990).

Tribe 8, *Indigofereae*

A stigmatic membrane occurs in members of this tribe (SHIVANNA & OWENS 1989: 174).

Indigofera L.

I. filiformis Thunb. (KNUTH 1904, based on SCOTT ELLIOT 1891). **STR c, f? ISS expl depos. SEQU stg. APPL v. RWD p. SYN** melitto. **VIS** –.

Little information is available about the flowers of this plant beyond the fact that it carries out explosive deposition of pollen that is triggered when an insect separates the thickened edges of the keel. Probably the method of pollen retention before the explosion is as in *Desmodium* (next tribe), in which case pollen presentation is secondary. However, in at least some species of the genus presentation is primary (CHR. WESTERKAMP, pers. comm.).

Tribe 9, *Desmodieae*

Desmodium DESV.

D. canadense DC. and *D. sessilifolium* TORR. (KNUTH 1904, citing BESSEY, FOERSTE & ROBERTSON). **STR c, f. ISS expl depos. SEQU stg. APPL a. RWD p. SYN** melitto. **VIS** large bees (*Hymenoptera-Apoidea*) (for *D. canadense*).

In these N American species the wings are firmly united to the keel and the base of the latter has projections connecting it with the standard. The keel and the column formed by the sexual organs develop contrary tensions. Pollen is shed into the tip of the keel before the flower opens and the tips of the filaments become dilated and curved outwards, forming a small basket within the keel keeping the pollen together. When a bee presses its head against the base of the standard and depresses the wings, the wings and keel are freed from the standard and spring downwards; at the same time the sexual column springs upwards and strikes the underside of the insect. The flower is nectarless. The flower of *D. canadense* is large and strong and can only be tripped by large bees; at first it is light purple with two white, purple-edged, marks at the base of the standard, but after tripping it becomes blue.

Tribe 10, *Phaseoleae*

The pollen-presenting stylar brush has evolved repeatedly in this tribe (see 'Discussion of Leguminosae') and secondary pollen presentation is widespread. The presence of the pollen-presenting brush is mentioned by LAVIN & DELGADO (1990) for *Adenodolichos* HARMS in subtribe *Cajaninae*, for *Clitoria* L. in subtribe *Clitoriinae* and for 17 genera of subtribe *Phaseolinae*. In addition it is mentioned for *Canavalia* DC. and *Rhynchosia* LOUR. by SCOTT ELLIOT (1891) and for the resupinate flowers of *Centrosema virginianum* (L.) BENTH. by KNUTH (1904, citing LINDMAN). A stigmatic membrane occurs (SHIVANNA & OWENS 1989: 174).

Phaseolus L.

P. coccineus L. (*P. multiflorus* LAM.). (MÜLLER, 1883, illus., citing also other authors). **STR c, s. ISS rel(dose)-poll. SEQU sim? APPL l. RWD n, p? SYN** melitto. **VIS** large bees (*Hymenoptera-Apoidea*).

The style presents pollen on a brush of hairs similar to that of *Lathyrus* (tribe 21). The tip of the keel is drawn out and coiled like a spring at one side of the body of the keel, so that the flower is asymmetric. The style is likewise coiled, and during insect visits to the flower it emerges with its load of pollen and touches the base of the insect's proboscis on one side.

Vigna SAVI

V. caracalla (L.) VERDC. (*Phaseolus caracalla* L.). (KNUTH 1904; FAEGRI & VAN DER PIJL 1979, illus.). **STR c, s. ISS rel(dose)-poll. SEQU sim? APPL d. RWD n, SYN** melitto. **VIS** large bees (*Hymenoptera-Apoidea*).

Pollen is presented on a stylar brush. Like *Phaseolus coccineus* this plant also has a coiled keel-tip and style, but in this bizarre, strongly asymmetric flower it has four complete turns. The style emerges from the keel on the left side of the insect and its tip then moves through an arc under the body and finally touches the insect dorsally. The fragrant flowers are white, flushed with rose, and have a yellow, black-striped guide-mark on the standard. Another species with a bizarre flower, *V. candida* (VELL.) MARÉCHAL & al. (*Phaseolus appendiculatus* BENTH.), has a keel with a single gyre and a style that contacts the insect on the right-hand side (KNUTH 1904, citing LINDMAN).

V. vexillata (L.) A. RICH. (HEDSTRÖM & THULIN 1986, illus.). **STR c, s. ISS rel(dose)-poll. SEQU stg. APPL l. RWD n, p? SYN** melitto. **VIS** the large bee *Xylocopa* (*Hymenoptera-Apoidea*).

The asymmetric flowers are purplish with a yellow guide-mark. The rather narrow keel-tip reacts to probing by an insect by moving across its back, hugging it, picking up available pollen on the near-apical stigma and then brushing more pollen onto it from the massive unilateral stylar pollen brush. A photograph shows the style-tip between the head and thorax.

Tribe 14, *Aeschynomeneae*

A ring of hairs around the stigma occurs in some but not all members of this tribe (SHIVANNA & OWENS 1989: 174).

Stylosanthes SWARTZ

The flowers in this genus are remarkable for having a long narrow hypanthium, the significance of which in relation to pollination is obscure.

S. gracilis HBK., *S. guianensis* (AUBLET) SWARTZ (PEREIRA-NORONHA, SILBERBAUER-GOTTSBERGER & GOTTSBERGER 1982, illus.). **STR c, (a), s. ISS rel(dose)-poll. SEQU sim. APPLIC a,v. RWD n, p. SYN** melitto. **VIS** mainly bees (*Hymenoptera-Apoidea*: nine genera are named, belonging to families *Andrenidae*, *Anthophoridae*, *Apidae*, *Halictidae* and *Megachilidae*). (Fig. 11).

Apart from the hypanthium (see above) the rest of this small flower (only 1 cm long overall) is typical of the subfamily in appearance. The stamens are monadelphous but the flower is one of the rare ones of this type that secrete nectar; this is produced in small quantity inside the base of the staminal tube. The staminal tube is relatively wide and is presumably penetrated by the proboscis of visiting insects. The flowers last about 4 hours, opening in the morning. About an hour before the flower opens the anthers of the inner whorl, which are small and rounded, are positioned below those of the outer whorl, which are large and elliptic and have dehisced. Soon after the flower has opened the inner anthers are found to be above the outer, their filaments having elongated rapidly, and they also then dehisce. During an insect visit the depression of the wings and keel causes the anthers of the inner whorl, in co-operation with the style, to expel a small portion of pollen. After the visit the organs return to their original position, and emission of further pollen is possible during subsequent visits. The bee visitors may forage for nectar or pollen exclusively or take both. Nectar-gathering is uncommon, evidently because the nectar supply is poor. Bees that take both foods probe for nectar and receive pollen on their undersides, combing it into their scopae after leaving the flower. Those foraging only for pollen gather it actively. Because of this statement I have indicated that pollen is applied to the vector anteriorly as well as ventrally. It is not clear whether bees engaged in this activity can cause

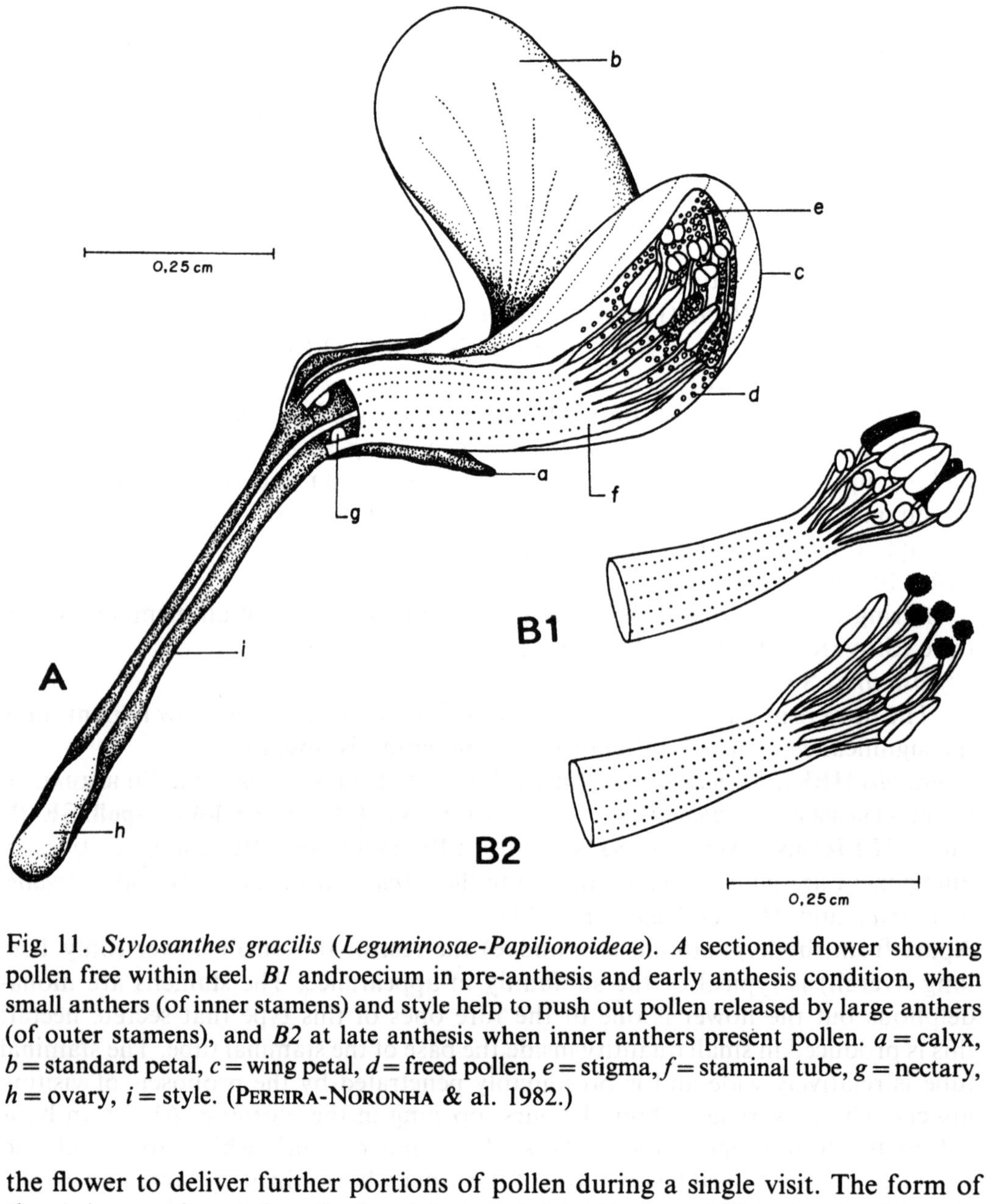

Fig. 11. *Stylosanthes gracilis* (*Leguminosae-Papilionoideae*). *A* sectioned flower showing pollen free within keel. *B1* androecium in pre-anthesis and early anthesis condition, when small anthers (of inner stamens) and style help to push out pollen released by large anthers (of outer stamens), and *B2* at late anthesis when inner anthers present pollen. *a* = calyx, *b* = standard petal, *c* = wing petal, *d* = freed pollen, *e* = stigma, *f* = staminal tube, *g* = nectary, *h* = ovary, *i* = style. (PEREIRA-NORONHA & al. 1982.)

the flower to deliver further portions of pollen during a single visit. The form of the style-apex is not described and is not clearly illustrated; the details of its role in pollen delivery are not supplied. Plants were found to be self-compatible.

S. biflora (L.) BRITT., STERNS & POGG. (KNUTH 1904, citing FOERSTE & ROBINSON). **STR c, (a), s. ISS rel(dose)-poll. SEQU sim. APPLIC ad, v. RWD n, p? SYN** melitto. VIS small bees, including *Calliopsis andreniformis* (*Hymenoptera-Apoidea*).

In this species the flowers are inverted; one observer found that because of this pollen was placed on the head of the bee, whereas the others found that the bees (different species) inverted themselves and so received pollen ventrally. It is mentioned that the large anthers are basifixed and the small dorsifixed. Pollen is issued in a ribbon by a pump mechanism, doubtless similar to that of the preceding species.

Tribe 16, *Galegeae*

A pollen-presenting brush occurs in seven genera in addition to those described below (LAVIN & DELGADO 1990).

Colutea L.

C. arborescens L. (KNUTH 1908, citing work of KIRCHNER & LOEW). **STR c, s. ISS rel(dose)-poll. SEQU pln. APPL v. RWD n. SYN** melitto. **VIS** bumblebees *Bombus* spp. and the honeybee *Apis* (but the latter cannot operate the stiff mechanism and takes nectar from the side of the flower without causing pollination) (*Hymenoptera-Apoidea*).

Pollen is presented on a stylar brush but the style is coiled in such a way that the pollen-laden part touches the visitor before the stigma, which is very small. The anthers themselves emerge during pollinator visits and they present part of the pollen directly. There is a ring of hairs round the stigma.

Clianthus LINDL.

C. puniceus (G. DON) LINDL. (KNUTH 1904, illus., citing G. M. THOMSON; MCCANN 1952). **STR c, s. ISS rel(dose)-poll. SEQU sim? APPLIC ad. RWD n. SYN ornitho. VIS** the bird *Prosthemadera novaeseelandiae* ('tui', *Meliphagidae*) and possibly birds of the family *Trichoglossidae*.

The floral parts are elongated and, while the keel hangs down, the standard is strongly reflexed, so that the total length of the flower is 8–10 cm. The plant is a shrub of coastal New Zealand and the flowers are scarlet apart from a shining blue-black area on the bulging part of the standard where it is reflexed. The wings are much shorter than the beak-like keel. Before the flower is fully open part of the style protrudes from the keel; it is slender and has a punctiform stigma. The exposed part is pubescent, but proximally, where it is within the keel and level with the anthers, it has longer hairs along its upper edge into which the flower's own pollen is deposited in the bud stage. The upper edges of the keel petals gape slightly at the base, giving access to the nectar, but otherwise are firmly overlapped. The stamens are diadelphous, the free one being arched at the base in the usual way, while the remainder form an expanded basal bowl holding a copious nectar supply. THOMSON thought that pollen was explosively scattered from the stylar brush but MCCANN did not confirm this. (Dr CHR. WESTERKAMP has found that a pollen cloud thrown from a brush does occur in *Sutherlandia* R. BR. of this tribe.) MCCANN found that the pollen is viscid and that after deposition it gradually moves towards the extremity of the style, apparently as a result of disturbance by visitors or wind causing movement of the style to and fro within the keel. The slightest pressure on the keel forces the style, together with some pollen, further out of the keel. This process can be repeated. Visiting birds rotate the flower and insert the bill into the base of the keel, the tip of which then rubs their heads (occipital region).

The punctiform stigma and the use of a pollen-holding brush positioned well back from the tip of the style are points in common with the entomophilous *Colutea* (above).

Tribe 19, *Loteae*

Lotus L.

Here also the androecium is diadelphous and again it presents two basal pores giving access to the nectar, as described on p. 34.

Lotus corniculatus L. (MÜLLER 1883, illus.). **STR c, f. ISS rel(dose)-poll. SEQU sim or pln. APPL v. RWD n, p. SYN** melitto. **VIS** mainly long-tongued bees (*Hymenoptera-Apoidea*), also *Lepidoptera* which are ineffective as pollinators (Fig. 12).

All the pollen (from the 10 stamens) is released into the tip of the keel well before anthesis and the anthers shrivel. During the considerable further growth of the

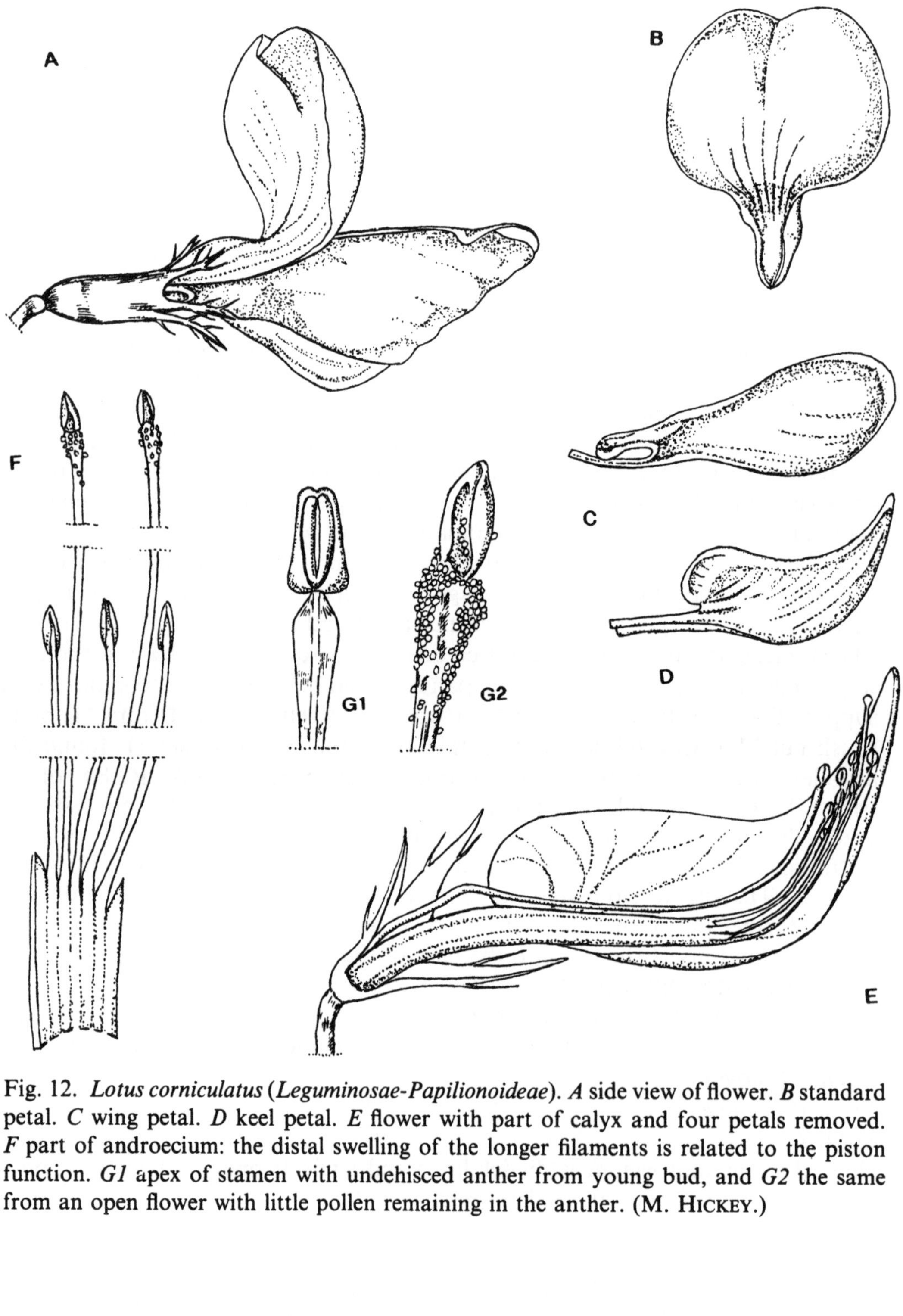

Fig. 12. *Lotus corniculatus* (*Leguminosae-Papilionoideae*). *A* side view of flower. *B* standard petal. *C* wing petal. *D* keel petal. *E* flower with part of calyx and four petals removed. *F* part of androecium: the distal swelling of the longer filaments is related to the piston function. *G1* apex of stamen with undehisced anther from young bud, and *G2* the same from an open flower with little pollen remaining in the anther. (M. HICKEY.)

flower the filaments of the five outer stamens keep pace with its growth and become clavately thickened, while those of the inner cease growth. During an insect visit to the mature flower a dose of pollen is extruded in a ribbon, propelled by the outer filaments acting as a piston. If downward pressure on the keel is sufficient the stigma emerges from the keel. DELPINO is quoted as saying that the stigma does not become receptive until its papillae have been slightly rubbed, by which the stigmatic surface is made sticky. This observation is confirmed by BUBAR (1958) who described a pliable membrane covering the stigma, jagged pieces of which could be seen after it was ruptured. He quotes an unpublished observation by W. F. GILES that breakage of the membrane releases a lipidic (fatty) material to which pollen sticks and in which it germinates.

Dorycnium Mill.

D. hirsutum (L.) SER. (*Lotus hirsutus* L.). (KUGLER 1977). **STR c, f, (a). ISS rel(dose)-poll. SEQU sim or pln. APPL av. RWD n, p. SYN** melitto. **VIS** large bees (*Hymenoptera-Apoidea*).

Pollen is expelled by the thickened tips of five of the filaments as in *Lotus*. However, because the keel is less beaked than in *Lotus* the tips of the stamens emerge from it during visits by insects to older flowers. KUGLER states that at this stage pollen is presented directly to the visitor by the stamens ('valvular presentation'). However, any pollen presented in this way will have been released by the anthers long before, and will be borne on the thickened filaments as well as perhaps resting within the shrivelled anther walls.

Anthyllis L.

A. vulneraria L. (MÜLLER, 1883, illus.). **STR c, f. ISS rel(dose)-poll. SEQU pln. APPL v. RWD n, p. SYN** melitto. **VIS** mainly long-tongued bees (*Hymenoptera-Apoidea*: *Apidae* and *Megachilidae*).

The shape and mechanical details of the flower are considerably different from those of *Lotus* in that all 10 filaments are clavately thickened and the stigma does not emerge until most of the pollen has been removed. As in *Lotus*, pollen does not adhere to the stigma until the latter has been rubbed, whereupon it becomes moist.

Tribe 20, *Coronilleae*

Hippocrepis L.

In *Hippocrepis* the androecium is as in *Lotus* (p. 39) and nectar is secreted.

H. comosa L. (MÜLLER, 1883, illus.). **STR c, f. ISS rel(dose)-poll. SEQU stg. APPL v. RWD n, p. SYN** melitto. **VIS** various bees (*Hymenoptera-Apoidea*).

The system of pollen delivery is the same as that of *Lotus corniculatus*. MÜLLER's illustration shows that there is some degree of thickening of the smaller filaments.

Coronilla L.

In this genus pollen delivery is similar to that in *Lotus corniculatus* but species differ, some having all 10 filaments thickened, others having them thickened alternately (HUTCHINSON 1964). Accounts consulted are incomplete.

C. emerus L. (treated taxonomically as *Hippocrepis emerus* (L.) LASSEN by LASSEN 1989). MÜLLER (1883) states that DELPINO found the piston apparatus to be the same as that in *Lotus corniculatus*. The visitors are large bees (*Hymenoptera-Apoidea*). One of the characters in which this species agrees with *Hippocrepis* rather than *Coronilla* and *Securigera* (q. v.) is the base of the androecium (LASSEN 1989).

However, FAEGRI & VAN DER PIJL (1971, 1979) depict the flower with no pores at the base of the staminal tube, and they state that there is no nectar; possibly their material was misidentified.

C. vaginalis LAM. (KNUTH 1908, citing illustration from H. MÜLLER). **STR c, f. ISS rel(dose)-poll. SEQU pln? APPL v. RWD –. SYN** melitto. **VIS** –.

KNUTH notes that insect visits to this plant are very rare.

C. valentina L. (YEO, unpublished 1983). Apparently six filaments are strongly dilated apically and the remaining four less so; the androecium is diadelphous but the free stamen is not arched at the base; the base of the staminal tube fits tightly round the base of the ovary, leaving no room for nectar, which appears to be absent. This is a character of *Coronilla* as redefined by LASSEN (1989).

C. varia L. (treated taxonomically as *Securigera varia* (L.) LASSEN by LASSEN 1989). (MÜLLER 1883, partly based on report by T. H. FARRER). **STR c, f. ISS rel(dose)-poll. SEQU stg. APPL v. RWD n?, p? SYN** melitto. **VIS** bees (*Hymenoptera-Apoidea*).

FARRER observed that nectar secreted on the outside of the calyx is taken by bees. The proboscis of the visiting bee passes under the standard and then between the unusually long and narrow claws of the petals to the outside of the calyx. This behaviour was also observed in three other species of *Coronilla*. It seems unlikely that this is an essential part of the pollination system and it is not mentioned by FAEGRI & VAN DER PIJL (1979). The base of the staminal tube is as as described here for *C. valentina*.

Tribe 21, *Fabeae* (*Vicieae*)

A pollen-presenting brush occurs in all four genera of this tribe.

Vicia L.

V. cracca L., *V. ervilia* (L.) WILLD., *V. pisiformis* L. and *V. sepium* L. (MÜLLER 1883, illus.; KNUTH 1908, partly based on MÜLLER and KIRCHNER; KUPICHA 1981). **STR c, s. ISS rel(dose?)-poll. SEQU sim? APPL v. RWD n, p. SYN** melitto. **VIS** bees, mostly long-tongued, especially *Bombus* (*Hymenoptera-Apoidea*).

The pollen is shed before the flower opens and is lodged in a brush of hairs on the style immediately below the stigma. This happens in *V. cracca* when the flower is only half its final size, according to MÜLLER. The hairs encircle the style but are better developed on the abaxial side. The brush emerges from the keel during insect visits; it seems unlikely that all pollen is taken on a single visit but observers have not reported on this point. MÜLLER stated that in *V. cracca* and *V. sepium* the stigma is not receptive to pollen until it has been rubbed. KUPICHA showed that the bearded part of the style is laterally compressed in *V. cracca* (it is dorsiventrally compressed in *V. sylvatica* L).

V. faba L. (STODDARD & BOND 1987, illus.). **STR c, s. ISS rel(dose)-poll. SEQU pln. APPL v. RWD n, p. SYN** melitto. **VIS** social bees *Bombus* spp. and *Apis mellifera* (*Hymenoptera-Apoidea: Apidae*) and solitary bees (*Hymenoptera-Apoidea*: *Anthophoridae* and *Andrenidae*).

When the flower opens the pollen is contained in a pouch formed by the keel petals and situated just above the stigma which itself is borne on an erect style that makes a right-angled bend at its junction with the ovary. It can be inferred from the illustration that at the time the pollen was deposited the style was much shorter relative to the stamens than in the mature flower. Bees depress the wing/keel complex, while the style is not displaced, so that its tip becomes exposed. The stigmatic surface and its brush sweep some of the pollen out of the pouch and into contact with the bee. The stigma is then rubbed by the bee and the stigmatic membrane is broken. Pollen already present on the bee may then get on to the stigma. Since it is stated that only some of the pollen is presented, it may be assumed that the process is repeatable. Unvisited flowers are said to self-pollinate. Some cultivars differ in that the stigma is fully receptive by the time the flower opens and self-pollination is the rule (LORD & HESLOP-HARRISON 1984).

Lens MILL.

L. culinaris MEDIKUS (*L. esculenta* MOENCH) (KNUTH 1908, citing work of KIRCHNER). Functionally this is said to be like *Vicia ervilia* but the pollen-presenting brush is on the adaxial side of the style.

Lathyrus L.

L. pratensis L. (MÜLLER 1883, illus.). **STR c, f, s. ISS rel(dose)-poll. SEQU pln?. APPL v. RWD n, p. SYN** melitto. **VIS** large long-tongued bees (*Hymenoptera-Apoidea*).

Pollen is deposited before the flower opens into an elongated brush of hairs on the adaxial side of the style (a difference from *Vicia*). The part of the style which bears the brush is swollen; it emerges from the keel during insect visits. (From its shape and position it seems likely that removal of pollen will result mainly from active scraping by the insect with its legs). Some of the pollen lies in the pouched tip of the keel and is delivered in a different way: when the keel is strongly depressed the filaments scour out the pouch and deliver it themselves.

L. japonicus WILLD. subsp. *maritimus* (L.) P. W. BALL (*L. maritimus* BIGELOW). (KNUTH 1908). It is stated that the mechanism agrees essentially with that of the preceding species.

L. grandiflorus SIBTH. & SMITH (TEPPNER 1988, illus.). **STR c, s. ISS rel(dose)-poll. SEQU stg. APPL l, d and legs. RWD n, p. SYN** melitto. **VIS** the large bee *Xylocopa violacea* (*Hymenoptera-Apoidea*).

The large, mainly purplish, flowers of this sub-Mediterranean species are so strongly constructed that only *Xylocopa* (in this case observed in Austria) can reach the nectar. The basal parts of the petals are asymmetric (full details are given by TEPPNER), so that, as the bee forces its proboscis under the standard the wing-and-keel complex is pressed downwards and to the left, while the bee is tilted to the right. The sickle-shaped style, which lies in its own groove formed by folds in the keel (cf. *Podalyria calyptrata*, below), emerges on the right side of the bee between the middle and hind legs and passes upwards around the bee's waist and over to the left side. Thus the style grips the waist like an arm in a similar manner to that of *Vigna vexillata* (see above); it is so twisted that the pollen-laden brush is directed towards the insect's body. The pollen from the stylar brush is deposited mainly on the fore-dorsal edge of the abdomen and to a lesser extent on the hind-dorsal edge of the thorax, but it may also 'overflow' onto the dorsal surfaces. In its passage, the stigma, which has a stigmatic membrane, picks up pollen from a previously visited flower while at the same time some pollen is stripped from the stylar brush on to the right side of the bee and the right hind-leg and middle femur. During the bee's withdrawal it sometimes manages to use the fore-legs to scrape some pollen off the stylar brush before it is concealed again in the keel.

Pollen from the fore-legs is passed to the crop, whereas that on the other legs and sides of the body is transferred to the scopae on the hind-legs for use also by the bee. It was not established to what extent, if any, the bee could gather up pollen deposited on its waist region but pollen can accumulate there to such an extent that as the stigma advances around the bee crumbs of pollen fall away.

L. sylvestris L. (MÜLLER 1883, citing also DELPINO and F. DARWIN) and *L. latifolius* L. (CHR. WESTERKAMP, personal communication) both show asymmetry of the flower so that the style emerges on the right-hand side of the insect. In *L. latifolius* the style is twisted so that the brush is turned inwards towards the visitor's body. Furthermore the style arches over the body between the thorax and abdomen (photograph by WESTERKAMP 1989), much as in *L. grandiflorus*. Both species are much visited by *Megachile* (*Hymenoptera-Apoidea*: *Megachilidae*) (YEO, unpublished).

Pisum L.

P. sativum L. (MÜLLER 1883, illus.; KUPICHA 1981). **STR c, f, s. ISS rel(dose)-poll. SEQU pln? APPL v. RWD n?, p? SYN** melitto. **VIS** a few species of large and strong bees (*Hymenoptera-Apoidea*). (Fig. 13).

The style bears a pollen brush similar to that of *Lathyrus*, but both the tip of the keel and the style are reflexed. As the keel returns to the neutral position after a visit the swollen filaments push pollen towards its tip, thereby recharging the brush. KUPICHA showed that the bearded part of the style is dorsiventrally compressed and reduplicate.

Tribe 23, *Trifolieae*

A stigmatic membrane is sometimes present, sometimes absent, in this tribe (SHIVANNA & OWENS 1989: 174).

Ononis L.

O. spinosa L. (MÜLLER 1883, illus.). **STR c, f. ISS rel(dose)-poll. SEQU pln. APPL v. RWD p. SYN** melitto. **VIS** bees (*Hymenoptera-Apoidea*), especially those with ventral pollen-carrying brushes (*Megachilidae*).

The coherent upper borders of the keel petals leave a small opening at the tip. During insect visits to young flowers doses of pollen emerge through this opening, pushed out by the thickened apices of the filaments. Later the opening extends proximally and the tip of the androecium and the style apex emerge during visits.

The androecium is monadelphous and all filaments are thickened at the tip. However, the outer stamens have larger thickenings of the filaments and smaller anthers than the inner. Male bees were often observed to visit several flowers in succession, apparently seeking nectar.

Medicago L.

M. sativa L. (KNUTH, citing several authors, illus.; ARMSTRONG & WHITE 1935, illus.). **STR c?, (a), s. ISS expl depos. SEQU sim. APPL v. RWD n, p?. SYN** melitto. **VIS** bees (*Hymenoptera-Apoidea*), *Lepidoptera*.

The flowers of this species are well known to require tripping; pollination is by bees that enter the flower centrally; honeybees and *Lepidoptera* can reach the nectar by probing from the side and these do not cause pollination. It is the description and illustrations of ARMSTRONG & WHITE that show that there is an element of secondary pollen presentation in this species. The anthers dehisce in the bud stage and although they shrink, they appear still to contain the pollen. The stigma is mushroom-shaped and it seems inescapable that it plays a part in

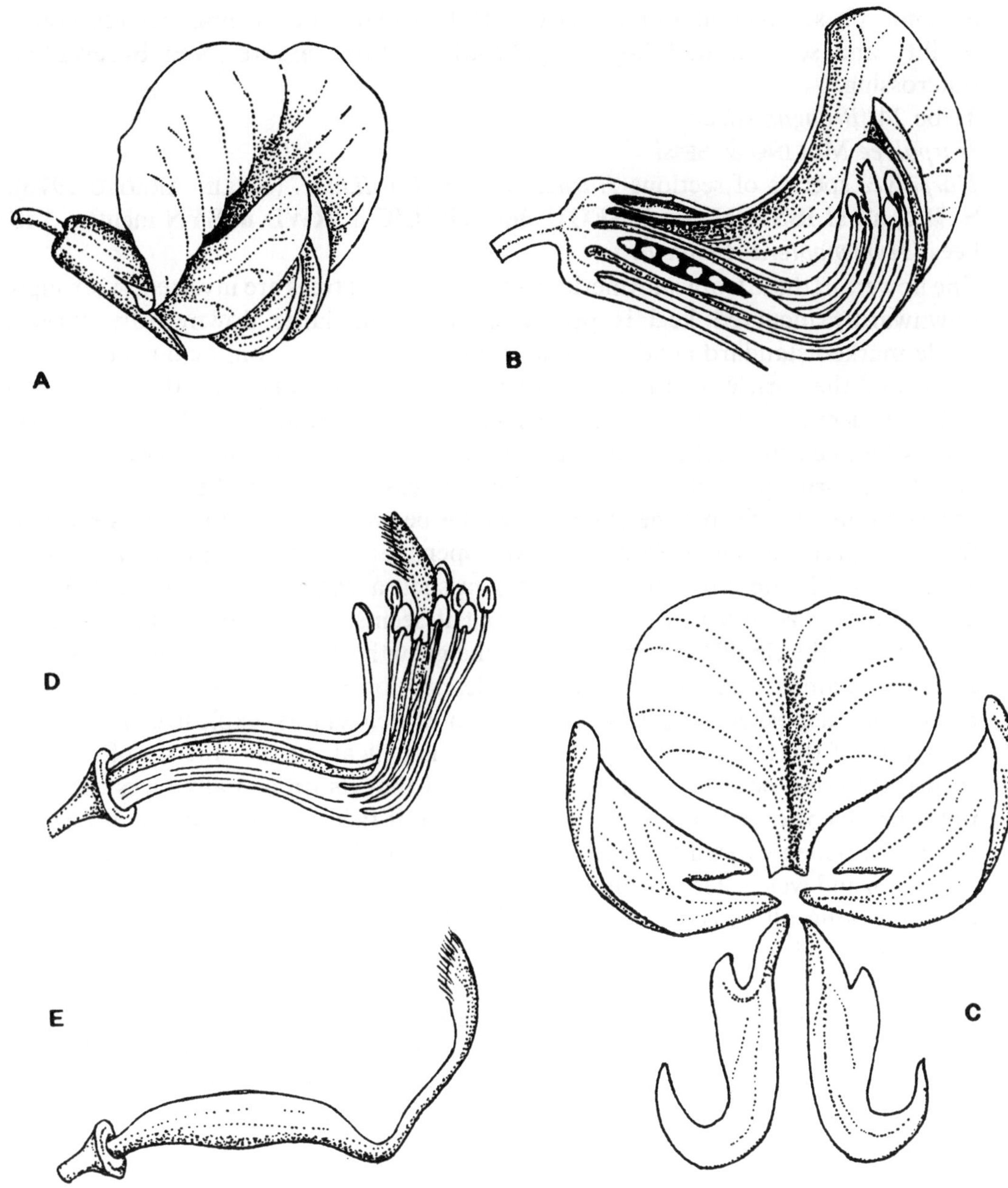

Fig. 13. *Pisum sativum* (*Leguminosae-Papilionoideae*). *A* oblique front view of flower. *B* sagittal section of flower. *C* separated petals. *D* androecium and gynaecium. *E* gynaecium. 'C' and 'D' show stylar pollen-presenting brush. (M. HICKEY, after WARMING 1895.)

expelling the pollen when the flower is tripped since until this moment it is situated at the base of the pollen mass. ARMSTRONG & WHITE also say that some pollen is forced into the tip of the keel before tripping; because the tips of the gynaecium and androecium are bent sharply upwards it is not evident that they could clear this pollen from the keel on tripping. The possibility remains that the edges of the keel-tip could be weak enough to bend back on meeting the pollinator's body and so place this pollen on to the insect. There is apparently no multipleallelomorph self-incompatibility system in this plant, but selfing reduces pollen fertility and seed-set in following generations; these are restored by eventual outcrossing.

Tribe 24, *Brongniartieae*

Harpalyce Mociño & Sessé

Harpalyce species of sections *Harpalyce* and *Brasilianae*. (Kalin Arroyo 1976). **STR c, a?, s? ISS expl cloud. SEQU sim. APPLIC d. RWD n. SYN** melitto. **VIS** bees (*Hymenoptera-Apoidea*).

The flowers in these tropical American shrubs or small trees are inverted. A strongly downwardly incurved keel is positioned over the large downwardly curved, guide-marked standard petal. Tension is built up between – on the one hand – the wings and the auricles of the keel, which press upwards, and – on the other hand – the gynaecium and androecium, which press downwards; additional tension occurs between the keel itself and the staminal tube. During insect visits the keel petals are torn apart, the staminal tube is released from the keel and pollen is sprayed onto the visitor; the style strikes the bee dorsally. The pollen is released from the large stamens before the flower opens, and that from the small anthers afterwards. The description is presented in such a way as not to answer clearly all the questions posed in this review, but it seems to be implied that pollen is pressed into the tip of the keel partly by the growth of the style, though probably the small stamens play a part too. The description of the pollen being sprayed, rather than deposited, suggests classification of delivery as 'explosive cloud'. The curvature of the keel attains up to 180° and is helical, so that the tip of the keel is turned to one side. The style is so far coiled that its tip actually faces upwards until the explosion has taken place, after which it faces downwards 'in a position to receive pollen carried by the bee' dorsally. If it did not hit the bee during the explosion and was instead pollinated only on later visits this would be an exception unique among all explosive flowers. Whether this is the case is not made clear, but I have provisionally assumed that pollen and stigma presentation are effectively simultaneous.

A few species of section *Brasilienses* are nectarless pollen flowers that do not explode; instead the bee tears open the tip of the keel to extract the pollen. The third section of this genus, *Cubenses*, consists mainly of species that exhibit the syndrome of ornithophily and have primary pollen presentation.

Tribe 27, *Podalyrieae*

Podalyria Willd.

P. calyptrata Willd. (Scott Elliot 1891, illus.). **STR c, a. ISS rel(dose)-poll. SEQU pln. APPL v. RWD n? SYN** melitto. **VIS** –.

The stamens lie along the outer border of the keel in a sort of tubular cavity (cf. *Lathyrus grandiflorus*, above). The younger anthers push up the pollen of both whorls. A gentle depression of the keel causes a thin ribbon of pollen to emerge but a violent depression brings out the style and anthers. In *P. cuneifolia* Vent. the edges of the keel are folded one within the other to form a more perfect tubular cavity for the style and pollen.

Tribe 28, *Liparieae*

The stamens are unequal and there is secondary pollen presentation by a piston mechanism as in the closely related following tribe (Polhill 1976: 165; Kalin Arroyo 1981: 732–736).

Liparia L.
L. spherica L. (SCOTT ELLIOT 1891, illus.). **STR c, a. ISS rel(dose)-poll. SEQU pln. APPL v. RWD –. SYN** melitto. **VIS** –.
The flower is highly modified. On depression the keel gets squeezed by the wings; this at first allows pollen to emerge in a thin thread and finally the stigma to appear. Pollinators were not seen, but *Xylocopa* is suggested as a likely one.
Tribe 29, *Crotalarieae*

In this tribe of 16 genera the anthers are almost always dimorphic (4 or 5 being longer and basifixed, 5 or 6 shorter and dorsifixed) and there is a ring of hairs on the style around the stigma. Secondary presentation of pollen deposited in the beak-like tip of the keel is usual, probably by the mechanism described below, involving both the smaller anthers and the style (POLHILL 1976). According to SHIVANNA & OWENS (1989: 174) a stylar brush occurs but in the examples described below pollen expulsion is mainly by the stamens.
Crotalaria L.

Although this genus is mainly bee-pollinated, adaptation to bird-pollination is found in some African species, either with little modification or occasionally with reduction of the wings and extra strengthening in the sepals (as in *C. rosenii* (PAX) POLHILL [POLHILL 1976: 166 and fig. 1(3)]).
C. micans LINK (YANAGIZAWA & GOTTSBERGER 1982–3, illus., as *C. anagyroides* HBK.). **STR c, a, s. ISS rel(dose)-poll. SEQU pln, sim? APPL v. RWD n, p. SYN** melitto. **VIS** the bees *Bombus*, *Centris*, *Megachile*, *Xylocopa* (*Hymenoptera-Apoidea*).
The five outer stamens have large anthers which dehisce some hours before anthesis. They form a cylinder containing the pollen. During insect visits the tip of the style, with its ring of small hairs, emerges from the keel, bringing a small proportion of the available pollen. At the end of the first day and the beginning of the second, the large anthers wither and the filaments of the remaining, much smaller, anthers elongate, bringing them to the tip of the keel where they dehisce. On the second day their pollen is issued in the same way as that of the large anthers. Although it appears that on the first day the large anthers form the cylinder in which the style acts as a piston I have assumed that the petals of the keel play a part in retaining the pollen, as described by POLHILL for the tribe. It is not made clear whether the stigma is receptive throughout anthesis or only some part of it, but it appears that pollen has slight priority of presentation at least during the first insect-visit.

This South American species (as well as at least some in Africa–POLHILL 1968: 173) is peculiar in supplying nectar as a reward while being monadelphous; the staminal tube has an opening on the upper side at the base giving access to the nectar. (Whether this entails a departure from symmetry I do not know.)
C. capensis JACQ. (SCOTT ELLIOT 1891, illus.). **STR c, a. ISS rel(dose)-poll. SEQU pln. APPL v. RWD n. SYN** melitto. **VIS** –.
The stamen whorls are differentiated and those with the shorter anthers push the pollen forward. On depression of the keel a thin streak of pollen exudes from its drawn-out tip; subsequently the stigma appears.
Aspalathus L.
A. chenopoda L. (SCOTT ELLIOT 1891, illus.). Emergence of a stream of pollen and later of the stigma occurs, as in *Crotalaria capensis*.

A. cordata (L.) DAHLGR. (*Borbonia cordata* L.). (SCOTT ELLIOT 1891, illus.). The five outer anthers push up the pollen and the odd stamen grows with them. On depression of the keel the stigma emerges first followed by the pollen in a conical mass round the style.

Bolusia BENTH.

Pollen is presented on a stylar brush; the flower recalls *Phaseolus* and *Vigna* (Tribe 10. *Phaseoleae*) in having a much-coiled keel and style (LAVIN & DELGADO 1990).

Tribe 32, *Genisteae*

Lupinus L.

In this genus the androecium is monadelphous and the flowers are nectarless. A ring of hairs is present below the stigma (POLHILL 1981: 166).

L. luteus L. (MÜLLER 1883, illus.). **STR c, a. ISS rel(dose?)-poll. SEQU? APPL v. RWD p. SYN** melitto. **VIS** the social bees *Apis* and *Bombus*, and the solitary bee *Megachile* (*Hymenoptera-Apoidea*), probably observed outside the natural range of the plant.

There are five large outer (antesepalous) anthers which deposit pollen in the tip of the keel before anthesis and then become shrivelled and retracted. The five inner stamens have small anthers which produce a little pollen; the filaments of four of them elongate later than those of the outer stamens and their anthers act as a piston. It is clear that the pollen of the five outer anthers (the greater part of the pollen which is produced) is secondarily presented; apparently the five smaller anthers do not emerge from the keel, in which case their pollen too may be regarded as secondarily presented. It is in any case enclosed by the keel petals.

L. nanus DOUGLAS. (DUNN 1956, illus.; JUNCOSA & WEBSTER 1989, illus.). **STR c, f, a. ISS rel(dose)-poll. SEQU pln. APPL v. RWD p. SYN** melitto. **VIS** bees (*Hymenoptera-Apoidea*).

A number of subspecies have been recognized within *L. nanus;* information quoted here from DUNN applies to some larger-flowered subspecies, among which is subsp. *latifolius* (BENTH.) DUNN, on which JUNCOSA & WEBSTER worked. The flowers are blue with at first a white guide-mark on the standard; the white area later changes to purple (DUNN said this occurred 20–24 hours after pollination, whereas JUNCOSA & WEBSTER found it to occur on the fourth day of flowering and independently of pollination). The piston arrangement for pollen delivery is the same as that of *L. luteus*. In the last 3 or 4 days before they open the flowers lengthen by 1 mm per day to a final length of 10–13 mm. The anthers of the outer (antesepalous) stamens are four times as long as those of the inner. The small anthers are dorsally medifixed but not versatile. During the 24 hours before anthesis begins the anthers dehisce introrsely and the inner stamens elongate rapidly, forcing the pollen into the tip of the keel. JUNCOSA & WEBSTER attribute the piston function of the open flowers to the 'atypical insertion of the anthers', the inflation of the distal portion of the filaments and the fact that they bear epidermal papillae. The reference to filaments here is to the inner ones, because the outer have by now shrunk back away from the pollen mass. The species is self-compatible but the stigma is not receptive to pollen on the first day; however, good seed set occurs with manipulation (imitation of insect visits) 30–96 hours after flower-opening. First-day and second-day pollen germinates readily on stigmas in their second and third days. DUNN noted that the ring of hairs round the stigma was less developed in inbreeding species of Lupinus that he studied, supporting the suggestion that it has a role in preventing self-pollination, while JUNCOSA & WEBSTER found that it became weaker in older flowers. This species is reported to show high levels of outbreeding but unvisited flowers in a greenhouse did often set seed. Thus the hair fringe functions mainly early in the life of the flower, limiting access to pollen before the flower opens (see later) and also during insect visits in the first period of anthesis. Therefore its importance is probably in reducing interference by self-pollen with the arrival of foreign pollen on the stigma.

The stigmatic surface is covered with unicellular papillae that are slightly inflated distally. The stigmatic fluid appears in the spaces between the narrower proximal parts of the papillae on the second day of anthesis; it appears to emerge from weak areas on the sides of the papillae where they are in contact. The cuticle of the papillae never separates from the underlying cell wall, as it does in species where the stigma becomes receptive when rubbed by the visitor. The exudate becomes copious on day 3. Thus the release of stigmatic fluid is not dependent on abrasion by insect visitors although its appearance on the stigmatic surface may be assisted by the disturbance of the papillae during insect visits (see also SHIVANNA & OWENS 1989).

Closely similar observations on two further species (*L. arizonicus* (WATS.) WATS. and *L. sparsiflorus* BENTH.) were made by WAINWRIGHT (1978). In this case, however, the stages of flower development were timed from the completion of meiosis. Like DUNN and JUNCOSA & WEBSTER, WAINWRIGHT thought that the ring of hairs below the stigma was significant in preventing self-pollination. Although DUNN's drawing shows the style well clear of the pollen mass in the keel, a photograph taken by WAINWRIGHT 20 hours before the flower opens shows the style-tip level with the tips of the large anthers, thus supporting JUNCOSA & WEBSTER in their view that the hair ring is important before anthesis. On the other hand WAINWRIGHT found that the pollen is apparently not capable of germinating until about the time the flower opens and that it never germinates if it reaches the stigma before the time the latter is receptive, despite the existence of considerable *in vitro* pollen longevity. This again is consistent with the idea that the hair-fringe is important because it reduces interference with the arrival of foreign pollen on the stigma.

In a self-incompatible species, *L. texensis* HOOK., the customary change of colour of the white area on the standard petal took place after five days, and was not influenced by pollination, fertilization or the manipulation of the flowers by insects; instead it corresponds with the loss of viability of the pollen (SCHAAL & LEVERICH 1980). As with other species there is an interval between flower opening and stigma receptivity; in this case pollination begins to be effective only on the fourth day and is most effective on the fifth. These results show that the onset of stigma receptivity is developmental, as shown by the direct observations of JUNCOSA & WEBSTER on *L. nanus*, and not related to abrasion by pollinators.

The only peculiarity of *L. confertus* KELL., from California, is that its racemes remain in flower for two to three months. It has strongly scented violet flowers, that later turn red (KNUTH 1904, citing MERRITT).

Genista L.

G. tinctoria L. (MÜLLER 1883, illus.; FAEGRI & VAN DER PIJL 1979). **STR c, s. ISS expl depos. SEQU stg. APPL v. RWD p. SYN** melitto. **VIS** varied, especially bees, including *Apis*, *Bombus*, *Megachile*, *Anthidium*, *Halictus*, *Lasioglossum* and *Andrena* (*Hymenoptera-Apoidea*).

At anthesis the style and staminal tube are in tension, opposed by the keel. The pollen has already been deposited in the narrow keel, on top of the style. When the flower is tripped the keel opens explosively and the tip of the style strikes the underside of the visitor, while the stem of the style forces the mass of pollen against the same place. The narrowness of the keel ensures that the style is effective in pushing the pollen upwards. During development four of the stamens of the outer whorl are much ahead of the remainingstamen of this whorl and those of the inner whorl. The growth of the latter helps to pack the pollen into the tip of the keel.

Ulex L.

U. europaeus L. (MÜLLER 1883, quoting OGLE). The system is the same as in *Genista tinctoria*.

Cytisus L.

C. scoparius (L.) LINK. (MÜLLER 1883, illus.; FAEGRI & VAN DER PIJL 1979; POLHILL 1976: 167–168). **STR c, a, s. ISS expl depos** and **expl cloud. SEQU stg. APPL general. RWD p. SYN** melitto. **VIS** the honeybee *Apis* and bumble-bees *Bombus* spp. (*Hymenoptera-Apoidea*).

During the explosive opening of this flower, triggered by the visit of an insect, the upper edge of the keel opens progressively from base to apex. There are five short stamens with basifixed anthers that are released from the keel when it has split half-way, striking the insect ventrally. Their anthers have dehisced in the bud stage and therefore the keel-petals have played a part in retaining the pollen until a visit takes place. Of the five remaining stamens, which have dorsifixed anthers, four are very long; their pollen is massed against the clavate tip of the style which helps to fling it out when it emerges explosively from the tip of the keel. Thus for this pollen both petals and style play a part in its presentation. The fifth stamen becomes folded downwards (CHR. WESTERKAMP, pers. comm.). The style finishes up in a coil of slightly more than one gyre. Dr WESTERKAMP also tells me that the tension which leads to the explosion is set up only as the bee forces its way into the flower; thus the bee's energy is used, rather than the plant's. Further, weak bees may release only the lower five stamens and thieve the pollen. *Apis* cannot trip the flower and the observed insects were presumably gleaning.

Lembotropis GRISEB.

This genus is retained in *Cytisus* by POLHILL (1976).

L. nigricans (L.) GRISEB. (*Cytisus nigricans* L.). (KNUTH 1908, citing work of H. MÜLLER) **STR c, f, a. ISS rel(dose)-poll. SEQU pln. APPL v. RWD p. SYN** melitto. **VIS** the bees *Megachile*, *Andrena* and *Bombus* (*Hymenoptera-Apoidea*).

The five outer stamens have large anthers which dehisce in bud and deposit their pollen in the keel. The filaments of the five inner stamens, which have very small anthers, elongate later, pushing the pollen of the large anthers further into the tip of the keel. During depression of the keel by an insect-visitor the small anthers act as pistons, delivering a dose of pollen; however, it is said to be the robust filaments of the outer stamens which act as piston-rods, pressing against the anthers of the inner stamens. In older flowers the keel gapes more easily and the stamens and stigma protrude.

Spartium L.

The flower is functionally similar to that of *Genista* but the details are sufficiently different to suggest a separate origin within the *Cytisus* alliance (POLHILL 1976).

Discussion of Leguminosae

The pollination system of the *Papilionoideae* typically involves a close juxtaposition of the anthers and stigma through their confinement within the keel. This arrangement is presumably for the protection of the pollen, either against its use as food by animals not likely to cause pollination (YEO 1972) or against its appropriation as brood-food by pollinating bees (see conclusion of this discussion). The concealment of the stamens and their exposure in response only to the visitor's exertion is typical of flowers adapted to bees. Evidently the papilionate flower (in this and other families) is primarily adapted to these insects and this adaptation is preserved in the vast majority of the Leguminosae-Papilionoideae despite some radiation in other directions (LEPPIK 1966; KALIN ARROYO 1981), for example, into bird pollination. Primary and secondary pollen presentation occur in association with the syndromes both of melittophily and ornithophily (ornithophilous secondary pollen presentation is described here for *Clianthus*, tribe 16. *Galegeae*, and *Crotalaria*, tribe 29. *Crotalarieae* but LAVIN & DELGADO (1990) state that it also occurs in some species of *Phaseolus*, tribe 10. *Phaseoleae*).

The floral structure of the typical *Papilionoideae* demands that there be arrangements to prevent self-pollination unless autogamy is to prevail, and clearly most plants with highly organized flowers are not prevailingly autogamous. Growth of pollen on the stigma in this subfamily may be restricted by the unavailability of stigmatic fluid or by physiological self-incompatibility. The stigmatic fluid can be withheld in two ways; one is simple dichogamy, in which the stigma releases stigmatic fluid some time after pollen is shed (and usually some time after the flower opens) and the other is containment of stigmatic fluid within a stigmatic membrane which is easily ruptured during the first visit of a pollinator to the flower.

The presence of such a stigmatic membrane was noted in a number of genera by MÜLLER (1883); the structure of the stigma and the nature of its secretion were first described by JOST (1907) and have been further investigated recently with modern techniques (see SHIVANNA & OWENS 1989). The membrane actually represents the cuticle of the stigmatic epidermis which has become detached from the cells that formed it. This type of membrane is reported by SHIVANNA & OWENS (1989: 174) to occur in seven tribes, all but one of which appear in my account of secondary pollen presentation, and to be absent in seven (according to the classification used here), of which four have secondary pollen presentation. The membrane is highly impermeable and encloses the partly lipidic stigmatic secretion which accumulates in the intercellular spaces that develop in the stigmatic tissue. The constitution of this fluid is typical of that of 'wet' stigmas which in turn are associated with binucleate pollen grains and self-incompatibility systems (if present) of the gametophytic type (HESLOP-HARRISON & SHIVANNA 1977; Y. HESLOP-HARRISON 1984).

In some cases the membrane breaks down spontaneously at or before the beginning of anthesis and the plants are autogamous. It is reasonable to assume that this is a secondary development. Examples are some species of *Lotus* (BUBAR 1958), some cultivars of *Vicia faba* (LORD & HESLOP-HARRISON 1984) and *Lupinus albus* (JOST 1907 – but note that in the species of *Lupinus* described above there is no membrane).

As is clear from our description of *Papilionoideae* with secondary pollen presentation, and from MÜLLER's illustrations of *Genista tinctoria*, *Cytisus scoparius*, *Lotus corniculatus*, *Anthyllis vulneraria* and *Hippocrepis comosa*, (MÜLLER 1883), the stigma is in many cases embedded before anthesis in the precociously deposited pollen mass. Interestingly, although it seems clear that in *Trifolium pratense* the anthers present the pollen, they too dehisce before flower-opening (J. & Y. HESLOP-HARRISON, 1983). It might be surmised for all such plants, regardless of whether or not they have secondary pollen presentation, that if the stigmatic fluid were not at this stage contained, it would get absorbed into the pollen mass by capillary action. This would remove most of the fluid from its proper field of action and modify the mutual adhesiveness of the pollen. Alternatively, if the pollen is unwettable by the stigmatic fluid, it might move about on the surface film.

The role of the stigmatic membrane, with a function that ceases at the moment of the first insect visit, is thus independent of whether pollen-presentation is primary or secondary or whether a physiological incompatibility system is present. It is a concomitant of the adopted system of bee-pollination and protection of the pollen against robbery.

What happens to the flower's own pollen at the time of the first insect visit to a flower with a stigmatic membrane is not entirely clear. Many *Papilionoideae* are adapted to repeated visits (as described in some of our examples). In *Lotus* the stigmatic secretion lasts until 20–24 hours after the application of 'adequate' quantities of pollen to the stigma (BUBAR 1958), suggesting that there would continue to be a danger that self-pollen would make contact with the stigmatic fluid. However, MÜLLER (1883) thought it likely that pollen-gathering insects would succeed in emptying the flower of *Lotus corniculatus* in the course of only a few visits and these might easily take place in a very short period of time. Thus the problem of separating self-pollen from the stigmatic secretion after the first insect visit may be less serious than seems likely at first sight. If this situation were verified it would imply that a flower was available to receive pollen over a very much longer period than that during which it was actively donating it. This period may be even more extended in outbreeding species of *Lotus*, where a supply of fresh fertilizable ovules within the flower is sustained over a period of 8–10 days (BUBAR 1958) (one would expect stigmatic secretion to be present throughout this time, despite the statement from BUBAR quoted above). [Incidentally, this observation is particularly interesting in view of the significance currently being attached to the relative durations of male and female phases of the flower as expressions of the different selective forces operating on the male and female functions (LLOYD & YATES 1982, DEVLIN & STEPHENSON 1984, HARDER & THOMSON 1989)]. The peristigmatic hair-fringe is discussed below.

In the *Papilionoideae* multi-allelic self-incompatibility systems are apparently widespread; they are gametophytic and monofactorial (KALIN ARROYO 1981: 724, SHIVANNA & OWENS 1989). They have been investigated mainly in a few economically important species, where the breeding system may have been influenced by artificial selection and may vary as between cultivars. However, a multi-allelic self-incompatibility system has been fully demonstrated in *Trifolium pratense* (references in J. & Y. HESLOP-HARRISON 1982); there is also one in *Phaseolus coccineus* (J. & Y. HESLOP-HARRISON 1984) and apparently one in some species of *Lotus* (BUBAR 1958). In addition to the above-mentioned examples it is known in the following genera which have secondary pollen presentation: *Indigofera, Desmodium, Clianthus, Coronilla, Lathyrus, Medicago, Crotalaria* and *Ulex* (KALIN ARROYO 1981, Table 1: 725–6). It also appears to be present in *Lupinus* species. The systems are not so strong in their operation as to completely rule out selfing (J. & Y. HESLOP-HARRISON 1984: 94). Indeed, LORD & HESLOP-HARRISON (1984) doubt the soundness of earlier claims (e. g. ROWLANDS 1958) that *Vicia faba* has a partly effective self-incompatibility, but this probably reflects variation within this important crop species. [Other possible factors in the breeding system of this species are zygotic lethality (ROWLANDS 1961) and a feed-back mechanism based on a positive correlation among individuals between parental inbreeding and the persistence of the stigmatic membrane (HOLDEN & BOND 1960; STODDARD & BOND 1987)]. The utility of the stigmatic membrane could be related to the prevalence of multi-allelic self-incompatibility systems in the group. Such a system enables the flower to combine the roles of pollen donor and pollen receptor simultaneously without risking self-fertilization, and where this is possible there could be disadvantages in not doing so (perhaps through greater expenditure of resources and longer exposure to any hazards that beset the flower during anthesis). In addition, the possibility should not be overlooked that there is some other advantage from the point of view of the breeding system in the presentation of both pollen and stigma to the pollinator on the occasion of the flower's first receiving a visit (see Chapters 3, 7., final paragraph).

BUBAR points out that if foreign pollen is in fact merely prepotent (as in some species of *Lotus*), it will preferentially effect fertilization if delivered to the flower on the occasion of the first insect visit, since self-pollen will have been prevented by the stigmatic membrane from germinating until this moment. If the supply of foreign pollen is inadequate, the deficit will be made up by self-pollen. The combination of a stigmatic membrane and prepotency of foreign pollen thus forms a nicely integrated system.

The alternative to the stigmatic membrane is dichogamy, which occurs in *Lupinus* (tribe 32. *Genisteae*), a genus showing secondary pollen presentation. Here the stigmatic fluid, instead of being held by a membrane, is secreted by the stigmatic papillae, but not until one or more days after the flower has opened. This prevents contamination of the flower's own pollen with stigmatic fluid from the time of anther-dehiscence until there has been an opportunity for visitors to remove the pollen. This type of stigma is a fairly widespread alternative to the membraned type and occurs in at least one other tribe dealt with here, 29. *Crotalarieae*. Whereas the secretion of membraned stigmas is copious, that of the membrane-less type is very sparse, but the pollen grains have the ability to hydrate and germinate without contact with it (SHIVANNA & OWENS 1989). It is not clear what is the significance of the difference between the two types of stigma. A peristigmatic hair-fringe occurs in both self-incompatible and self-compatible species of *Lupinus*. The fact that it is less well developed in inbreeding than outbreeding self-compatible species is evidence that its role is to keep self-pollen away from the stigma and to keep the stigma clear for the arrival of foreign pollen. Such a hair-fringe also occurs in tribes 6. *Millettieae*, 7. *Robinieae*, 8. *Indigofereae*, 10. *Phaseoleae*, 16. *Galegeae*, 20. *Coronilleae*, 23. *Trifolieae* and 29. *Crotalarieae*. In 32. *Genisteae* it occurs in a genus which does not have secondary pollen presentation [*Laburnum* (KNUTH 1908, under *Cytisus laburnum*)]. A second function of the hair-fringe is suggested by SHIVANNA & OWENS (1989: 163), namely as a barrier against the escape of a copious stigmatic secretion from the stigmatic surface. This would not apply to *Lupinus*.

We now have to consider possible roles for secondary pollen presentation. One that it certainly takes is that of forming part of the system of explosive deposition (*Genista tinctoria*, *Ulex europaeus*) or explosive release of a cloud of pollen (*Cytisus scoparius*, *Harpalyce* spp.). *Genista tinctoria* has achieved a rather high degree of specialization to leaf-cutter bees (*Megachile*) from which each flower requires only a single visit, and which it rewards with pollen only. *Cytisus scoparius* may be regarded as having a more refined arrangement, scattering pollen so that some is deposited where it cannot be removed by grooming, and sending the tip of the style scouring the insect's back and probably one or other of the grooves between the divisions of the body. Explosive systems give complete protection against loss of pollen to non-pollinating pollen-gathering insects (until after the explosion, when gleaners may arrive). Where nectar is not offered the bait has to be pollen and the reward must be adequate, in which case there is apparently selection in favour of arrangements which ensure that at least some pollen will remain on a part of the insect's body which can be reached by the stigma of the next flower visited, as in *Cytisus*. An alternative possibility, suggested to me by Dr CHR. WESTERKAMP, is that some of the nectarless flowers are deceit flowers and offer no reward, in which case any removal of pollen by bees has to be regarded as theft. Species for which this seems likely are some in *Coronilla* (in the sense adopted here, see p. 41).

A second method of placing pollen where the insect finds it difficult to collect it by grooming is to combine possession of a stylar brush with asymmetry of the flower, so that the style rubs the insect on the side of the proboscis-base or the side or back of the body, as in tribe 7. *Robinieae*, in *Phaseolus* and *Vigna* in tribe 10. *Phaseoleae*, and *Lathyrus* in tribe 21. *Fabeae*. In these examples nectar is present.

Regarding the three other floral functions in which secondary pollen presentation has a putative role, namely the exact positioning of pollen and stigma, the protection of pollen, and the issuance of pollen in doses, the first two do not depend on secondary pollen presentation. They are achieved in the other *Papilionoideae* as well; we could, however, postulate that the job is done better by a system of secondary pollen presentation, and this could be tested by exact studies of floral movements during real or simulated pollinator-visits. The occurrence of pollen-dosing, on the other hand, is probably very much a matter of chance in *Papilionoideae* unless they have secondary pollen presentation. Pollen-dosing is described for *Stylosanthes*, *Crotalaria*, *Lembotropis*, *Ononis*, *Clianthus*, *Lotus*, *Lathyrus* and *Pisum*, and implied for *Anthyllis*, *Hippocrepis* and

Coronilla; it could well occur in all remaining non-explosive secondary pollen presentation systems in *Papilionoideae*. That strong pressure exists for evolution of systems carrying out these functions is clear from the existence of three different systems (piston-arrangement, stylar brush and explosive) which presumably evolved independently, and from the evidence, to be mentioned shortly, that the piston arrangement and the brush system have each arisen more than once.

It is, however, part of the purpose of this review to consider whether the occurrence of secondary pollen presentation gives ground for reconsidering existing classifications. In an introductory review to Advances in Legume Systematics (POLHILL & RAVEN 1981), which I have followed for the classification of the subfamily, it is made clear that floral developments have occurred in parallel in various evolutionary lines within the family (POLHILL, RAVEN & STIRTON 1981) and this is also emphasized by KALIN ARROYO (1981) in the same publication. These developments include the papilionate flower itself as well as specialized features of the latter such as those involved in secondary pollen presentation. A framework of informally designated groups above the rank of tribe is outlined within the subfamily *Papilionoideae* by POLHILL, RAVEN & STIRTON (1981: 7, diagrams p. 5 and 199). Firstly there are the *Sophoreae*, a diverse basal group (with free stamens). Then there is a major dichotomy into the 'epulvinate alliance' of temperate herbaceous tribes (such as *Loteae*, *Fabeae* and *Trifolieae*), centred on tribe 16. *Galegeae*, and the 'pulvinate alliance', centred on tribe 6. *Millettieae* (*Tephrosieae*) and consisting mainly of tropical and woody groups, one of which is tribe 7. *Robinieae*, another the herbaceous tribe 10. *Phaseoleae*. Finally there is a separate and smaller 'genistoid alliance' including the tribes 29. *Genisteae* and 32. *Crotalarieae*. Despite their proximity to the *Sophoreae* in this evolutionary scheme these are placed at the end of the taxonomic sequence.

Thankfully, this classification is generously annotated. Looking at it in detail we find that at least the morphological features associated with secondary pollen presentation have been taken into account. Consequently, our survey of the phenomenon is unlikely to demand amendment of the classification. The only hint of an anomaly comes with the reported similarity of the pollen presentation mechanisms in *Lupinus* (*Genisteae*) and *Crotalaria* (*Crotalarieae*). These two tribes are not far apart and in fact were treated as one by BENTHAM. Apart from this, we may therefore confine ourselves to noting the occurrence of the different types of secondary pollen presentation in the tribal classification, and the homogeneity or otherwise of the tribes with respect to this character.

In the pulvinate tribes, secondary pollen presentation using a stylar brush is present in 6. *Milletieae* (syn. *Tephrosieae*), 7. *Robinieae*, 10. *Phaseoleae* and 16. *Galegeae* (this tribe shows a transition between the pulvinate syndrome and the epulvinate). Tribe *Robinieae* falls into two parts, one of which has beardless styles; in *Galegeae* the bearded style is similarly confined to the nine genera of the *Coluteinae*, one of four subtribes. In *Phaseoleae* the bearded style is considered to have arisen at least three times (LACKEY 1981; LAVIN & DELGADO 1990); it is general in two subtribes but of isolated occurrence in the remaining one. [An explosive system occurs in *Mucuna*, belonging to subtribe *Erythrininae* of this tribe (KALIN ARROYO 1981: 755) but this apparently does not involve secondary presentation (VOGEL 1969, bat-pollinated species)]. The pulvinate tribes 8. *Indigofereae* and 9. *Desmodieae* tend to develop mechanisms for an explosive release of pollen (Polhill, Raven & Stirton 1981), while a piston system using both stamens and style occurs in tribe 14. *Aeschynomeneae*.

Among the epulvinate tribes 19. *Loteae* and 20. *Coronilleae* are considered to be so closely related that Polhill (1981) discusses them and keys them as one. He says "a dilation of the filaments, at least of those which push the pollen into the keel tip, seems to be almost consistent throughout the two tribes." Thus a single origin will suffice for the two tribes, although genera do differ in the number of thickened filaments. Nevertheless the same system re-appears in *Ononis* which, in this and other characters, stands somewhat apart from the other genera of its tribe (23. *Trifolieae*). Tribe 24. *Brongniartieae*, on the other hand, shows explosive tendencies.

The genistoid alliance is also diverse. Of the 16 genera of tribe 29. *Crotalarieae*, probably only *Crotalaria* and *Bolusia* have a bearded style, though the alternation of large basifixed and small dorsifixed anthers is the rule in the tribe. In *Lupinus*, which occupies an isolated position in tribe 32. *Genisteae*, a version of the androecial piston system occurs. The rest of the tribe *Genisteae* is made up of the "*Cytisus-Genista* complex" where there are various pollen-delivery systems, including simple primary presentation, as well as the various forcible and explosive types involving secondary pollen presentation (Polhill 1976; Bisby 1981). Another tribe where piston mechanisms occur is 28. Liparieae.

It is clear that the *Papilionoideae* provide in their systems of pollen presentation a quite remarkable case of plural parallel evolution. This implies a pressure for the selection of pollen-protecting systems that is both strong and consistent. It is worth calling attention at this point to Yeo's (1972) suggestion that pollen which is presented from below may be taken more easily by non-pollinating insects than that which is presented from above (in the nototribic flowers of the *Lamiaceae*, for example). This would necessitate stronger protective arrangements in sternotribic flowers. However, another and probably more important factor is the prevalent adaptation to bee-pollination; it seems likely that the bees' need for pollen as brood-food has necessitated measures to regulate its release and hence its loss through appropriation by the pollinators.

For additional references on some of the aspects of legume floral biology considered here see Schrire (1989: 211–220).

Only binucleate pollen is recorded for *Leguminosae-Papilionoideae*, and it is known in many genera with secondary pollen presentation (Brewbaker 1967).

References for Leguminosae

Armstrong, J. M., White, W. J., 1935: Factors influencing seed-setting in alfalfa. – J. Agric. Sci. (Cambridge) **25**: 161–179.

Bentham, G., 1865: Leguminosae. – In Bentham, G., Hooker, J. D., (eds.): Genera Plantarum **1(2)**. – London: Reeve and Williams & Norgate.

Bisby, F. A., 1981: Tribe 32. *Genisteae* (Adans.) Benth., 1865. – In Polhill & Raven (1981): 409–425.

Brewbaker, J. L., 1967.

Bubar, J. S., 1958: An association between variability in ovule development within ovaries and self-incompatibility. – Canad. J. Bot. **36**: 65–72.

DELPINO, F., 1873–4 ['1870'].

DEVLIN, B., STEPHENSON, A. G., 1984: Factors that influence the duration of the staminate and pistillate phases of *Lobelia cardinalis* flowers. – Bot. Gaz. **145**: 323–328.

DUNN, D. B., 1956: The breeding systems of *Lupinus*, group *Micranthi*. – Amer. Midl. Naturalist **55**: 443–472.

FAEGRI, K., VAN DER PIJL, L., 1971.

FAEGRI, K., VAN DER PIJL, L., 1979.

HARDER, L. D., THOMSON, J. D., 1989.

HEDSTRÖM, I., THULIN, M., 1986: Pollination by a hugging mechanism in *Vigna vexillata* (*Leguminosae-Papilionoideae*). – Pl. Syst. Evol. **154**: 275–283.

HESLOP-HARRISON, J. & Y., 1982: Pollen-stigma interaction in the *Leguminosae*: constituents of the stylar fluid and stigma secretion of *Trifolium pratense* L. – Ann. Bot. (Oxford) **49**: 729–735.

HESLOP-HARRISON, J. & Y., 1983: Pollen-stigma interaction in the *Leguminosae*: the organization of the stigma in *Trifolium pratense* L. – Ann. Bot. (Oxford) **51**: 571–583.

HESLOP-HARRISON, J. & Y., 1984: Stigma organisation and the control of fertilisation in *Phaseolus*. – In REIMANN-PHILLIPP, R., (ed.): Proceedings of the Eucarpia Meeting on *Phaseolus* Bean Breeding, Held in Hamburg, July, 1983: 88–96.

HESLOP-HARRISON, Y., 1984: Organisation and function of the Angiosperm stigma: some features of significance for plant breeding. – In HERVÉ, Y., DUMAS, C., (eds.): Incompatibilité Pollinique et Amélioration des Plantes: 27–39. – Rennes: Dépt. de Formation de l'École Nat. Sup. Agronomique.

HESLOP-HARRISON, Y., SHIVANNA, K. R., 1977.

HEYWOOD, V. H. (ed.), 1978.

HOLDEN, J. H. W., BOND, D. A., 1960: Studies on the breeding system of the field bean, *Vicia faba* L. – Heredity **15**: 175–192.

JOST, L., 1907: Über die Selbststerilität einiger Blüten. – Bot. Zeitung (Berlin) **65**: 77–117.

JUNCOSA, A. M., WEBSTER, B. D., 1989: Pollination in *Lupinus nanus* subsp. *nanus* (*Leguminosae*). – Amer. J. Bot. **76**: 59–66.

KALIN ARROYO, M. T., 1976: The systematics of the Legume genus *Harpalyce* (*Leguminosae*: *Lotoideae*). – Mem. New York Bot. Gard. **26**: 1–80.

KALIN ARROYO, M. T., 1981: Breeding systems and pollination biology in *Leguminosae*. – In POLHILL & RAVEN (1981): 723–769.

KNUTH, P., 1904.

KNUTH, P., 1908.

KUPICHA, F. K., 1981: Tribe 21. *Vicieae* (ADANS.) DC., 1825. – In POLHILL & RAVEN (1981): 377–381.

LACKEY, J. A., 1981: Tribe 10. *Phaseoleae* DC., 1825. – In POLHILL & RAVEN (1981): 301–327.

LASSEN, P., 1989: A new delimitation of the genera *Coronilla*, *Hippocrepis*, and *Securigera* (*Fabaceae*). – Willdenowia **9**: 49–62.

LAVIN, M., DELGADO, A., 1990: Pollen brush of *Papilionoideae* (*Leguminosae*): morphological variation and systematic utility. – Amer. J. Bot. **77**: 1294–1312.

LEPPIK, E. E., 1966: Floral evolution and pollination in the *Leguminosae*. – Ann. Bot. Fenn. **3**: 299–308.

LLOYD, D. G., YATES, M. A., 1982.

LORD, E. M., HESLOP-HARRISON, Y., 1984: Pollen-stigma interaction in the *Leguminosae*: stigma organization and the breeding system in *Vicia faba* L. – Ann. Bot. (Oxford) **54**: 827–836.

MCCANN, C., 1952: The tui and its food plants. – Notornis **1952**: 6–14.

MABBERLEY, D. J., 1987: The Plant-book. – Cambridge: Cambridge University Press.

MÜLLER, H., 1883.
PEREIRA-NORONHA, M. R., SILBERBAUER-GOTTSBERGER, I., GOTTSBERGER, G., 1982: Biologia floral de *Stylosanthes* (*Fabaceae*) no cerrado de Botucatu, Estado de São Paulo. – Revista Brasil. Biol. **42**: 595–605.
POLHILL, R. M., 1968: Miscellaneous notes on African species of *Crotalaria* L.: II. – Kew Bull. **22**: 167–348.
POLHILL, R. M., 1976: *Genisteae* (ADANS.) BENTH. and related tribes (*Leguminosae*). Bot. Syst. **1**: 143–368.
POLHILL, R. M., 1981: Tribe 19. *Loteae* DC., 1825, and Tribe 20. *Coronilleae* (ADANS.) BOISS., 1872. – In POLHILL & RAVEN (1981): 371–375.
POLHILL, R. M., RAVEN, P. H., (eds.), 1981: Advances in Legume Systematics, Parts 1 & 2. – Kew: Ministry of Agriculture Fisheries and Food, London.
POLHILL, R. M., RAVEN, P. H., STIRTON, C. H., 1981: Evolution and systematics of the *Leguminosae* – In POLHILL & RAVEN (1981): 1–26.
RENDLE, A. B., 1925: The Classification of Flowering Plants **2**. – London: Cambridge University Press.
ROWLANDS, D. G., 1958: The nature of the breeding system in the field bean (*V. faba* L.) and its relationship to breeding for yield. – Heredity **12**: 113–126.
SCHAAL, B. A., LEVERICH, W. J., 1980: Pollination and banner markings in *Lupinus texensis* (*Leguminosae*). – SouthW. Naturalist **25**: 280–282.
SCHRIRE, B. D., 1989: A multidisciplinary approach to pollination biology in the *Leguminosae*. – In STIRTON, C. H., ZARUCCHI, J. L. (1989): 183–242.
SHIVANNA, K. R., OWENS, S. J., 1989: Pollen-pistil interactions. – In STIRTON, C. H., ZARUCCHI, J. L. (1989): 157–182.
STODDARD, F. L., BOND, D. A., 1987: The pollination requirements of the faba bean. – Bee World **68**: 144–152.
TEPPNER, H., 1988: *Lathyrus grandiflorus* (*Fabaceae-Vicieae*): Blüten-bau, -Funktion und *Xylocopa violacea*. – Phyton (Austria) **28**: 321–336.
VOGEL, S., 1969: Chiropterophilie in der neotropischen Flora: neue Mitteilung II. – Flora, Abt. B **158**: 185–222.
WAINWRIGHT, C. M., 1978: The floral biology and pollination ecology of two desert lupines. – Bull. Torrey Bot. Club **105**: 24–38.
WESTERKAMP, C., 1989.
YANAGIZAWA, Y., GOTTSBERGER, G., 1982–3: Competição entre *Distictella elongata* (*Bignoniaceae*) e *Crotalaria anagyroides* (*Fabaceae*) com relação às abelhas polinizadoras no cerrado de Botucatu, Estado de São Paulo, Brasil. – Portugaliae Acta Biol., Sér. A **17**: 149–166.
YEO, P. F., 1972.

Rhizophoraceae

Systematic position: *Rosidae, Myrtales*
Restriction of occurrence within the family: Tribe Rhizophoreae

Distribution within the tribe. Rhizophoraceae are woody plants and the tribe *Rhizophoreae* (one of two or three tribes recognized – MACNAE & FOSBERG 1981) comprises four genera of mangrove plants. Secondary pollen presentation is known in all six species of *Bruguiera* and one of *Ceriops*.
Floral features. Flower actinomorphic with inferior ovary, above which is a shortly campanulate hypanthium. Sepals in *Bruguiera* usually 8–16 (varying within the species), in *Ceriops* 5 or 6, longer than the petals. Petals free, equal in number to

the sepals, elongate in shape and involute so that each enfolds two stamens. Stamens twice the number of the petals, those of a pair unequal, the filament of the antesepalous one being twisted to bring the anther opposite the petal; anthers introrsely dehiscent. Style simple. Nectar is secreted within the hypanthium.
Type of pollen presentation. Pollen-discharge is explosive, with the petals playing a part in its presentation ('Pollenhaufen').

Examples of Secondary Pollen Presentation in Rhizophoraceae

Bruguiera LAM.
B. gymnorhiza (L.) SAVIGNY. (DAVEY 1975, as *B. gymnorrhiza*, illus.; TOMLINSON & al. 1979, illus.; TOMLINSON 1986, illus.; KONDO & al. 1988; TANAKA 1989). **STR c,a. ISS expl cloud. SEQU pln. APPL general. SYN** ornitho. **VIS** unnamed bees (*Hymenoptera-Apoidea*) and the following birds (*Aves*): sunbirds (*Nectariniidae*), honey-eaters (*Meliphagidae*), white-eyes (*Zosteropidae*) and a bulbul (*Pycnonotidae*). (Fig. 14).
The flowers are about 3.5cm long, with the ovary and hypanthium accounting for about half the length, and are finally about 2.5cm wide. The calyx and hypanthium are pinkish or reddish brown or scarlet and the petals white, soon becoming brown. The flower-opening is directed vertically downwards and somewhat inwards towards the crown of the tree, which is thought to facilitate access by perching birds. In the mature bud the stamen filaments are sinuous and the anthers have dehisced and shed some of their pollen into the chamber provided by the adjacent petal. The bilobed petals are well-vascularized and heavily cuticularized but there are ridges along their edges with thinner-walled cells, and as these dry out tension is set up. There is also some internal pressure from the curling of the filaments and from hairs on the inner surface of the petals pressing on the backs of the anthers. On the furled base of each petal are two tufts of stiff hairs directed obliquely downwards into the cavity of the hypanthium. Pressure on these can trigger an explosive unfurling of the petal and a scattering of the already freed pollen. Tripping can also occur when a glabrous area just above these hair-tufts is pressed. The two sensitive areas are likely to be touched when a bird probes the hypanthium, where a large amount of nectar is secreted. Not all the petals are discharged during a single pollinator-visit; the process may require three to six days. The petals and stamens are then shed. In Japan it was found that the flowers opened in the morning and were much visited at that time. Only one third of the flowers produced nectar on the first day, but the others did so later. Tests showed that plants were self-fertile.

In *B. exaristata* DING HOU, which is also pollinated by sunbirds and honey-eaters, the stigmas become receptive one or two days after the flowers open; nectar is present and pollinator visits may continue for up to eight days (TOMLINSON & al.).

The floral details of *B. sexangula* (LOUR.) POIR. (*B. eriopetala* ARN.), described by GEHRMANN (1911) from Buitenzorg (Bogor), Indonesia, are closely similar to those of *B. gymnorhiza*, even though the account gives no firm evidence that any of the pollen is presented secondarily. The explosion in this species is remarkably violent, sending up a cloud of pollen over 20 cm high in a still room (presumably

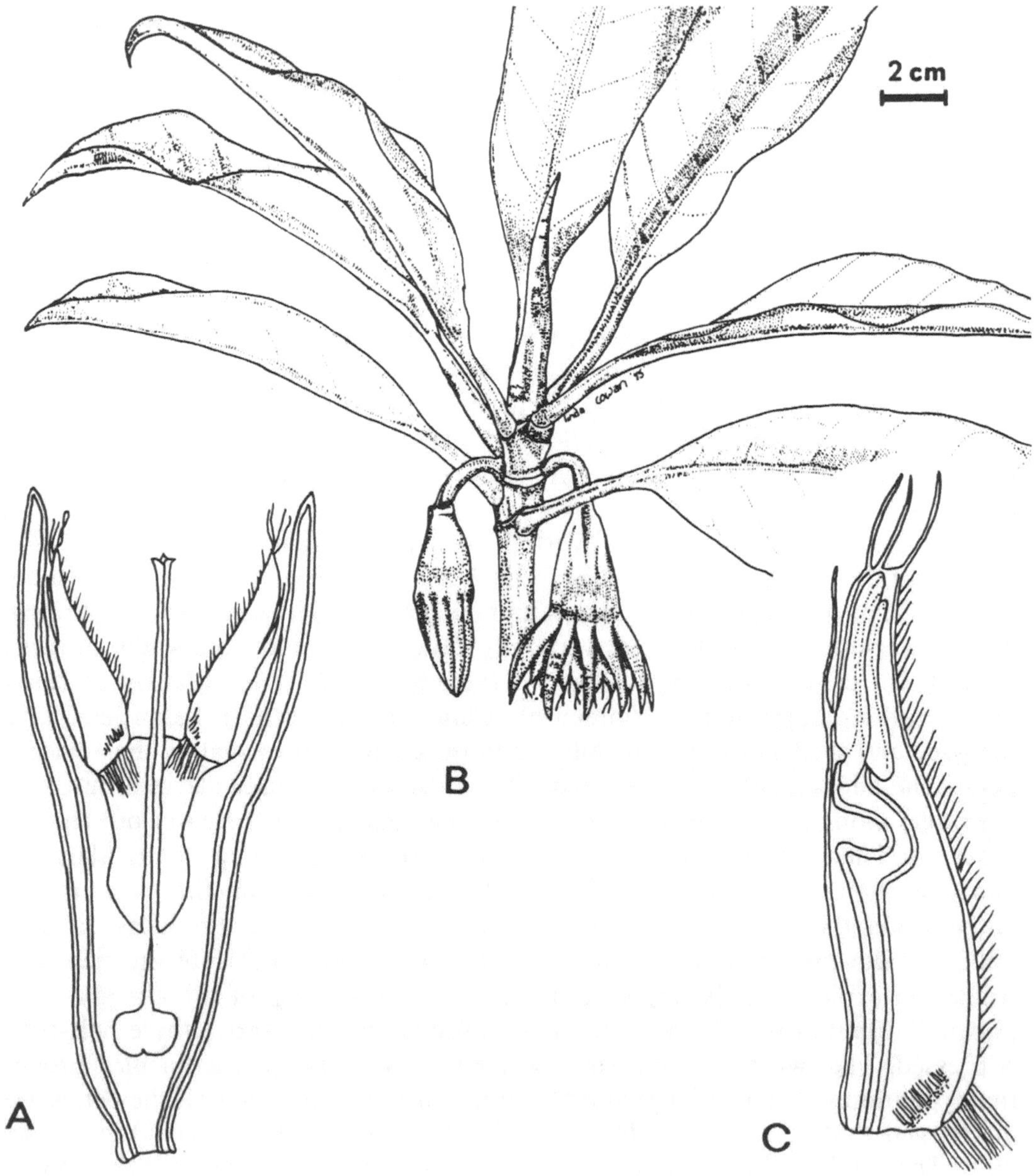

Fig. 14. *Bruguiera gymnorhiza* (*Rhizophoraceae*). *A* flowering shoot. *B* longisection of flower with two petals shown complete (the hairy structures). *C* petal as if taken from left-hand side of 'B', cut open to show the two stamens which it enwraps (it has a deep apical slit with a bristle in it and there are bristles at the tip of each lobe); there are two tufts of sensitive hairs at the base. (L. COWAN, in DAVEY 1975, B and C re-drawn.)

the flower was tripped in an upright position). The flowers were visited by sunbirds and hawkmoths (*Lepidoptera-Sphingidae*).
B. parviflora GRIFFITH. (TOMLINSON & al. 1979, illus.). **STR c,a. ISS expl depos?. SEQU pln. APPL** proboscis? **SYN** psycho. **VIS** large butterflies (*Lepidoptera*). The flower is much smaller than in the bird-pollinated species, having a length of about 15mm, most of which is made up of ovary and a very short hypanthium.

The flower is erect, facing outwards on the plant, apparently scentless and with a meagre nectar production. Tripping requires only a delicate touch on the distal parts of the petals and is not violent, there being no visible cloud of pollen. It is presumed that bees could also pollinate the flowers.

Ceriops ARN.

C. tagal (PERR.) C. B. ROBINS. (TOMLINSON & al. 1979, illus.; TOMLINSON 1986, illus.). **STR c,a. ISS expl depos?. SEQU pln. APPL** proboscis? **SYN** phalaeno. **VIS** moths (*Lepidoptera*).

Only two species of *Ceriops* are now recognized, and *C. tagal* is the only one with explosive pollen release. The flowers are more or less nodding and only 5 mm long. The petal and stamen construction and their response to touch are as in *Bruguiera parviflora*. However, the flowers open in the late afternoon, produce a distinctive perfume and have white petals, points consistent with the syndrome of moth pollination.

Discussion of Rhizophoraceae

The bird-pollinated species of *Bruguiera*, in having an explosive mechanism that birds alone can trip, have effectively protected their pollen against removal by insects. Their secondary pollen presentation is incidental to the explosive pollen release system which enhances their commitment to ornithophily. The protective value of secondary pollen presentation in the insect-pollinated members of the genus seems doubtful, as the flowers are small and not obviously specialized to particular insect groups. This suggests that the genus is primarily bird-pollinated and that secondary pollen presentation is merely residual in the insect-pollinated species. However, it is then difficult to account for the occurrence of secondary pollen presentation in one of the only two species in the genus *Ceriops*.

Binucleate pollen is reported from *Bruguiera* (BREWBAKER 1967).

References for Rhizophoraceae

BREWBAKER, J. L., 1967.

DAVEY, J. E., 1975: Note on the mechanism of pollen release in *Bruguiera gymnorrhiza*. – J. S. African Bot. **41**: 269–272.

GEHRMANN, K., 1911: Zur Blütenbiologie der *Rhizophoraceae*. – Ber. Deutsch. Bot. Ges. **29**: 308–318.

KONDO, K., NAKAMURA, T., TSURUDA, K., SAITA, N., YAGUCHI, Y., 1988: Pollination in *Bruguiera gymnorrhiza* and *Rhizophora mucronata* (*Rhizophoraceae*) in Ishigaki Island, The Ryukyu Islands, Japan. – Biotropica **19**: 377–380.

MACNAE, W., FOSBERG, F. R., 1981: *Rhizophoraceae*. – In DASSANAYAKE, M. D., FOSBERG, F. R. (eds.): A Revised Handbook to the Flora of Ceylon **2**: 487–500. – New Delhi: Amerind.

TANAKA, H., 1989: Field notes on pollination of two species of *Rhizophoraceae*. – J. Phytogeogr. Taxon. **37**: 65–67.

TOMLINSON, P. B., 1986: The Botany of Mangroves. – Cambridge: Cambridge University Press.

TOMLINSON, P. B., PRIMACK, R. B., BUNT, J. S., 1979: Preliminary observations on floral biology in mangrove *Rhizophoraceae*. – Biotropica **11**: 256–277.

Myrtaceae

Systematic position: *Rosidae, Myrtales*
Restriction of occurrence within the family: Subfamily *Leptospermoideae*

Distribution within the subfamily. Leptospermoideae, one of two subfamilies that make up the *Myrtaceae*, has been informally divided into several 'alliances' by JOHNSON & BRIGGS (in MORLEY & TOELKEN 1983). Secondary pollen presentation occurs only in a few genera of the *Chamelaucium* alliance, which is mainly developed in the south-west of Western Australia.
Floral features. Flowers actinomorphic or slightly zygomorphic with five sepals and petals; stamens five or a multiple of five; ovary usually inferior, syncarpous with a single style terminated by a small stigma.
Type of pollen presentation. The pollen is released suspended in a viscid oily liquid and flows on to a brush on the style or sometimes on to staminodes ('Pseudo-Staubblatt'). The liquid originates from a lysigenous terpene gland in the connective, whence it saturates the contents of the thecae of the anthers (VOGEL 1984). A condensed account of secondary pollen presentation in this family by HOLM (1978) has been superseded by a detailed illustrated account in his book (HOLM 1988).

Examples of Secondary Pollen Presentation in Myrtaceae

Darwinia RUDGE

All eight species studied by HOLM (1988) release their pollen in an oily fluid on to a stylar brush as described by WERTH (1915). The sepals are partly united to form a flloral tube above the ovary in which nectar is stored. KEIGHERY (1975) stated that the flowers are protandrous but HOLM (1988: 76) could find no changes with age in the stigma.
D. vestita BENTH. (KEIGHERY 1975, illus.; HOLM 1988, illus.). **STR s. ISS exposed. SEQU pln. APPL a? RWD n. SYN** entomo. **VIS** solitary bees and *Thynnidae* (*Hymenoptera-Apoidea* and -*Scolioidea* respectively). (Fig. 15).
The small (c. 2mm diameter) purplish and/or white flowers are borne in dense capitula at the tips of the branches. The sepals are minute and the petals stand erect. The anthers of the ten very short stamens dehisce in bud and extrude their pollen in an oily liquid on to the pollen presenter located near the tip of the style (see 'Type of pollen presentation' above). This is a zone of knobbed hairs shaped like a bottle-brush and situated immediately below the punctiform stigma. When the flower first opens the pollen is presented on the well-exserted style, while the stamens remain deep within the flower but are rendered conspicuous by their purple connectives. The flower also includes 10 filiform staminodes longer than the stamens. HOLM, who saw no visitors, concluded that the flower structure would force nectar-drinking insects to stand nearly vertically.
D. macrostegia (TURCZ.) BENTH. (KEIGHERY 1975, illus.; HOLM 1988, illus.). **STR s. ISS exposed. SEQU pln. APPL a? RWD n. SYN** ornitho. **VIS** western spinebill *Acanthorhynchus superciliosus* (*Aves-Meliphagidae*). (Fig. 16A–C).
The process of secondary pollen presentation in this species is the same as in *D. vestita*. Adaptation to bird-pollination is expressed by the nodding involucrate

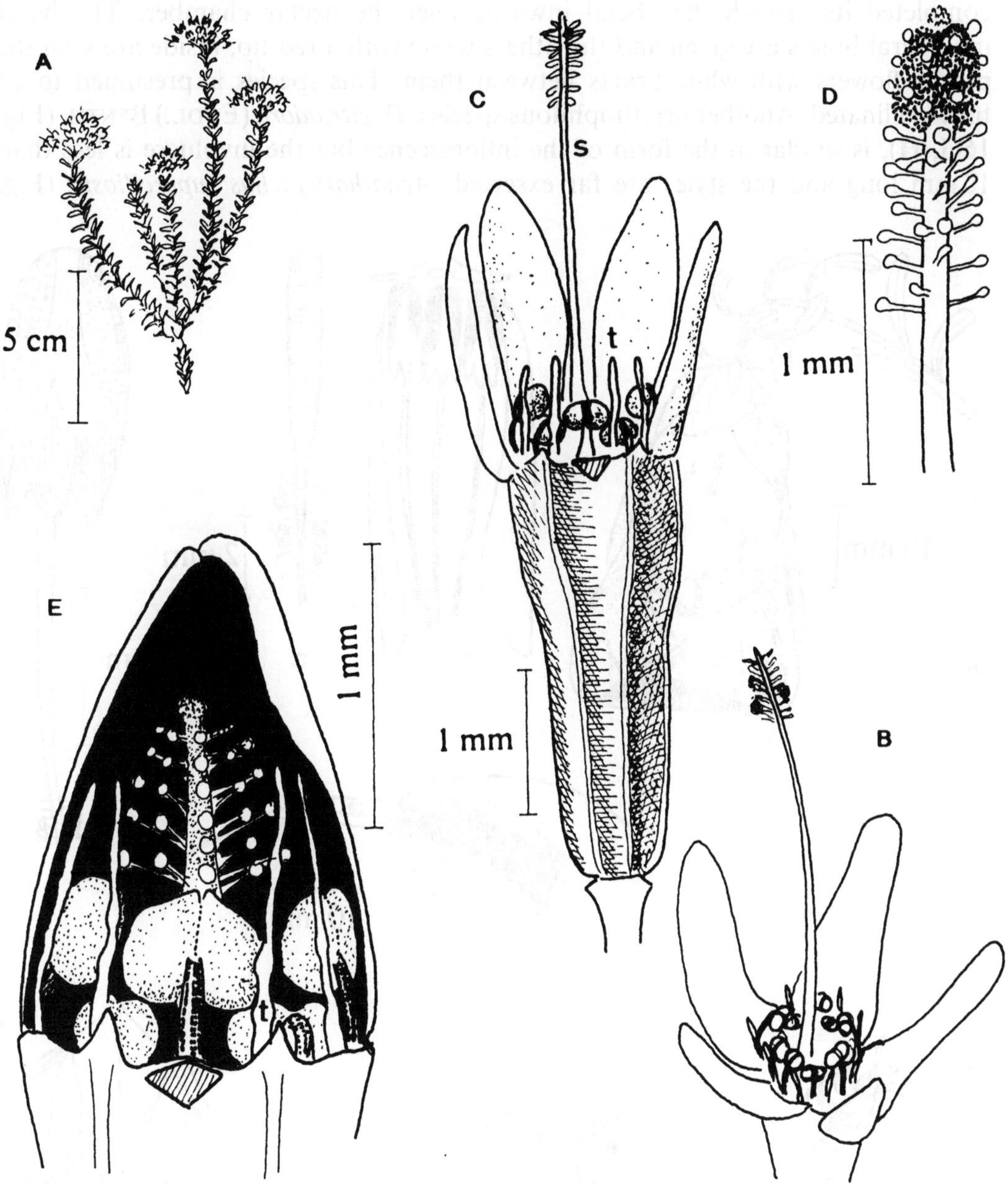

Fig. 15. *Darwinia vestita* (*Myrtaceae*), an insect-pollinated species. *A* habit of plant. *B* view of flower (sepals are vestigial). *C* side view of flower with one petal removed. *D* style with pollen grains among knobbed lateral hairs and small stigma at apex. *E* upper part of flower bud with one petal removed, before release of pollen. *s* = style, *t* = staminode. (Holm 1988.)

flower-cluster, the shorter perianth, more exserted style and red and white colouring of the involucre. The involucre is 4cm long and the styles reach nearly to its mouth.

D. meeboldii C. A. Gardner (Keighery 1975; Holm 1988, illus.) is organized similarly to *D. macrostegia* but the involucre is only about 2.5 cm long and the styles emerge slightly from its mouth. After pollen deposition, and during the growth of the style, the stamens and staminodes are erect but when the style has completed its growth they bend inwards over the nectar chamber. The basal involucral bracts are green and the others white with a red tip; inside are seen the purple flowers with white bracts between them. This species is presumed to be bird-pollinated. Another ornithophilous species, *D. citriodora* (ENDL.) BENTH. (Fig. 16G, H), is similar in the form of the inflorescence but the involucre is less than 10 mm long and the styles are far exserted. *Acanthorhynchus superciliosus* (Fig.

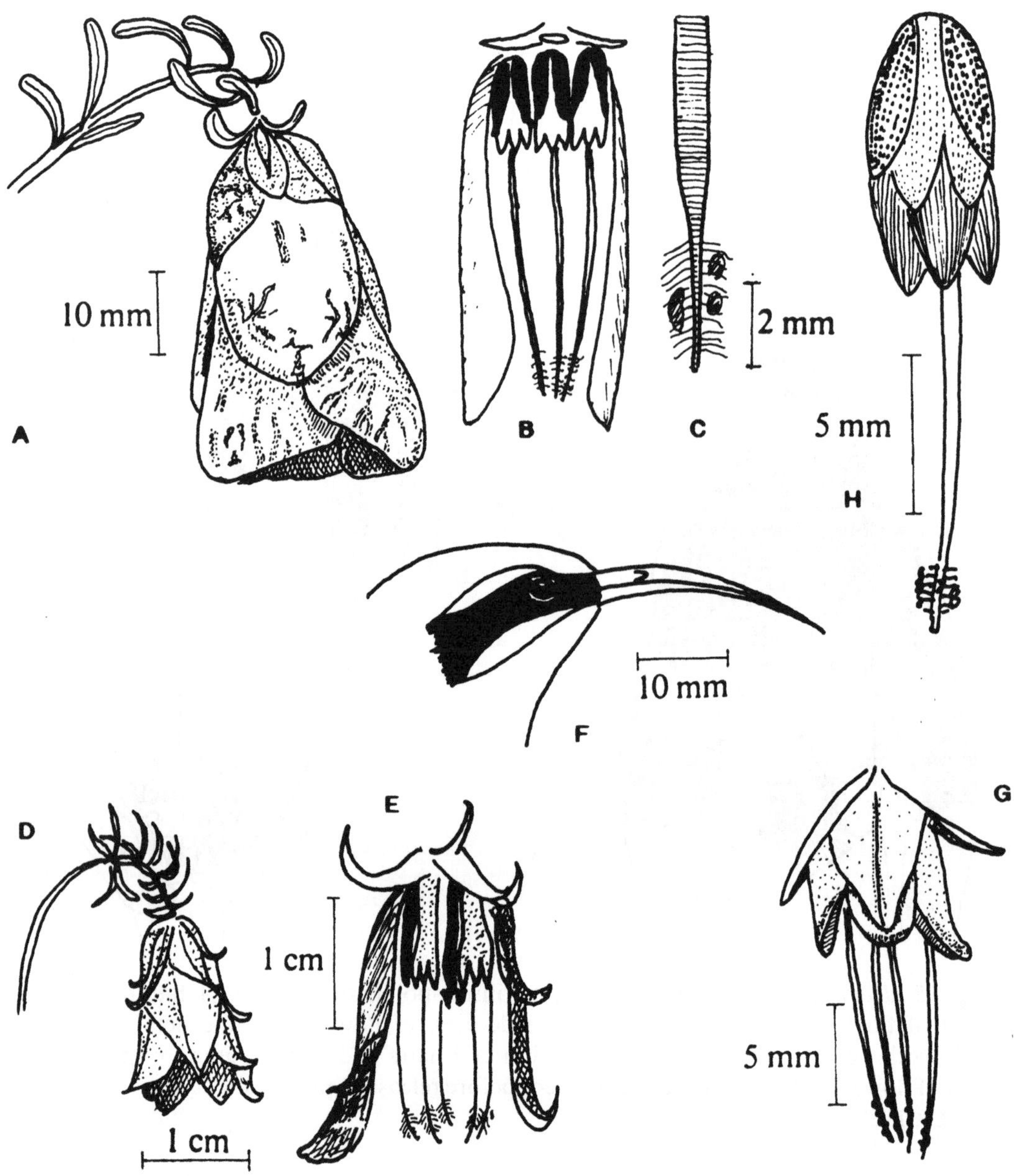

Fig. 16. *Darwinia* (*Myrtaceae*), bird-pollinated species with involucrate inflorescences. *D. macrostegia*: *A* side view of involucre; *B* section of same, showing three flowers; *C* style-apex with pollen grains among hairs. *D. leiostyla*: *D* side view of involucre; *E* section of same, showing five flowers; *F* head of western spinebill, *Acanthorhynchus superciliosus* (known pollinator of *D. macrostegia* and *D. citriodora*). *D. citriodora*: *G* side view of involucre; *H* single flower. (HOLM 1988.)

16F) visits it and may be the only bird that can pollinate it (Keighery; Holm). Four further ornithophil species with nodding involucres were described by Holm (e.g. Fig. 16D, E).

Verticordia DC.

Eight *Verticordia* species were studied by Holm (1988) and secondary pollen presentation was observed in five of them. *Verticordia* has received the popular name 'feather-flower' on account of the fine dissection of the sepals and commonly other floral parts. They show fascinating variation in their detailed floral structure which is not yet understood and their flowers frequently show colour polymorphism or colour-changes with age. The plants are small or medium-sized prolifically flowering shrubs of Western Australia. Their sepals are usually partly united to form a floral tube above the ovary and some of the stamens are staminodial.

V. huegelii Endl. (Holm 1988, illus.). **STR s. ISS exposed. SEQU sim? APPL d? RWD n. SYN** entomo. **VIS** *Coleoptera*. (Fig. 17A–D).

The flowers vary from red to white and are 12 – 15 mm in diameter. The calyx-lobes are divided into long, loosely plumose segments far exceeding the fringed petals, which form a bowl. There are 10 small erect fimbriated staminodes and 10 short incurved stamens with knob-like anthers. There are spaces between these through which the nectar in the base of the floral cup can be reached. The style is hairy distally and has a peltate stigma; in bud the hairs hang down and touch the anthers. At this stage the pollen is released and clings to the hairs in sticky masses (see 'Type of pollen presentation' for the family). When the flower is open the style has elongated considerably so that the stigma stands slightly above the rim of the petaline bowl. Holm found beetles visiting the flowers but of ten specimens that were examined, none was carrying pollen.

V. picta Endl. (Holm 1988, illus.).

Secondary pollen presentation in this species conforms with the pattern in the preceding. However, the stamens are relatively long and erect and they alternate with clusters of three basally coherent flattened staminodes so as to form a short wide tube above the shallow calyx-tube. In addition every other detail of the flower is different. Nectar is provided. The only insect visitors seen by Holm were again beetles which ate the floral parts and did not carry pollen.

V. grandis Drumm. (Holm 1988, illus.). **STR s. ISS exposed. SEQU sim? APPL a. RWD n. SYN** ornitho. **VIS –.** (Fig. 18).

The flowers are about 27mm in diameter and slightly zygomorphic. Two scarious bracteoles embrace the base of the ribbed and turbinate ovary and calyx-tube. Opposite the plumose calyx-lobes are reflexed and gland-dotted appendages, and alternating with them are reflexed fimbriate epicalyx-lobes. The calyx-tube is continued upwards from the ovary by a broadly cylindric epigynous zone consisting of corolla and androecium; above this the petals and the ten stamens stand erect, while the ten narrowly triangular staminodes bend inwards, forming a complete cover over the nectar. The anthers are small but in the mature flower their shining globular connectives make them conspicuous. The style lies to one side of the tube, loosely embraced by a petal that is less incurved than the others; it has a punctiform stigma and a zone of hairs just below this. Pollen is released into the hairs of the style before the flower opens; at this stage the style is coiled at the tip.

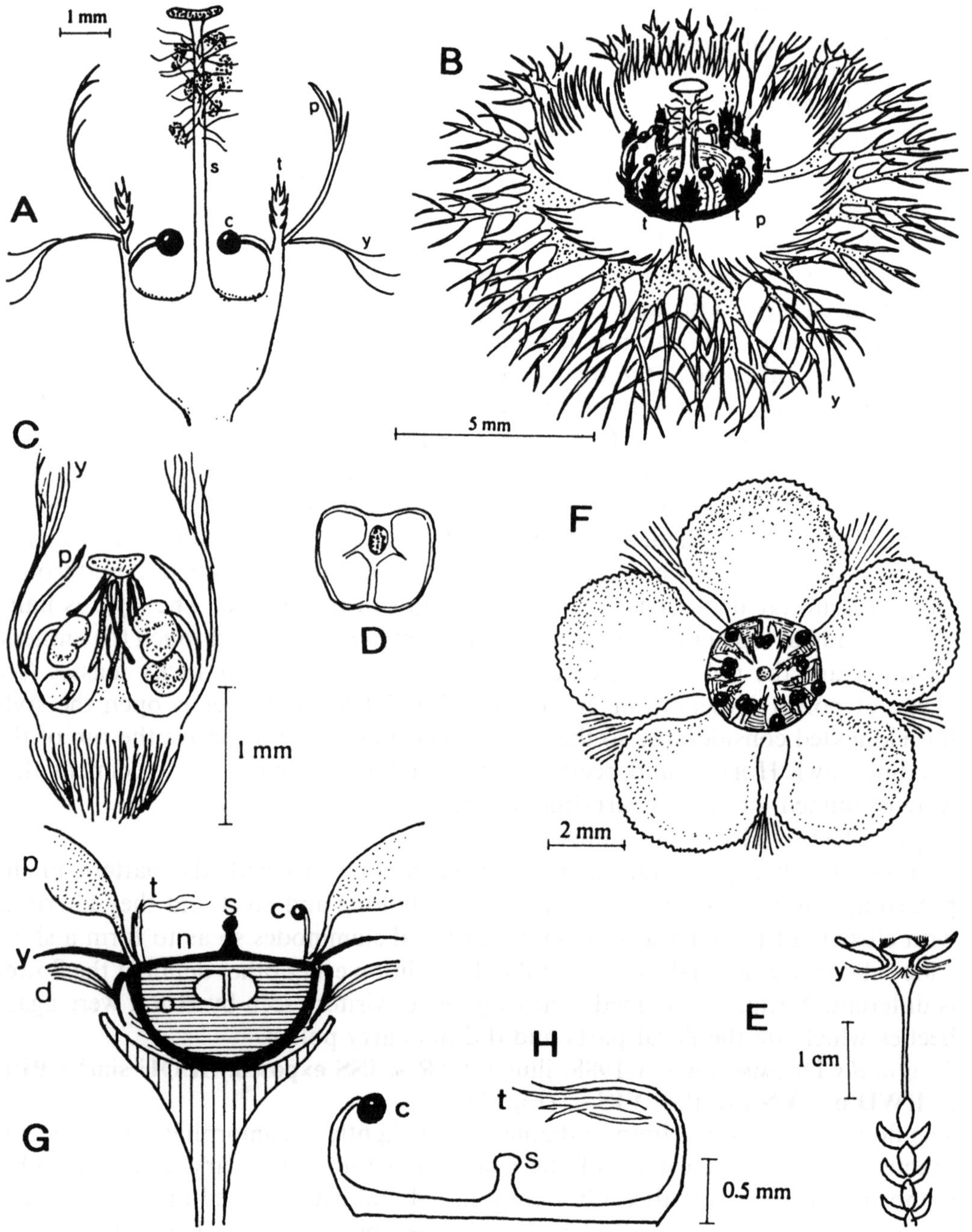

Fig. 17. *Verticordia* species (*Myrtaceae*) pollinated by insects. *V. huegelii*: *A* general view of flower; *B* longisection of flower, showing pollen among stylar hairs; *C* the same in bud before dehiscence of anthers; *D* transection of anther showing gland in connective. *V. habrantha*: *E* side view of flower on its leafy shoot; *F* general view of flower; *G* transection of flower; *H* longisection of top of ovary with one stamen with swollen connective and one staminode. *c* = anther, *d* = epicalyx, *n* = nectary, *p* = petal, *s* = style, *t* = staminode, *y* = sepal. (HOLM 1988.)

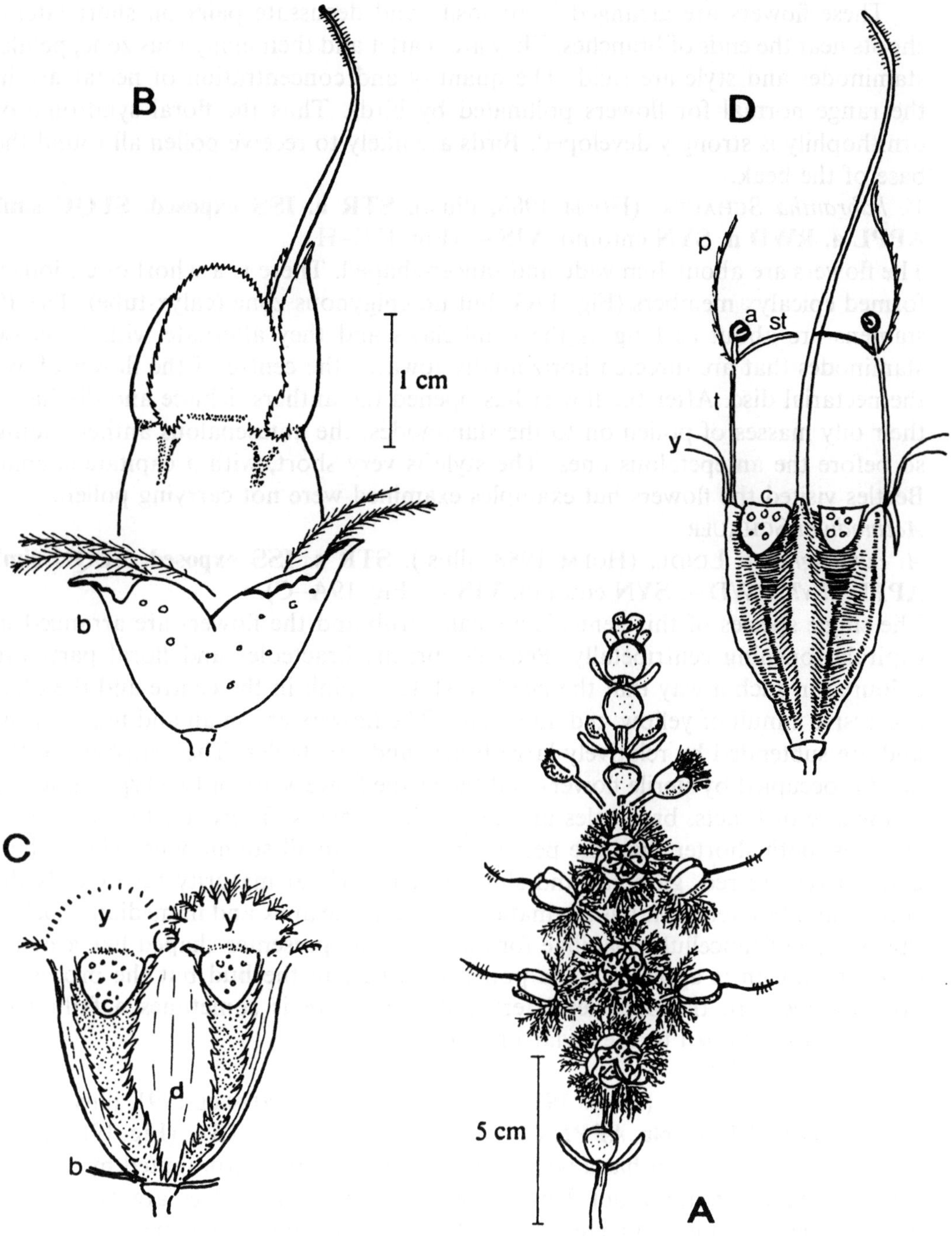

Fig. 18. *Verticordia grandis* (*Myrtaceae*), a bird-pollinated species. *A* inflorescence. *B* side view of flower. *C* side view of calyx with appendages and epicalyx. *D* longisection of flower-tube above intact ovary, epicalyx lobes removed (except on right). *b* = bracteole, *c* = herbaceous gland-dotted appendage (reflexed), *d* = epicalyx lobe (reflexed), *p* = petal, *st* = staminode, *t* = tubular epigynous zone, *y* = calyx lobe. (HOLM 1988.)

These flowers are arranged in opposite and decussate pairs on short lateral shoots near the ends of branches. They are scarlet and their epigynous zone, petals, staminodes and style are rigid. The quantity and concentration of nectar are in the range normal for flowers pollinated by birds. Thus the floral syndrome of ornithophily is strongly developed. Birds are likely to receive pollen all round the base of the beak.

V. habrantha SCHAUER. (HOLM 1988, illus.). **STR a. ISS exposed. SEQU sim? APPL a. RWD n. SYN** entomo. **VIS** –. (Fig. 17E–H).

The flowers are about 1cm wide and saucer-shaped. There is a whorl of curiously formed epicalyx members (Fig. 17G) but no epigynous zone (calyx-tube). The 10 stamens are about as long as the petal-claws and they alternate with 10 erose staminodes that are directed horizontally towards the centre of the flower above the nectarial disc. After the flower has opened the anthers dehisce and discharge their oily masses of pollen on to the staminodes, the antesepalous anthers doing so before the antepetalous ones. The style is very short, with a capitate stigma. Beetles visited the flowers but examples examined were not carrying pollen.

Actinodium SCHAUER

A. cunninghamii LINDL. (HOLM 1988, illus.). **STR s. ISS exposed. SEQU sim? APPL a?,v?. RWD –. SYN** entomo. **VIS** –. (Fig. 19A–C).

The single species of this genus is a small shrub and the flowers are arranged in capitula, opening centripetally. Pedicels, bracts, bracteoles and floral parts are coloured in such a way that the head is white or pink in the centre and then has successive annuli of yellow, red and white. The flowers are small and tetramerous and are subtended by relatively large bracts and bracteoles. The periphery of the head is occupied by sterile flowers and the fertile flowers are of two types differing in the size of bracts, bracteoles and perianth members. There are two whorls of stamens, both shorter than the petals, and 8 very small staminodes. The anther-connectives are red, globular and shining. The style at maturity far exceeds the corolla and bracteoles; it has stigmatic papillae at the apex and immediately below this a zone of unicellular papillae forming a pollen presenter shaped like a maize cob. The pollen is applied to the pollen presenter in the bud but the process is not described. In the peripheral fertile flowers there is sometimes a failure of transfer of the pollen to the stylar presenter.

Chamelaucium DESF.

C. uncinatum SCHAUER. (HOLM 1988, illus.; SLATER & BEARDSELL 1991, illus.). **STR s. ISS exposed. SEQU pln. APPL a?. RWD n. SYN** entomo. **VIS** –. (Fig. 19D–F).

The pentamerous flower has a wide ovary and nectar-disc surmounted by a calyx-tube (epigynous zone) about 2mm deep. The sepals themselves are obsolescent but the flower bud is completely encased in a pair of bracts. The petals are white or (YEO, unpublished) bright purplish pink, orbicular, spreading and non-contiguous. There are 10 short stamens in two whorls and 10 knobbed staminodes. In bud the style reaches the level of the rim of the calyx-tube and bears spreading hairs immediately below the stigmatic apex. The stamens at this stage are curled inwards so that the swollen red-brown oil gland, which is placed at the base of the connective on the morphologically outer side, is nearest the style-head. Then the filaments move upwards so that the gland brushes the stylar hairs and releases its fluid into them; with further uncoiling the morphological inner side of the

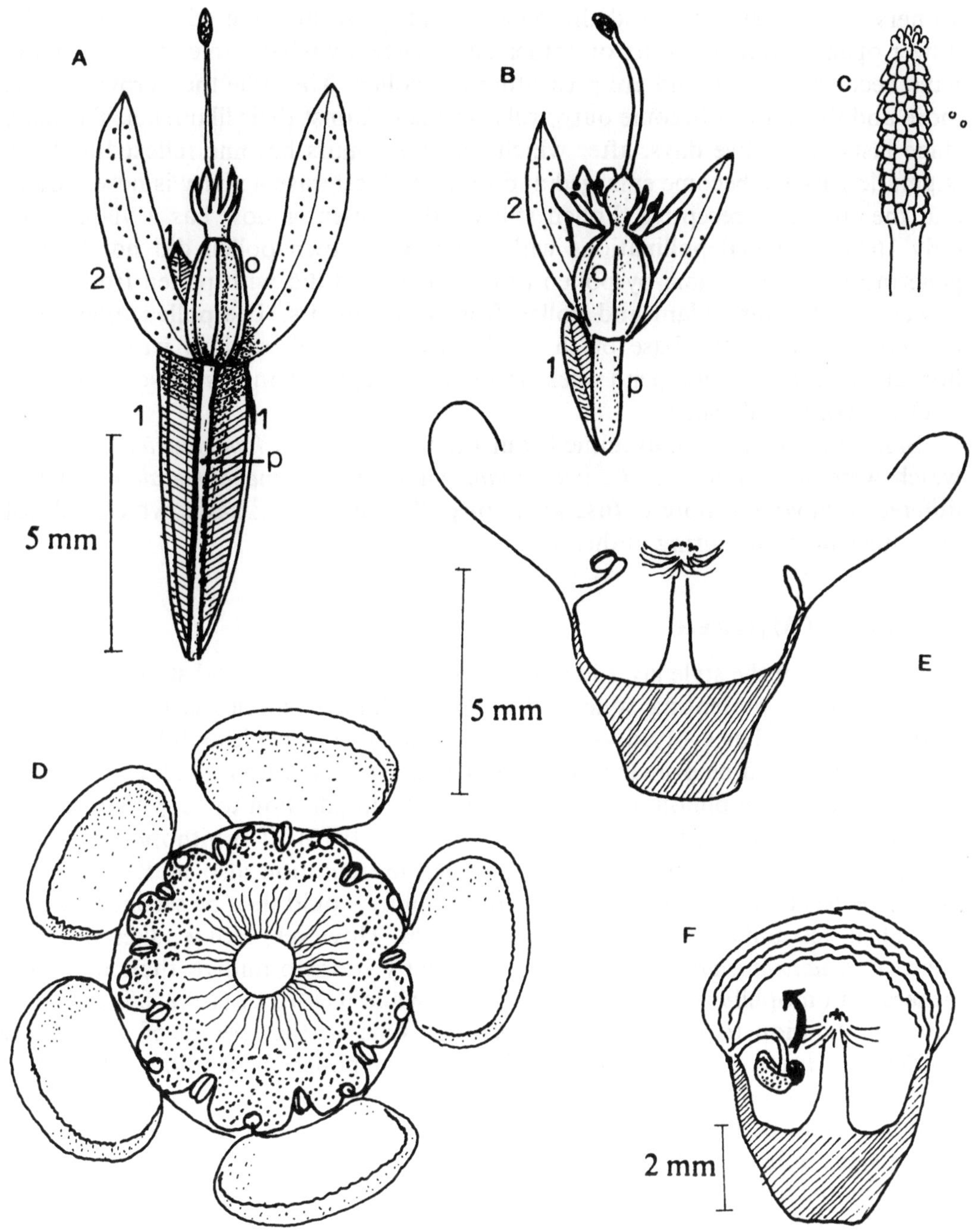

Fig. 19. Further *Myrtaceae*. *Actinodium cunninghamii*: *A* side view of outer fertile flower; *B* side view of inner fertile flower; *C* style-head. *Chamelaucium uncinatum*: *D* face view of flower showing petals, stamens, staminodes and style; *E* longisection of flower; *F* longisection of flower-bud (arrow shows direction of movement of stamens). *1* = bract, *2* = bracteole, *o* = ovary (surmounted by reduced sepals and petals), *p* = pedicel. (HOLM 1988.)

anthers touches the hairs and the pollen is taken up into the sticky liquid. The flower opens within 24 hours of anther-dehiscence, by which time the stylar hairs have become elevated and are presenting the pollen. The style then grows a little more and the anthers become outwardly reflexed against their filaments. The male stage lasts about five days, after which the stylar hairs become reflexed and the stigmatic papillae become enlarged and separated; esterase activity is now present and the stigma is receptive. Pollen still on the presenter now has a viability of only 16%. Artificial pollinations with the flower's own pollen did not lead to penetration of the stigma by pollen tubes, whereas it did with pollen from other flowers on the same plant and pollen from other plants. The pollen tubes were observed to reach the base of the style. This is the Geraldton wax-flower of horticulture; the flowers spend weeks in the bud stage and anthesis also lasts many weeks (YEO, unpublished).

SLATER & BEARDSELL examined four other species of *Chamelaucium*, three of which were very similar to *C. uncinatum*, while one, *C. megalopetalum* BENTH., differed in having a more diffuse array of pollen presenting hairs, which did not alter their position during anthesis.

Discussion of Myrtaceae

Presentation on the style gives identical positioning of pollen and stigma. The stiff exserted pollen-presenting style of the bird-pollinated species is reminiscent of *Proteaceae*. If it has evolved as part of the syndrome of ornithophily its presence in *Darwinia* would be primary (which is the view of KEIGHERY 1982: 85). There is at least one ornithophilous species of *Verticordia* in addition to *V. grandis* (as can be seen from the illustrations in ERICKSON & al. 1973). If ornithophily was the reason for the evolution of secondary pollen presentation also in *Verticordia* it would be basic in this genus too, and entomophily would thus be secondary, even though it accounts for the majority of species. However, HOLM takes the view that, as *Chamelaucium* appears to be distinctly more primitive than the other genera, entomophily is primitive in the group while KEIGHERY (1982: 85) considers insect-pollination to be basic in *Verticordia*. One of the species of *Verticordia* studied by HOLM, *V. acerosa* LINDL., has primary pollen presentation; it receives visits from beetles, like several other species. If entomophily is primitive in the group it could be that the use of the oily adhesive is connected with the relative hairlessness of beetles, though it is curious that HOLM, although frequently finding beetles on the flowers, never found pollen on the beetles. In *V. grandis* it is thought that pollen would be placed around the base of a bird's beak; here again, adhesive would be useful. It would be interesting to know how this very viscous liquid behaves, both on coming into contact with the bodies of birds and insects and again when these touch the receptive stigmas. The alternative arrangement of presenting pollen on staminodes (*V. habrantha*) is interesting and more information on it would be welcome.

EAST (1940) knew of no occurrences of self-incompatibility in this large family. Binucleate pollen is recorded for six of its genera by BREWBAKER (1967) but none of these has secondary pollen presentation.

References for Myrtaceae

BREWBAKER, J. L., 1967.

EAST, E. M., 1940.

ERICKSON, R., GEORGE, A. S., MARCHANT, N. G., MORCOMBE, M. K., 1973: Flowers and Plants of Western Australia. – Sydney: Reed.

HOLM, E., 1978: Some unusual pollination mechanisms in western Australian wildflowers. – W. Austral. Naturalist **14**: 60–62.

HOLM, E., 1988.

HOOKER, W. J., 1861: *Verticordia nitens*. – Bot. Mag. **87**: t. 5286.

KEIGHERY, G. J., 1975: Parallel evolution of floral structures in *Darwinia* (*Myrtaceae*) and *Pimelea* (*Thymeleaceae*). – W. Austral. Naturalist **13**: 46–50.

KEIGHERY, G. J., 1982: Bird-pollinated plants in Western Australia. – In ARMSTRONG, J. A., POWELL, J. M., RICHARDS, A. J., (eds.): Pollination and Evolution: 77–89. – Sydney: Royal Botanic Gardens.

MORLEY, B. D., TOELKEN, H. R., 1983: Flowering Plants in Australia. – Adelaide etc.: Rigby.

SARGENT, O. H., 1918: Fragments of the flower biology of Westralian plants. – Ann. Bot. (Oxford) **32**: 215–231.

SLATER, A. T., BEARDSELL, D. V., 1991: Secondary pollen presentation in the *Chamelaucium* alliance of the *Myrtaceae*: a compact substigmatic ring in *Chamelaucium*. – Austral. J. Bot. **39**: 229–239.

VOGEL, S., 1984: Blütensekrete als akzessorischer Pollenkitt. Mitteilungsband einer Botaniker-Tagung, Wien, 9–14 Sept. 1984: 123. – Wien: Institut für Botanik der Universität Wien.

WERTH, E., 1915: Kurzer Überblick über die Gesamtfrage der Ornithophilie. – Bot. Jahrb. **53**, Beiblatt **116**: 314–378.

Proteaceae

Systematic position: *Rosidae, Proteales*
Restriction of occurrence within the family: Subfamilies *Proteoideae* and *Grevilleoideae* (absent from the three other subfamilies)

Distribution within the subfamilies. The classification of the *Proteaceae* used here is that of JOHNSON & BRIGGS (1975), which is also available in MORLEY & TOELKEN (1983: 238–244), though in a sequence slightly modified by the fact that the taxa are presented in a key. Secondary pollen presentation (as evidenced by the development of a pollen-presenter on the style) occurs in tribe *Proteeae*, in one subtribe of each of the two remaining tribes of the subfamily *Proteoideae* and in all seven tribes of subfamily *Grevilleoideae* (JOHNSON & BRIGGS 1975).
Floral features. Usually zygomorphic; tepals 4, usually strap-shaped, rarely fused basally, often extensively coherent in the congenitally free part, but ultimately separating from below upwards; stamens 4, usually epitepalous; gynaecium monocarpellary with a simple style and two to many ovules. Flowers protandrous. The tip of the perianth is frequently recurved before and after anthesis.
Type of pollen presentation. Usually the filaments are adnate to the tepals to a high level and the anthers introrse. They deposit their pollen on the pollen-presenter at an early stage ('Pseudo-Staubblatt') but after the flower has opened they are

distanced from it by the growth of the style that has taken place in the interim and sometimes by the recurvature of the tepals. JOHNSON & BRIGGS (1975) state that the term 'pollen-presenter' was proposed in 1950 by GUTHRIE and SALTER and they describe it as a slightly to markedly thickened, often oblique or disc-like, upper region of the style which is in contact with the anthers at the time of their dehiscence and to which the pollen adheres (Fig. 20A, C, 22B). The stigma, on the other hand, is often very small and may take the form of a recess or furrow, rather than a knob. The style is often indurated and longer than the perianth; as its tip in such cases is, at the bud stage, held by the coherent tepal apices its middle part forms a loop which protrudes between two of the free tepals whose edges have ceased to cohere ('hairpin' effect, Fig. 20D). After release it is usually either less sharply bent or straight. The pollen-presenter is sometimes sticky or provided with fine hairs among which the pollen is retained. (Dr P. S. WESTON, Royal Botanic Gardens, Sydney, has kindly amended this section.)

Examples of Secondary Pollen Presentation in Proteaceae (Sequence of tribes as in JOHNSON & BRIGGS 1975, re-numbered consecutively)

Subfamily *Proteoideae*

Tribe 3, *Conospermeae*
Isopogon KNIGHT

Species of the genus *Isopogon* differ in the extent to which the functional florets are erect and in the colour of the florets, which may be whitish or purplish pink, and their external hairiness (ERICKSON & al. 1973).

I. anethifolius (SALISB.) KNIGHT. (CAROLIN 1961, illus.). **STR s. ISS exposed. SEQU pln. APPL l?, v? RWD –. SYN** melitto. **VIS** –.
The blossom is of capitulum-like appearance, containing numerous slender yellow flowers. The pollen-presenter is sticky when it receives the pollen; it is not itself swollen, but the style is swollen immediately below it. The appearance of the loaded pollen-presenter in the open flower is spear-like; the stigmatic region is at the extreme tip. The flowers in the head open centripetally.

I. dubius DRUCE (LAMONT 1985).
This species was found to have nectarless flowers and to be visited and pollinated by bees (two species of the short-tongued genus *Leioproctus*). The swollen part of the style just below the stigma is treated as composed of a pollen presenter and a landing knob. The bees alighted on the knob and gathered the pollen. They usually took all the pollen on a single visit but denuded pollen presenters regularly received further visits. The colour of the stylar swelling changed from yellow to orange-red over a period of two days but the reason for this is not apparent.

Petrophile KNIGHT (= *Petrophila* R. BR.)

Petrophile is closely related to *Isopogon* but the flowers are arranged in a shortly cylindrical head and the pollen-presenter is clothed with fine hairs, which receive the flower's own pollen in the usual way. *P. biloba* R. BR. was found by LAMONT (1985) to resemble *Isopogon dubius* in lacking nectar, in having a landing knob and pollen presenter that changed colour and in being visited by pollen-gathering bees (one species of *Leioproctus* and the honeybee, *Apis mellifera*). (See also *Grevillea pilulifera*, Tribe 11, *Grevilleeae*).

P. longifolia R. BR. (BENTHAM 1873, illus.). **STR s. ISS exposed. SEQU pln. APPL l?, v? RWD –. SYN** melitto. **VIS** –. (Fig. 20C).

The hairs of the pollen-presenter are well-developed in this species but may be greatly reduced in others. In all species, however, the extent of the hair tract was found to match exactly the size of the thecae of the anthers at the time of their dehiscence. BENTHAM was unable to be sure what is the function of the swelling immediately below the presenter; it is shown to be coloured red in an otherwise pale straw-coloured flower by ERICKSON & al. (1973: 61) and may, therefore, be a landing knob.

P. pulchella (SCHRAD.) R. BR. (*P. fucifolia* (SALISB.) KNIGHT). (CAROLIN 1961, illus.).

The functional arrangements are as in *P. longifolia* and the general appearance of a floret is similar to that of *Isopogon*.

Tribe 4, *Franklandieae*

Adenanthos LABILL.

A. obovatus LABILL. (BENTHAM 1873, illus.). **STR s. ISS exposed. SEQU pln. APPL ad? RWD n? SYN** ornitho. **VIS** – (see below). (Fig. 20D, E).

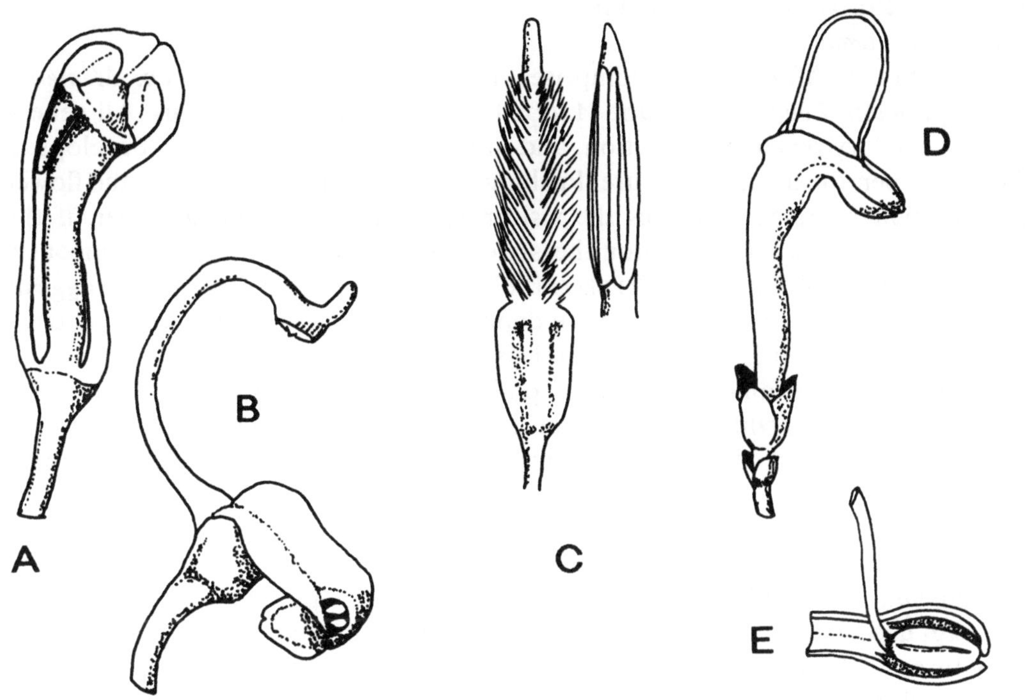

Fig. 20. *Proteaceae. Grevillea buxifolia*: *A* bud opened to show two anthers close to pollen presenter and reflexed appendage of latter (dense hair-clothing of perianth and style not shown): *B* mature flower with repositioned pollen presenter and appendage. *Petrophile longifolia*: *C* style and anther, showing exact match in length of stylar pollen-presenting brush and theca of anther; the terminal part of the style carries the stigma and the swelling below the brush is probably an alighting knob. *Adenanthos obovatus*: *D* general view of flower showing style protruding between partly separated perianth segments; *E* apex of perianth opened to show stylar pollen presenter and anthers (stippled). (P. F. YEO, after W. H. FITCH in BENTHAM 1873.)

About twenty upwardly inclined scarlet flowers are arranged in a loose spike (ERICKSON & al. 1973: 29). Above its hairpin-like loop the style is sharply bent forward again where it is held, with the anthers, in the tightly encasing perianth tip. The pollen-presenter is somewhat dorsiventrally flattened and the anther opposed to its upward surface is barren. The other surface is moist and receives pollen from the fertile anthers. The style is finally released from the perianth with a jerk that, according to BENTHAM, shakes the pollen from its end. However, the photograph in ERICKSON & al. shows a number of exposed styles which appear still to be carrying pollen; as they are well placed to touch the crown of a small, long-billed bird, it seems possible that the system is not designed to scatter pollen, and that any that is scattered is wasted. NELSON (1978), writing of the whole genus, states that the tips of the tepals separate to release the style, implying that the process is spontaneous. The tip of the style has a cleft which opens in the later stage of flowering and presents the receptive stigmatic surface.

Other species of *Adenanthos* have four fertile anthers and a terete, more or less spear-shaped, pollen-presenter (BENTHAM; NELSON 1978). One of these, *A. pungens* MEISSN., is illustrated by BENTHAM, and the pollen presenters of several others by NELSON (l.c.). NELSON states that all species, with one possible exception, are pollinated by nectarivorous birds.

Tribe 5, *Proteeae*

This South African tribe comprises 14 genera including *Protea*, *Mimetes*, *Leucospermum*, *Leucadendron* and *Aulax*, the last two being dioecious. All genera show secondary pollen presentation, and in those named here tension develops in the style when the flower is mature. Probing by a strong pollinator trips the flower and the pollen presenter strikes the visitor and deposits pollen; however, if the flower is not tripped by a visitor the style is eventually released spontaneously (REBELO & al. 1984; COLLINS & REBELO 1987). This generalized description suggests that pollen application is by explosive deposition, but most of the accounts cited below suggest that it is by explosive release of a pollen cloud.

I have no individual accounts of pollen presentation in genera other than *Protea* and *Leucospermum*. In the dioecious genera the gynaecium is present in the male flowers and consists of a typical stylar pollen presenter surmounting an undeveloped ovary.

Protea L.

Protea species have dense, capitulum-like heads of flowers.

P. incompta R. BR. [SCOTT ELLIOT 1890, illus. (pollen presenter only)]. **STR s. ISS expl depos? SEQU pln. APPL a? RWD n? SYN** ornitho. **VIS** –.

Only three of the stamens are fertile and the anthers of these are appendaged. The stylar pollen presenter has horny ridges between which the pollen is deposited. The style becomes bowed and pushes the perianth segment carrying the sterile anther away from the others. There is a minute stigma at the tip of the style which is unable to receive the flower's own pollen. The separation of the perianth members is described as almost explosive (this could have been observed on flowers left untouched or artificially probed).

P. mellifera THUNB. and *P. lepidocarpon* R. BR. [SCOTT ELLIOT 1890, illus. (*P. mellifera*)].

These two species are very similar to *P. incompta*.

P. longiflora LAM. (SCOTT ELLIOT 1890, illus.). **STR s. ISS expl cloud? SEQU pln. APPL a? RWD n? SYN** ornitho. **VIS** –.
This differs from *P. incompta* in the greater elongation and consequent curvature of the style before its release. 'When the separation [of the perianth segments] finally takes place, the style straightens by its own elasticity, and a cloud of pollen is scattered'. Such a release of pollen is evidently what happens when all visitation is prevented but tripping by a visitor might also release the pollen in a cloud; it raises the possibility that wind pollination could occur if visitors were scarce.
P. kilimandscharica ENGLER. (KNUTH 1904, citing VOLKENS). **STR s. ISS expl cloud. SEQU pln. APPL a. RWD n. SYN** ornitho. **VIS** sunbirds (*Aves-Nectariniidae*).
The detailed description of this yellow-flowered species makes it clear that external pressure is the inducement to the opening of the flower and that when this happens the pollen is flung away and bird visitors acquire a pollen deposit on the bill, forehead and breast. In the bud stage the base and apex of the perianth are indurated, while the intervening length is elastic. The style becomes arched during growth and it then protrudes from one side of the flower, pushing outwards one of the four perianth segments, on either side of which a slit appears. Its apex bears a punctiform stigma and below this are eight ridges which become tightly clamped by the dehisced anthers. The style grows further and resembles a drawn bow with a compressed cross-section, and a string formed by the remaining three perianth segments. The flower is now ready to be tripped by a bird's beak entering one of the slits in the perianth. The disturbance suffices to make the indurated tip of the solitary displaced perianth segment break free from the others. The stigma does not take up pollen at the moment of the explosion.
P. amplexicaulis R. BR. and *P. humiflora* ANDREWS. (WIENS & ROURKE 1978, illus. [*P. amplexicaulis*]; WIENS & al. 1983, illus.). **STR s. ISS exposed. SEQU pln. APPL ad. RWD n. SYN** thero. **VIS** mainly rodents, particularly *Aethiomys namaquensis*, *Praomys verreauxi* and *Rhabdomys pumilio* (*Mammalia-Muridae*).
The features of these plants that represent the syndrome of therophily (ROURKE & WIENS 1977) are the low spreading habit, the concealment of the flowers below the foliage, the proximity to the ground of the flowers, the bowl-shaped heads up to 12 cm wide on stout peduncles, the reddish bracts and pale interior of the heads, the wiry styles, the large quantity of nectar (up to 2.4 ml per head) and a yeasty odour. Pollen is offered in longitudinal furrows on a fusiform stylar presenter. The styles are at first arched slightly outwards, but then become angled inwards so that the distal part is at about 45° to the vertical. The stigmatic groove, above the pollen-presenting part of the style, is microscopic. The emergence of the style from the perianth is mildly explosive but the pollen is sticky and is not dislodged by this movement. A curious feature is that the perianth, similarly bent but some distance below the pollen presenter, forms a little trough, about 10mm below the pollen presenter, in which the nectar is presented. Thus a mouse whose head is over the centre of the flower-head and facing outwards can reach the nectar in each floret, and as it does so, it receives pollen from the same floret on the upper side of its snout. There is at least some degree of self-compatibility in *P. humiflora*.

Leucospermum R. Br.
L. conocarpum R. Br. (Scott Elliot 1890, illus.; Knuth 1904, citing Marloth). **STR s. ISS expl cloud. SEQU pln. APPL a? RWD n? SYN** ornitho. **VIS** sunbirds (*Nectarinia chalybea, Nectariniidae*) and sugar-birds (*Promerops cafer, Meliphagidae*). The pollen presenter is pear-shaped and without grooves. The tip of the style is restrained by the perianth lobes in the usual way. When the style bursts free or the perianth is torn apart by a visiting bird the style is released with a small explosion and a scattering of pollen.
L. nutans R. Br. (Scott Elliot 1890, illus.).
In this species pollen is deposited on a flat-topped pollen presenter.

Subfamily *Grevilleoideae*

Tribe 10, *Embothrieae*
Lomatia R. Br.
L. silaifolia (Sm.) R. Br. [Carolin 1961, illus. (style head only)]. **STR s. ISS exposed? SEQU pln. APPL d? RWD n? SYN** melitto. **VIS** –.
The *Lomatia* species are shrubs or trees. *L. silaifolia* and at least its close allies have loose inflorescences of quite numerous whitish flowers not much more than a centimetre long. Before the flower opens the tip of the perianth is spherical, accommodating the disc-like pollen-presenter (Cochrane & al. 1968: 26 and 133, in *L. ilicifolia* R. Br. and *L. fraseri* R. Br.). Then the perianth splits along the lower side and a short loop of the style emerges. It is not clear whether the flower has to be tripped to make the pollen-presenter emerge but Dr P. S. Weston – pers. comm. – thinks this is unlikely, in which case pollen presentation is 'exposed'.
Telopea R. Br.
T. speciosissima (Sm.) R. Br. and *T. oreades* F. Muell. (Carolin 1961, without distinction of species). **STR s. ISS exposed?. SEQU pln. APPL av? RWD n? SYN** ornitho. **VIS** honey-eaters (*Aves-Meliphagidae*).
Both species have red flowers in a capitulum, the involucral bracts being coloured red in *T. speciosissima.* The perianth is arcuate-ascending and the style, although emerging on one side of the split perianth, is not looped. As in *Lomatia* the pollen presenter is disc-like. Probably the pollen-presenter emerges from the perianth spontaneously, in which case the method of pollen issue is 'exposed'.
Tribe 11, *Grevilleeae*
Grevillea R. Br.

In many species of *Grevillea* the perianth is more or less sharply decurved towards the tip and the style protrudes before anthesis from the slit on the upper side as a 'hairpin' (see description of family). Exceptions to this are noted below. The stigma is a hole fringed with fine papilloid hairs opening into a smooth cavity in the centre of the stylar disc (Hildebrand, cited by Knuth 1904, 239; Yeo, unpublished 1982; R. Makinson, pers. comm.). As the flower enters the female stage the lining of the cavity tends to evert, bringing the papillae to the exterior (Fig. 21; Collins & Rebelo 1987: 392b and fig. 5). Probably all species offer nectar as a reward, and most but not all species are bird-pollinated.
G. eriostachya Lindl. (Holm 1988, illus.). **STR s. ISS exposed. SEQU pln. APPL a. RWD n. SYN** ornitho. **VIS** –. (Fig. 22).

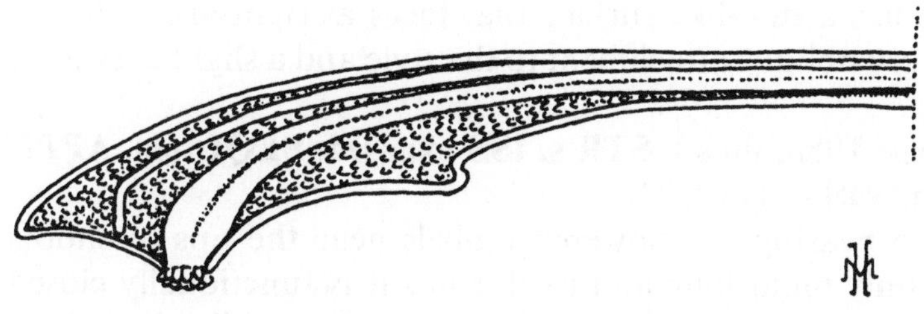

Fig. 21. *Grevillea × lanigera* (*Proteaceae*). Longisection of style showing stigmatic area which is initially hollow, and later is filled in and then protrudes. (M. HICKEY.)

This species is unusual in habit, forming a dense low shrub from which wand-like inflorescence stalks emerge to height of 3m. Near their tips these bear 300–400 flowers in shortly stalked spikes. The perianth is densely covered with hairs but colour in the inflorescence is provided by the yellow styles which straighten when the flower opens and end in a fairly substantial top-shaped pollen presenter. Birds can perch not only on the axes of the inflorescence but also on the flowers, the 'hairpins' of which are quite strong enough to support them.

G. wilsonii A. CUNN. (BENTHAM 1873, illus.). **STR s. ISS exposed. SEQU pln. APPL ad? RWD n** (almost certainly). **SYN** probably ornitho. **VIS** –.

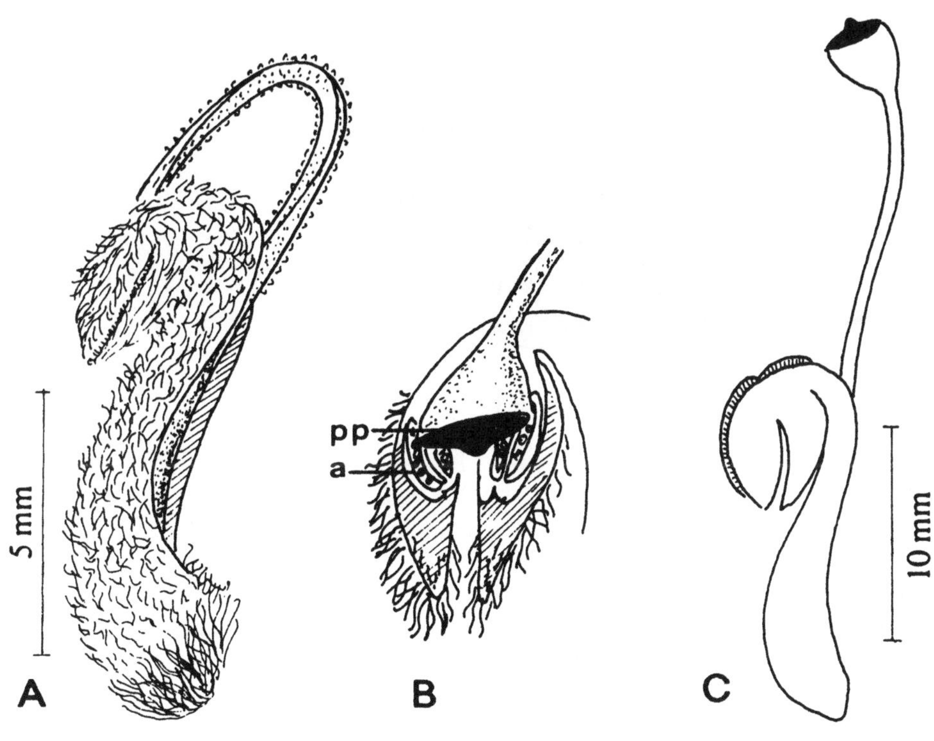

Fig. 22. *Grevillea eriostachya* (*Proteaceae*). *A* flower in advanced bud stage with looped style. *B* the same, section of tip of perianth with anthers (*a*) and pollen presenter (*pp*). *C* open flower (hairs omitted). (HOLM 1988.)

The pollen-presenter here has a disc-like surface that faces at right-angles to the floral axis as a result of a very oblique attachment to the style and a slight curvature of the latter.

G. fasciculata R. Br. (Holm 1988, illus.). **STR s. ISS exposed. SEQU pln. APPL ad. RWD n. SYN** ornitho. **VIS** –. (Fig. 23).

The plant is a small shrub bearing few-flowered umbels near the tips of shoots laterally. The perianth is only 6mm long and its dorsal slit is functionally closed by hairs. The gynaecium is stout and straight and is exposed only distally, where the perianth tips are recurved; the massive pollen presenter is lateral and downwardly-facing as in *G. wilsonii*. Ornithophily is inferred from the robust construction of the flower and the colouring, the perianth being red with orange lobes and the pollen presenter yellow. If, as suspected, the visitors are Western Spinebills (*Acanthorhynchus superciliosus*, Fig. 16F) the pollen would be carried on the distal part of the bill.

G. brownii Meissn. (Holm 1988, illus.)

The perianth is about half as long again as that of *G. fasciculata* and has a saccate nectar-store at the base. The style is again straight with a lateral pollen-presenter but this is exserted by several millimetres. If the flower is visited by the Western Spinebill its style-head will touch the bird's bill in its basal half.

G. buxifolia (Sm.) R. Br. (Bentham 1873, illus.). (Fig. 20A, B).

The style of this species is peculiar in having an appendage by which it is continued beyond the stigmatic disc, which accordingly appears lateral. However, in the bud

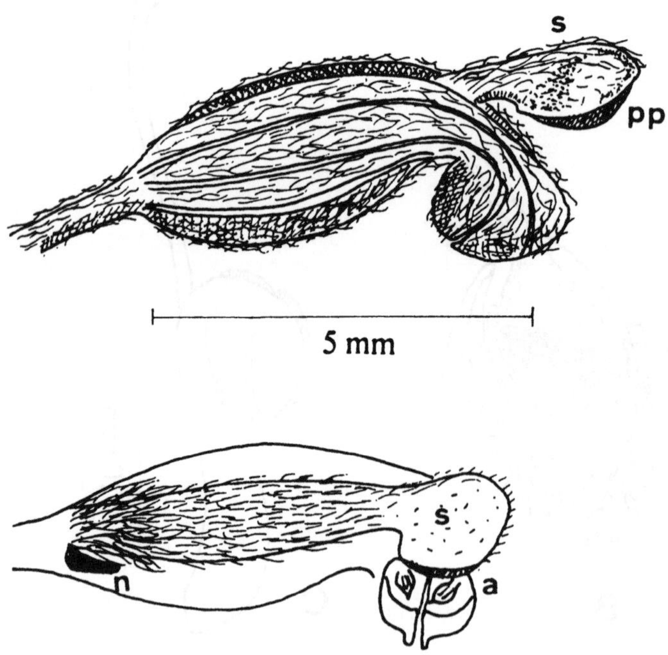

Fig. 23. *Grevillea fasciculata* (*Proteaceae*). Above, side view of flower. Below, side view of style in sectioned flower. *a* = anthers, *n* = nectary, *pp* = pollen presenter, *s* = style-head. (Holm 1988.)

the appendage is reflexed so that the pollen-presenter is more or less symmetrically appressed to the anthers in the usual way. At maturity the perianth is deflexed just above the base, while the arcuate style bends backwards, giving a wide separation between style-head and the nectarial region in the base of the perianth.

G. leucopteris MEISSN. (LAMONT 1982, illus.). **STR s. ISS expl depos. SEQU pln. APPL d? RWD n. SYN** cantharo. **VIS** 70 species of insects, belonging to 12 orders; *Pachytricha* (*Coleoptera*) and *Gryllacris* (*Orthoptera*) were effective as pollinators.

This south-west Australian species is a tall shrub growing as an emergent in scrub. It is quick-growing and behaves as a nomadic colonizer. Its inflorescences are grouped into panicles. The ovary is borne on a gynophore and, since it forms part of the gynaecial 'hairpin', it emerges through the slit in the perianth. When the bud is moderately advanced the tepals on each side of the slit twist outwards in their mid-region, forming a platform. This twisting also tends to advance the slit but this is resisted because the tips of these two tepals are lodged in a groove on either side of the pollen presenter. Pressure on the platform can lever these tepals out of the grooves; the same result is achieved by a blunt-nosed insect approaching the nectar, or by a large insect attempting to pass through the stylar loop. The effectively tubular base of the perianth is 5 mm long and its entrance is closed by hairs. Flowers become receptive to visitors about 18.00 hr and by midnight all flowers are tripped, because any not visited are self-tripping. The stigma is exposed by the recurvature of two lobes, which takes place 8–20 hr after the beginning of floral receptivity. The flowers are self-fertile and much insect-pollination is by 'own pollen'; in addition, unvisited flowers fertilize themselves, since the pollen either gets onto the stigmatic lobes or into overflowing stimgatic fluid as a result of self-tripping. New flowers open in one head over a period of 3 or 4 days; the heads in a panicle, however, are 5 or fewer days out of phase, so that overlap is common. The bagging of inflorescences on wild plants reduced fruit set only to 80% of that of controls; emasculation without bagging brought this down to 20%, showing that some outcrossing occurs.

Of the two effective pollinators mentioned above, the cricket was rare and the beetle common. The beetles took nectar but not pollen, and one of them could trip up to 30 flowers in an hour. The pollination syndrome is shown above as cantharophilous, but most of the components of this syndrome that are present are also those of phalaenophily. These are nocturnal anthesis, cream coloured flowers and strong scent. Absence of clear-cut floral shape and guide-marks indicates cantharophily rather than phalaenophily, but flowers of this type do attract moths [they came to this plant and are suspected of being legitimate pollinators of *Macadamia* (*Proteaceae*, Tribe 13, below)]. The nectar is abundant and dilute at night (13.5–26.5% sucrose equivalent), and sparse and very concentrated during the day. Each anther produces 280 pollen grains, a small number (but there is only one seed per flower).

G. pilulifera GARDNER. (LAMONT 1985).

This species was found to resemble *Isopogon dubius* and *Petrophile biloba* (Tribe 3, *Conospermeae*) in having a landing knob adjacent to the pollen presenter (both changing colour during anthesis), and in offering only pollen, which is collected by bees (two species of *Leioproctus*).

G. vestita MEISSN. (BENTHAM 1873, illus.).
This is a straight-flowered species in which the ovary is borne on a gynophore and passes immediately into a short urceolately thickened style with a conical tip. Only the latter acts as the pollen-presenter; the urceolate part is furrowed but has no role in pollen-presentation (it could be another occurrence of the alighting knob, found in the preceding species). The stigma is at the tip of the conical part and is apparently not receptive at first.

Tribe 13, *Macadamieae*

Lambertia SM.

L. formosa SM. (CAROLIN 1961, illus.). **STR s. ISS exposed. SEQU pln. APPL a. RWD n. SYN** ornitho. **VIS** honey-eaters (*Aves-Meliphagidae*).
The flowers are actinomorphic, with a perianth 5cm long but with the lobes revolute to varying degrees. They are cylindric and arranged seven to a head at the tips of branches of the shrubby plant. Both bracts and perianths are red. The pollen-presenter is slender, adhesive in bud and well-exserted during anthesis. The stigma is a cleft at the tip of the style, reaching down into the zone of pollen deposition; it opens and becomes receptive late in the day on which anthesis begins. Pollen is transported on the bills of the honey-eaters.

L. ericifolia R. BR. (HOLM 1988, illus.). **STR s. ISS exposed. SEQU pln. APPL d. RWD n. SYN** ornitho. **VIS** honey-eaters, White-cheeked – *Phylidonyris niger* – and Tawny-crowned – *Gliciphila melanops* (*Aves-Meliphagidae*).
The 7-flowered inflorescences are scattered through a very loosely growing shrub. The flowers are tubular but slightly zygomorphic. The bud has a slender tip encasing a fusiform pollen-presenter on to which the pollen is deposited. At anthesis the lobes forming the slender part of the bud separate and roll back into tight coils, leaving a fairly widely open throat. The style is 40–45 mm long; the pollen presenter is set at a slight angle to the rest of the style, so that it will rub the crown of a visiting bird. The zygomorphy of the flower is sufficient to ensure that the bird can only probe the flower from one direction. The perianth and style are stiff in the longitudinal direction but are easily bent. The perianth is orange-red and the style yellow.

L. inermis R. BR. (BURBIDGE, HOPPER & COATES 1979).
In a study of the pollen loads of New Holland Honeyeaters in Western Australia it was found that the pollen presenter rubs against the crown of the bird, and that pollen of *L. inermis* was present on the crowns of captured birds. The bills of the same birds bore pollen of two other *Proteaceae* that they were visiting, (*Banksia coccinea* and *Dryandra cuneata*). The same deposition of pollen was observed on the Red Wattlebird (*Anthochaera carunculata*) by HOLM (1988, illus.).

Macadamia F. MUELL.

M. integrifolia MAIDEN & BETCHE and *M. tetraphylla* L. A. S. JOHNSON (SEDGLEY 1982; SEDGLEY & al. 1985, illus.; CORBET 1986; VITHANAGE & IRONDSIDE 1986, illus.). **STR s. ISS exposed. SEQU pln. APPL a? RWD n(p). SYN** phalaeno. **VIS** moths (*Lepidoptera*) and the social bees *Trigona*, *Apis mellifera* – honeybee – and *Bombus* – bumblebees (*Hymenoptera-Apoidea*); the beetle *Metriorrhynchus rhipidius* (*Coleoptera-Lycidae*).
These two Australian species of a tropical and subtropical genus are the source of the *Macadamia* nuts of commerce, and were studied in cultivation by SEDGLEY & al. in Australia and CORBET in New Zealand. The flower is initially

actinomorphic and the pollen presenter is ellipsoidal and sticky. The style develops the usual hairpin-bend as the flower develops, while the perianth bends away from the side on which the style loop emerges. The flower opens in two stages: first the tips of the tepals turn back, exposing the anthers still embracing the pollen presenter; the style then elongates – pushing the anthers apart – and straightens, while the perianth with the anthers remains bent. The stigmatic papillae are confined to a small area at the tip of the style and are not receptive until one or two days after the flower has opened, by which time pollen is likely to have been removed from the pollen-presenting zone of the style by insects. The plant is a tree and its inflorescences are pendent and situated within the canopy. The flowers are white or cream (*M. integrifolia*) or pale pink (*M. tetraphylla*), with a sweet scent that is perhaps stronger in the evenings. These features, combined with the narrow opening between the tepals of the flower and the style-length (10–17 mm), suggest that moths are the pollinators and that they may hover while feeding. There is one report of moths visiting the flowers in New Zealand. The flower does not seem to be adapted to bees, but native and introduced social bees take the nectar and pollen and are effective pollinators when doing so; honeybees hover in order to take the pollen. *Metriorrhynchus* takes nectar; it spends at least 30 minutes in each inflorescence but then flies rapidly to an inflorescence on another tree. *Macadamia* has a self-incompatibility system that operates in the style and is therefore presumably gametophytic.

Tribe 14, *Banksieae*

Banksia L. f.

Banksias have very numerous slender flowers arranged in massive cylindric or somewhat conical heads. Usually the perianth is so slender that its members are little more than secondary stamen filaments. The stiff styles emerge through a slit in the perianth, and are bent like a bow or a hairpin. Pollen-presenters are small and slender, and either sticky or papillose; BENTHAM states (of the whole genus) that when liberated from the perianth they shake off their pollen, which implies that those that are adhesive must dry out. There seems to be wide variation in the ease of spontaneous tripping, leading in many cases to doubt about the method of issue of pollen; the problem is discussed at the end of this generic account.

The prevalent pollination syndrome in *Banksia* is that of ornithophily but that of pollination by non-flying mammals (therophily) is also present (ROURKE & WIENS 1977). Although the most famous flower-visiting Australian mammal, the Honey-Possum (*Tarsipes rostratus* [syn. *T. spencerae* or *spenserae*]) – a marsupial, is an arboreal feeder, the most distinctive expression of therophily in Australian plants is related to ground-living animals.

B. marginata CAV. (BENTHAM 1873, illus.). **STR s. ISS expl cloud. SEQU pln. APPL a? RWD n? SYN** ornitho or thero. **VIS –**.

BENTHAM's illustration of *B. marginata* shows no papillae on the pollen-presenter so it is presumably of the adhesive type. BENTHAM was unsure of the mechanism by which the pollen-presenter is released from the perianth. D. MACFARLAND (pers. comm. 1985) states that the style can be released by applying pressure. The flowers are yellow (BEADLE & al. 1982).

B. integrifolia L. f. (D. MACFARLAND, pers. comm. 1985, illus.; MACFARLAND 1985). **STR s. ISS exposed. SEQU pln. APPL a? RWD n** (n,p for bees). **SYN** ornitho. **VIS** honey-eaters (*Aves-Meliphagidae*).

The release of the style from the perianth can be induced by pressure. This does not scatter the pollen, which remains on the small spherical presenter (although some gets left behind in the depression in the tip of the tepal where the anther resides). The style is straight after release (in some species it remains hooked). In this species there are about 1000 flowers per inflorescence and they open from below upwards in a period of 6 to 12 days.

B. coccinea R. Br. (Burbidge, Hopper & Coates 1979; Holm 1988, illus.). **STR s. ISS exposed. SEQU pln. APPL general. RWD n. SYN** ornitho. **VIS** honey-eaters, Tawny-crowned – *Gliciphila melanops* – and New Holland – *Phylidonyris niger* (*Aves-Meliphagidae*). (Fig. 24).

The short dense heads, subtended by a ring of spreading leafy bracts, become red from below upwards as the flowers open. The colour is provided by the styles, which are sufficiently rigid in the hooked stage to support the birds that perch on them. At anthesis they are straight with a small swollen pollen presenter at the

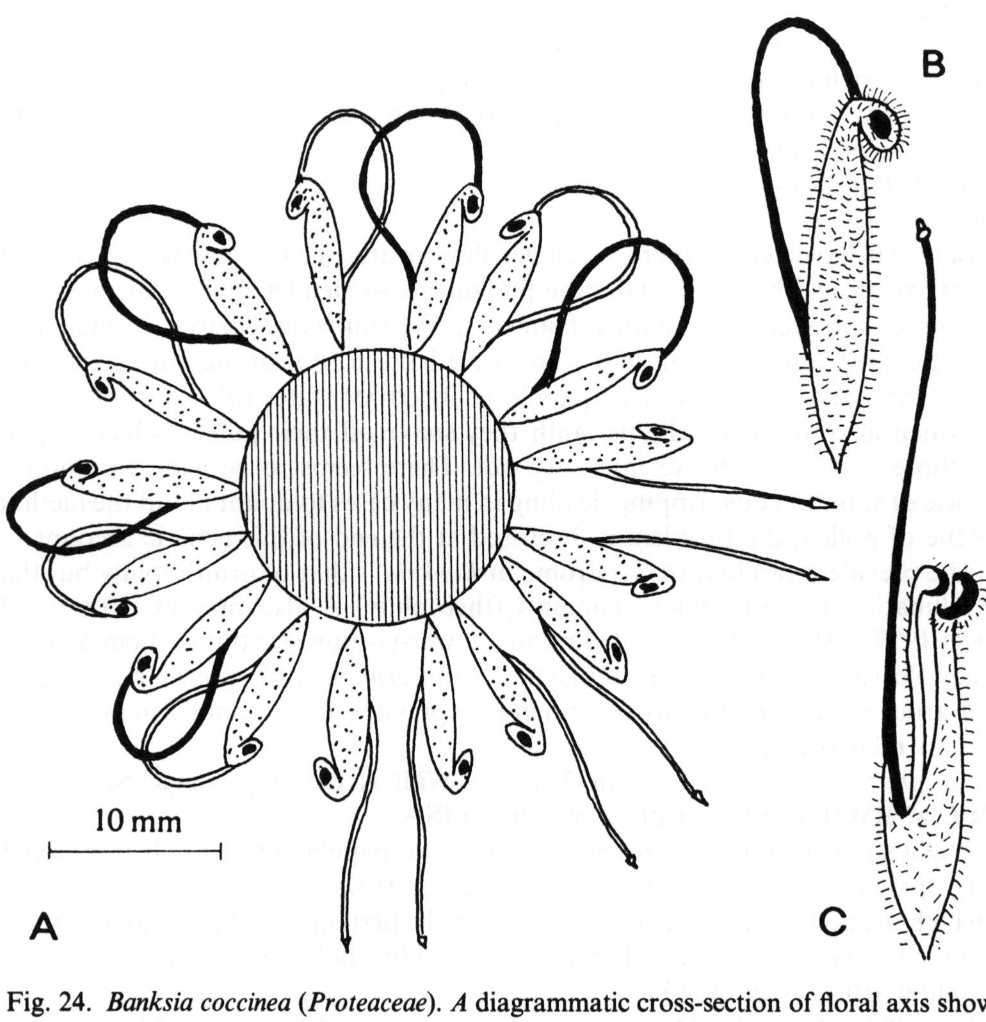

Fig. 24. *Banksia coccinea* (*Proteaceae*). *A* diagrammatic cross-section of floral axis showing radiating flowers in paired rows, younger styles hooked, older straight; alternate hooked styles blacked in for clarity. *B* flower in late bud stage. *C* open flower. (Holm 1988.)

tip. The grey-hairy perianths are slightly less reduced than in some other species. HOLM found no nectar in the flowers, but apparently the nectar runs out of the flowers and is absorbed by plumose bracts between them (secondary nectar presentation!). The distance from the nectar to the pollen presenter is 30–35 mm. In a study of the pollen loads of New Holland Honeyeaters in Western Australia it was found that the pollen of this species (together with that of *Dryandra cuneata* R. BR. [below]) was picked up by the bill of the bird, the pollen presenters reaching only as far as the base of the beak when the bird probed the inflorescence (cf. *Lambertia inermis* above). However, HOLM states that the inflorescence acts like a brush, dusting the bird in many places.

B. prionotes LINDL. (COLLINS & SPICE 1986, illus.). **STR s. ISS exposed. SEQU pln. APPL av. RWD n. SYN** ornitho. **VIS** honey-eaters, mainly *Lichmera indistincta* and *Phylidonyris niger* (*Aves-Meliphagidae*). (Fig. 25A, B).

The flower-head consists of several hundred florets, which appear grey-white when still closed, and orange when the tepals and anthers have separated from the pollen presenters. This separation can be induced by touch but it presumably occurs spontaneously in many cases. The floret is 40 mm long. As in most species of *Banksia* the florets of an inflorescence open from below upwards. The visiting birds perched on the top of the head and leaned down to probe the most recently opened flowers, thus touching the pollen presenters that were most likely to be carrying pollen. Although inter-plant pollinations were relatively infrequent they appeared to be adequate for normal seed-set, which was higher than that of the majority of other species for which figures are available. Some indications of the existence of self-incompatibility were observed.

B. grandis WILLD. (HOLM 1988).

A tree or tree-like shrub, this species produces cylindric flower-heads composed of some thousands of flowers. The buds are green, the styles yellow and the pollen orange, so that zones of colour appear as the flowers open from below upwards. In its flower form, including the arched styles, it resembles *B. prionotes*, but the perianth is much less hairy. HOLM quotes reports of visits by the Little and Red Wattlebirds and the White-eared Honey-eater (respectively *Anthochaera leucotis*, *A. carunculata* and *Meliphaga leucotis*, all *Meliphagidae*).

B. dryandroides BAXTER (HOLM 1978). **STR s. ISS exposed. SEQU pln. APPL ?. RWD n. SYN** thero. VIS –. (Fig. 25C, D).

This species forms a broad, extremely dense shrub 1–1.5 m high. The flower-heads are 4 cm long, axillary and hidden in the interior of the shrub. The perianth has annular folds at the base. The style is only slightly arched and does not project beyond the perianth; it remains bent just below the tip. The head is green with brown perianth-tips when in flower. There are hairy bracts at the base of the flowers, and nectar is found among them (cf. *B. coccinea*) as well as inside the perianth. At noon a faint sweet floral odour was observed but at 22.00 h it was strong and a little sour. Pollen presentation takes place in the usual way. The leaves of the plant are linear and pinnate, with contiguous triangular spine-tipped leaflets; these are horny and have revolute margins giving them further rigidity. If any attempt is made to push the leaves aside they interlock, and a man may have to use all his strength to reach the flowers, which are thus evidently protected from browsers.

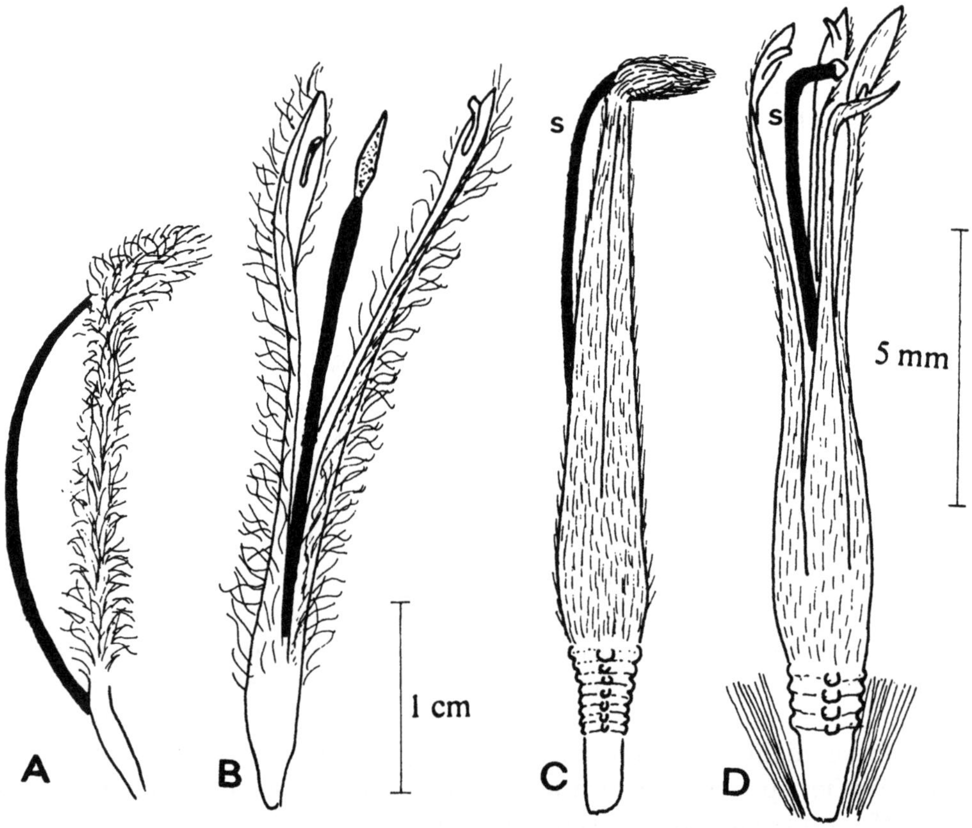

Fig. 25. *Banksia* species (*Proteaceae*) with scarcely looped styles. *A* unopened and *B* opened flower of *B. prionotes*, a bird-pollinated species. *C* opened and *D* unopened flower of *B. dryandroides*, a mammal-pollinated species. *s* = style. (HOLM 1988.)

A species of similar habit described by HOLM is *B. candolleana* MEISSN. HOLM also describes some other *Banksia* species which may be therophilous but show a different habit: these are geophytes with all stems underground and a central rosette; round the edges of the plant are produced cone-like inflorescences at soil level.

For the particular species of *Banksia* of which I have been able to give descriptions above there is uncertainty about the method of pollen issue, as only artificially stimulated release has been reported. However, a general statement for straight-styled species is now available from Dr V. TURNER (pers. comm. 1985). She tells me that in these plants she has seen several species of birds and non-flying mammals tripping the flowers. The pollen is sticky, and the whole mass of pollen on the presenter is usually deposited on the body of the pollinator as a single unit ('explosive deposition'). The dependence of *Banksia* flowers on tripping is further evidenced by marked variations in (apparent) age-structure of flowers in a head and in total time required for all the flowers to open, corresponding to variations in the tripping activity of vertebrates (COLLINS & REBELO 1987: 398a). *Banksia* cannot be tripped by insects, so the requirement for tripping may be important in preventing pollen losses to these animals in a type of flower which is adapted to vertebrate pollination (TURNER 1982, and pers. comm.).

Spontaneous tripping and pollen presentation in the event of a delay in pollinator visits is implicit in the statement of COLLINS & REBELO quoted in the previous paragraph; other publications (MACFARLAND 1985 [on *B. integrifolia* and *B. spinulosa*], PATON & TURNER.1985 [on *B. ericifolia* SM.] and COLLINS & SPICE 1986 [on *B. prionotes*]) also give the impression that the pollen may spontaneously become fully exposed. Further, a table supplied by Dr TURNER giving the number of days for an inflorescence to complete anthesis shows that whereas in some species this is very much longer without outside assistance than with, in *B. spinulosa* and *B. ericifolia* it is exactly the same in both circumstances. It seems probable, therefore, that in such cases there is a period of only mild tension in the flower followed shortly by a spontaneous release of the style. The prevalence of spontaneous tripping, either sooner or later, could account for a general ambiguity on the question of 'exposed' presentation versus explosive types of pollen release that has prevailed in the literature until very recently. There are no reports that support BENTHAM'S opinion that pollen is shaken from the presenter on its release from the perianth.

Dryandra R. BR.

Dryandra species show close parallels in habit, inflorescence form and flower structure with *Banksia*. HOLM (1988) describes *D. formosa* R. BR., a bird-pollinated shrub with conspicuous bronzy golden flower-heads, and *D. mucronulata* R. BR. which is almost identical with *B. dryandroides* and is presumably therophilous.

Discussion of Proteaceae

The taxonomic groups in which secondary pollen presentation is said to occur by JOHNSON & BRIGGS (1975) are shown by name in Table 2-2 together with the names of the genera for which some description of the process of pollen presentation is available.

Table 2-2. Distribution of Secondary Pollen Presentation in the Proteaceae

Taxa in which pollen presenter occurs (tribal numbering as in descriptive section, above)	Genera for which descriptions are given above
Subfam. 2 Proteoideae	
Tribe 3 Conospermeae	
Subtribe Petrophilinae	*Isopogon, Peterophile*
Tribe 4 Franklandieae	
Subtribe Adenanthinae	*Adenanthos*
Tribe 5 Proteeae	*Leucospermum, Protea, Mimetes, Aulax*
Subfam. 5 Grevilleoideae	
Tribe 8 Oriteae	
Tribe 9 Knightieae	
Tribe 10 Embothrieae	*Lomatia, Telopea*
Tribe 11 Grevilleeae	*Grevillea*
Tribe 12 Helicieae	
Tribe 13 Macadamieae	*Lambertia, Macadamia*
Tribe 14 Banksieae	*Banksia, Dryandra*

As the pollen presenter is absent in the simpler-flowered members of all subfamilies of the *Proteaceae* it must have arisen repeatedly in this family (JOHNSON & BRIGGS 1975: 125). These authors conclude that in subfamily *Proteoideae* it arose independently in at least the tribes *Conospermeae*, *Franklandieae* and *Proteeae*. (In *Conospermeae* it is found in only one of five subtribes, and in *Franklandieae* in one of two subtribes.) In subfamily *Grevilleoideae* it is considered to have arisen separately in the tribe *Oriteae* (where it is only feebly developed) while in the remaining six tribes it either arose separately or from a common incipient condition (see Table 2-2).

The massing of the flowers into heads is a common feature in the *Proteoideae* and *Grevilleoideae* but it is not an original or essential feature of the family (JOHNSON & BRIGGS 1975: 125–6). It is probably an adaptation to pollination by birds and small mammals and, in the less spectacular cases, insects (JOHNSON & BRIGGS 1975: 126). A large cone-like inflorescence would seem to be particularly appropriate as an adaptation to non-flying mammals that can crawl over it. However, fruits in *Banksia* are large and it would not be possible for more than a small proportion of the flowers in the head to produce fruits. Accordingly, the heads are thought often to be andromonoecious (with a mixture of bisexual and male flowers).

It is considered by JOHNSON & BRIGGS (1975: 125) that Proteaceous flowers are pre-adapted, by their conformation, to the development of a pollen presenter. The relevant aspects of this floral conformation are presumably the attachment of the stamens to the perianth and the introrse anthers. These two pre-adaptive features are not at all peculiar to the *Proteaceae*. Attachment of stamens to the perianth is a very common feature of Angiosperm flowers, and two possible roles for it seem fairly obvious; one is economy of material and the other is maintenance of the anthers in a particular relationship with the distal part of the perianth for functional reasons. Introrse anthers would be expected to be correlated both with adnation to the perianth and with a more or less tubular or campanulate form of the blossom, which requires axial entry by the pollinator. What is peculiar in many members of the *Proteaceae* is the high degree of adnation of the stamens to the perianth members, and the concomitant reduction of these, so that in many cases they are not much more than coloured stamen filaments, though their coherent tips do also have the role of constraining the apex of the style. This narrowing of the perianth parts is most pronounced in the forms with dense inflorescences of massed florets, where the attribute of conspicuousness resides in the head as a whole. One would not, however, expect that all types of *Proteaceae* had evolved via types with dense inflorescences and narrowed perianth parts; it is therefore difficult to avoid JOHNSON & BRIGGS's conclusion that constitutive factors in the floral structure of the *Proteaceae* predispose them to some rather special responses to selective pressures in floral ecology.

The possible roles of secondary pollen presentation in the *Proteaceae* must now be summarized. Presentation of pollen and stigma in exactly the same position in the flower is achieved in all genera though it is perhaps most significant in those cases where the flowers are large relative to the inflorescence and are presumbaly seen and probed individually. These are *Telopea*, *Lambertia*, *Adenanthos*, *Macadamia*, probably some species of *Grevillea*, and examples in *Isopogon*, *Petrophile* and

Banksia in which an alighting knob for insects is present. Where the flowers are densely massed and small in relation both to the size of the whole inflorescence and the size of the pollinators (birds and mammals) only the identical distance of pollen and stigma from the nectar would be important. Pollen protection until the moment of a pollinator's visit is provided in flowers that have to be tripped; these are found in *Protea*, *Leucospermum* and some species of *Banksia* (see discussion at end of description of these species). Tripping followed by explosive deposition is found in *Banksia* and *Grevillea leucopteris*. Tripping followed by release in an explosive cloud is found in *Protea* and *Leucospermum*. Thus identical placing of pollen and stigma and pollen protection seem to be its principal roles. In addition it has provided the base for the evolution of explosive systems. Another possible factor in secondary pollen presentation in *Proteaceae* is rigidity of the style. Firm construction is a well-known component of the syndromes of ornithophily and chiropterophily and it may also be expected in flowers pollinated by non-flying mammals. Secondary pollen presentation using the style means that strengthening is required in only this one organ instead of also in the perianth and androecium. Particularly in *Banksia*, with its massed flowers, a large perianth is not needed for display or alighting purposes, but stiffening of the style would be needed as an adaptation to vertebrate pollinators. Plants growing in poor soils in a warm dry climate are not likely to be short of photosynthate for stiffening the floral parts but, if there was no secondary pollen presentation and the filaments had to be indurated too, there could be interference with access to the nectar (see Chapter 3, 7.).

It has been suggested by Sussman & Raven (1978) that a pollination relationship involving non-flying mammals was widespread at the end of the Cretaceous, and that this relationship was supplanted by that involving bats during the Tertiary. It is postulated, therefore, that in areas where bats are absent, rare or only seasonally present, some of the present-day relationships between plants and non-flying mammals are relictual, and one of these areas is temperate Australia. *Banksia* might thus be a primitively 'mouse'-pollinated genus with a still further reduced need for a colourful display. Species which then turned to bird-pollination would be pre-adapted in the matter of strengthening and in the distance between the style apex and the nectar. The linear floral parts, being massed together, can provide an adequate visual display when brightly coloured. If the tripping mechanism in these flowers requires a strength greater than that of insects, this would provide protection against nectar robbery by them.

In bird-pollinated Proteaceous inflorescences with fewer flowers, such as those of *Telopea*, *Grevillea* and *Lambertia*, the perianth can have an important role in display and nectar-presentation, and it is therefore well developed and firm; here, as already mentioned, the role of secondary pollen presentation in ensuring identical positioning of pollen and stigmas must be more important. In nectarless species of *Grevillea*, *Isopogon* and *Petrophile* studied by Lamont (1985), visiting bees (including native species) alighted on the pollen presenter to gather pollen, showing that identical positioning is again important. However, secondary pollen presentation seems quite inessential here; it would not need to be present to serve the placement function were the styles to be shortened. Perhaps the flowers are secondarily entomophilous.

Though *Banksia* and similar genera might be primitively vertebrate-pollinated the *Proteaceae* apparently arose in the mid-Cretaceous (JOHNSON & BRIGGS 1975) when the only available pollinating animals were insects (SUSSMAN & RAVEN 1978). The most primitive members of the family were thus insect-pollinated and they probably had loose inflorescences. Not only does it seem certain that secondary pollen presentation has arisen quite independently on separate occasions in Proteaceae (as is implied by JOHNSON and BRIGGS's classification) but very likely that its primary role was not the same in each case.

It is interesting to find that there are *Proteaceae* that have flowers that have to be tripped but do not have secondary pollen presentation. This occurs in the subtribe *Conosperminae* (next to *Petrophilinae* in JOHNSON & BRIGGS 1975). The style, though variously modified, does not function as a pollen presenter (JOHNSON & BRIGGS 1975: 171). When flowers of *Conospermum* are tripped the explosive release of a cloud of pollen occurs (CAROLIN 1961). Before the flower is tripped the ripe pollen is retained in the thecae of the anther; these thecae are cup-shaped and wide open but the rim of each is appressed to the rim of the theca of the neighbouring anther. The flower is bilabiate and only the stamen that lies under the one-lobed upper lip is perfect; the stamen on either side of this has a single fertile theca and a sterile prong-like theca, while the fourth stamen has two sterile prong-like thecae (see HOLM 1988, for clear illustrations). The somewhat elongated style becomes hooked back under the upper lip as the flower develops and is under tension when the flower opens (BENTHAM 1873). The slightest touch on the style or the sterile thecae releases both style and anthers; the style flips downwards in such a way as to touch the object that released it. This will generally be a pollinator – often in fact a fly – from which it may brush off previously deposited pollen. The stigma moves fractionally ahead of the cloud of pollen released from the flower's own anthers at the moment of tripping (CAROLIN 1961; HOLM 1988). The form of the anthers and the way in which they retain pollen is similar in the related genus *Synaphea* but here it is the lower anther and the lower thecae of the lateral anthers that are fertile and the remaining thecae sterile; the completely sterile upper stamen holds the curiously shaped style against the upper lip of the perianth (BENTHAM 1873). Presumably this flower also has to be tripped and the sequence of events when this happens is the same as in *Conospermum*. In *Stirlingia* (subtribe *Stirlingiinae*, Tribe *Conospermeae*) the thecae are all fertile and are shaped and arranged like the fertile thecae of *Conospermum* and *Petrophile* (BENTHAM 1873). Although it is conceivable that secondary pollen presentation could have arisen from such a system, the discussion so far implies that this would not have been the only route to its acquisition. The mechanism in these *Conosperminae* evidently promotes pollen conservation, as is very often the case where secondary pollen presentation occurs. Possibly, therefore, the overriding factor in floral evolution in this family has been protection of the pollen from loss from one cause or another.

Proteaceae are normally protandrous by 24–36 hours (COLLINS & REBELO 1987) so that self-pollination resulting from the stylar system of secondary pollen presentation is minimized. Self-incompatibility occurs in *Macadamia* and it was found in ten species of *Protea* and three of *Leucospermum*, while two of the latter genus were self-compatible (HORN 1962). Indications of self-incompatibility were also found by Trelease in *Grevillea* (KNUTH 1904). A recent review (COLLINS & REBELO 1987) showed self-incompatibility in two species of *Dryandra* and partial or complete self-incompatibility in all seven species of *Banksia* for which data were available. Full self-compatibility was found in three species of *Grevillea* and one of *Hakea*. (COLLINS & REBELO 1987, suggest that various incompatibility systems might be operating.) As in other families, dichogamy is apparently related to secondary pollen presentation whereas self-incompatibility is probably not. Binucleate pollen is reported in *Grevillea*, *Leucadendron* and *Macadamia* (BREWBAKER 1967).

References for Proteaceae

BREWBAKER, J. L., 1967.

BEADLE, N. C. W., EVANS, O. D., CAROLIN, R. C., TINDALE, M. D., 1982: Flora of the Sydney Region (3rd. edn.). – Frenchs Forest and Welllington: Reed.

BENTHAM, G., 1873: Notes on the styles of Australian *Proteaceae*. – J. Linn. Soc. Bot. **13**: 58–65.

BURBIDGE, A. H., HOPPER, S. D., COATES, D. J.; 1979: Pollen loads on New Holland honeyeaters at Qualup, Western Australia. – W. Austral. Naturalist **14**: 126–128.

CAROLIN, R., 1961: Pollination of the *Proteaceae*. – Austral. Mus. Mag. **13**: 371–374.

COCHRANE, G. R., FUHRER, H. A., ROTHERHAM, E. R., WILLIS, J. H., 1968: Flowers and Plants of Victoria. – Sydney: Reed.

COLLINS, B. G., REBELO, A. G., 1987: Pollination biology of the *Proteaceae* in Australia and southern Africa. – Austral. J. Ecol. **12**: 387–421.

COLLINS, B. G., SPICE, J., 1986: Honeyeaters and the pollination biology of *Banksia prionotes* (*Proteaceae*). – Austral. J. Bot. **34**: 175–185.

CORBET, S. A., 1986: *Macadamia* pollination: some questions. – New Zealand Macadamia Nutgrowers Co-op. Co. Ltd News Bull. August 1986: 8–11.

ERICKSON, R., GEORGE, A. S., MARCHANT, N. G., MORCOMBE, M. K., 1973: Flowers and Plants of Western Australia. – Sydney: Reed.

HOLM, E., 1988.

HORN, W., 1962: Breeding research on South African plants: II. fertility of *Proteaceae*. – J. S. African Bot. **28**: 259–268.

JOHNSON, L. A. S., BRIGGS, B. G., 1975: On the *Proteaceae*–the evolution and classification of a southern family. – Bot. J. Linn. Soc. **70**: 83–182.

KNUTH, P., 1904.

LAMONT, B., 1982: The reproductive biology of *Grevillea leucopteris* (*Proteaceae*), including reference to its glandular hairs and colonising potential. – Flora **172**: 1–20.

LAMONT, B., 1985: The significance of flower colour change in eight co-occurring shrub species. – Bot. J. Linn. Soc. **90**: 145–155.

MORLEY, B. D., TOELKEN, H. R., 1983: Flowering Plants in Australia. – Adelaide etc.: Rigby.

MACFARLAND, D., 1985: Flowering biology and phenology of *Banksia integrifolia* and *B. spinulosa* (*Proteaceae*) in New England National Park, New South Wales. – Austral. J. Bot. **33**: 705–714.

NELSON, E. C., 1978: A taxonomic revision of the genus *Adenanthos* (*Proteaceae*). – Brunonia **1**: 303–406.

PATON, D. C., TURNER, V., 1985: Pollination of *Banksia ericifolia* SMITH: birds, mammals and insects as pollen vectors. – Austral. J. Bot. **33**: 271–286.

REBELO, A. G., SIEGFRIED, W. R., CROWE, A. A., 1984: Avian pollinators and the pollination syndromes of selected Mountain Fynbos plants. – S. African J. Bot. **3**: 285–296.

ROURKE, J., WIENS, D., 1977: Convergent floral evolution in South African and Australian *Proteaceae* and its possible bearing on pollination by nonflying mammals. – Ann. Missouri Bot. Gard. **64**: 1–17.

SCOTT ELLIOT, G. F., 1890.

SEDGLEY, M., 1982: Pollination biology of fruit and nut crops. – In WILLIAMS, E. G., KNOX, R. B., GILBERT J. H., BERNHARDT, P. (eds.): Pollination **'82**: 69–73. – Parkville: University of Melbourne.

SEDGLEY, M., BLESING, M. A., VITHANAGE, H. I. M. V., 1985: A developmental study of the structure and pollen receptivity of the *Macadamia* pistil in relation to protandry and self-incompatibility. – Bot. Gaz. **146**: 6–14.

SUSSMAN, R. W., RAVEN, P. H., 1978: Pollination by lemurs and marsupials: an archaic coevolutionary system. – Science **200**: 731–736.

TURNER, V., 1982: Marsupials as pollinators in Australia. – In ARMSTRONG, J. A., POWELL, J. M., RICHARDS, A. J. (eds.): Pollination and Evolution. – Sydney: Royal Botanic Gardens.

VITHANAGE, V., IRONSIDE, D. A., 1986: The insect pollinators of macadamia and their relative importance. – J. Austral. Inst. Agric. Sci. **52**: 155–160.

WIENS, D., ROURKE, J. P., 1978: Rodent pollination in southern African Protea spp. – Nature **276**: 71–73.

WIENS, D., ROURKE, J. P., CASPER, B., RICKART, E. A., LAPINE, T. R., PETERSON, C. J., CHANNING, A., 1983: Nonflying mammal pollination of southern African proteas: a non-coevolved system. – Ann. Missouri Bot. Gard. **70**: 1–31.

Meliaceae

Systematic position: *Rosidae, Sapindales*
Restriction of occurrence within the family: Subfamily *Melioideae* (one of five subfamilies – D. J. MABBERLEY in HEYWOOD 1978)

Distribution within the subfamily. Secondary pollen presentation has previously been reported only briefly in *Meliaceae*. It is known only in the southern African genus *Turraea*.
Floral features. Flowers with varying numbers of free or united sepals and the same number of usually free petals. Stamens usually 5 to 10, often more numerous than the petals; filaments normally united into a tube.
Type of pollen presentation. On the style-head ('Pseudo-Staubblatt').

Examples of Secondary Pollen Presentation in Meliaceae

Turraea L.

Mr F. WHITE tells me that the appearance of the style-head of most species of this genus suggests that it acts as a pollen presenter. Also, that under normal conditions virtually all the pollen is transferred to the style-head but failure of transfer may occur if temperature and humidity are not right. In addition, if the atmosphere is dry pollen may then fall from the style-head.

T. vogelioides BAGSHAWE. (WHITE & STYLES 1963). **STR s. ISS exposed. SEQU pln. APPL** face or proboscis?. **RWD n. SYN ? VIS** –.
Dehiscence of the anthers before the flower opens and deposition of the pollen on the proximal part of the style-head was seen on a plant in the University Botanic Garden, Oxford. A change in the colour of the stigmatic surface from green to yellow one or two days after the flower opened was taken to indicate the beginning of its receptivity. This account appears in a Flora which does not treat this species. However, the generic description states that the proximal part of the style-head functions as a receptaculum pollinis (the term used for 'pollen presenter' by BREMEKAMP in dealing with the *Rubiaceae*, q. v.). The authors considered that

this interpretation needed confirmation in the field for this and other species. They supposed that species with very long staminal tubes are pollinated by moths.

T. obtusifolia HOCHST. (YEO, unpublished 1984–5; MARLOTH 1925. illus.; WHITE & STYLES 1963, illus.; WHITE 1986 1988, illus.). **STR s. ISS exposed. SEQU sim. APPL** face or proboscis? **RWD n. SYN** sphingo. **VIS** moths and butterflies (*Lepidoptera*). (Fig. 26).

The insignificant calyx is a five-lobed cup 2–3 mm long. The corolla is white and before opening it is narrowly club-shaped; it consists of five narrowly spathulate petals that attain a length of about 3cm and are erect at the base and spreading or recurved distally. The androecium also attains a length of about 3 cm and its filamental tube is white and petaloid and also narrowly club-shaped. Just below the anthers it is somewhat constricted and bears a coronal fringe of numerous narrow, divergent segments. The anthers are introrse. The style terminates in a head which is urn-shaped in profile, the slightly expanded and nearly flat top representing the stigmatic surface. The head is green in colour and just before the flower opens the sides of the 'urn' are level with the anthers, which have by then dehisced. Stages representing the opening of the flower have not been seen by me, but once it is open there is always a gap between the pollen-caked stylar head and the anthers (which I found were retaining part of their pollen–but see introduction to genus). The gap is about 1mm in the young flower and 2 mm in older ones. The annular space between the top end of the filamental tube and the style is also only about 1mm. The stigma looks moist in the early stages; it does not normally receive any of the pollen that is released in the bud. Nectar is present in the staminal tube and has been found to fill it to a height of 6 mm. I could detect no scent in the flowers, nor could WHITE (pers. comm.). The report of visitors is due to MARLOTH.

These observations were made on a plant growing in the Temperate House at the University Botanic Garden, Cambridge, in the autumn and late summer of 1984 and 1985 respectively. The plant was obtained from the Royal Botanic Gardens, Kew, in 1981 and is an evergreen shrub, 50 cm tall, with spreading branches bearing (in season) numerous erect flowers. WHITE (1988) says that in forest it can scramble to a height of 5m. In the greenhouse pollen gradually disappeared from the style-head, perhaps by falling after loss of adhesion (see introduction to genus).

T. floribunda HOCHST. (WHITE & STYLES 1963, illus.; WHITE 1986). **STR s. ISS exposed. SEQU –. APPL** face or proboscis? **RWD n. SYN** sphingo. **VIS** hawkmoths (*Lepidoptera-Sphingidae*).

The flower is similar in form to that of *T. obtusifolia* but it has the petals and staminal tube still more elongated and the style exserted by at least 1 cm. The petals are greenish white and there is a strong vanilla scent (WHITE, pers. comm.). Hawkmoths were seen to arrive at the trees at dusk and stay for 15–30 minutes; they arrived and departed en masse. It is suggested that the fringe of appendages on the staminal tube helps the insects to locate visually the very narrow entrance.

Discussion of Meliaceae

The flowers of several species of *Turraea* present features typical of the hawkmoth-pollination syndrome (sphingophily). The proboscis must be slender and approxi-

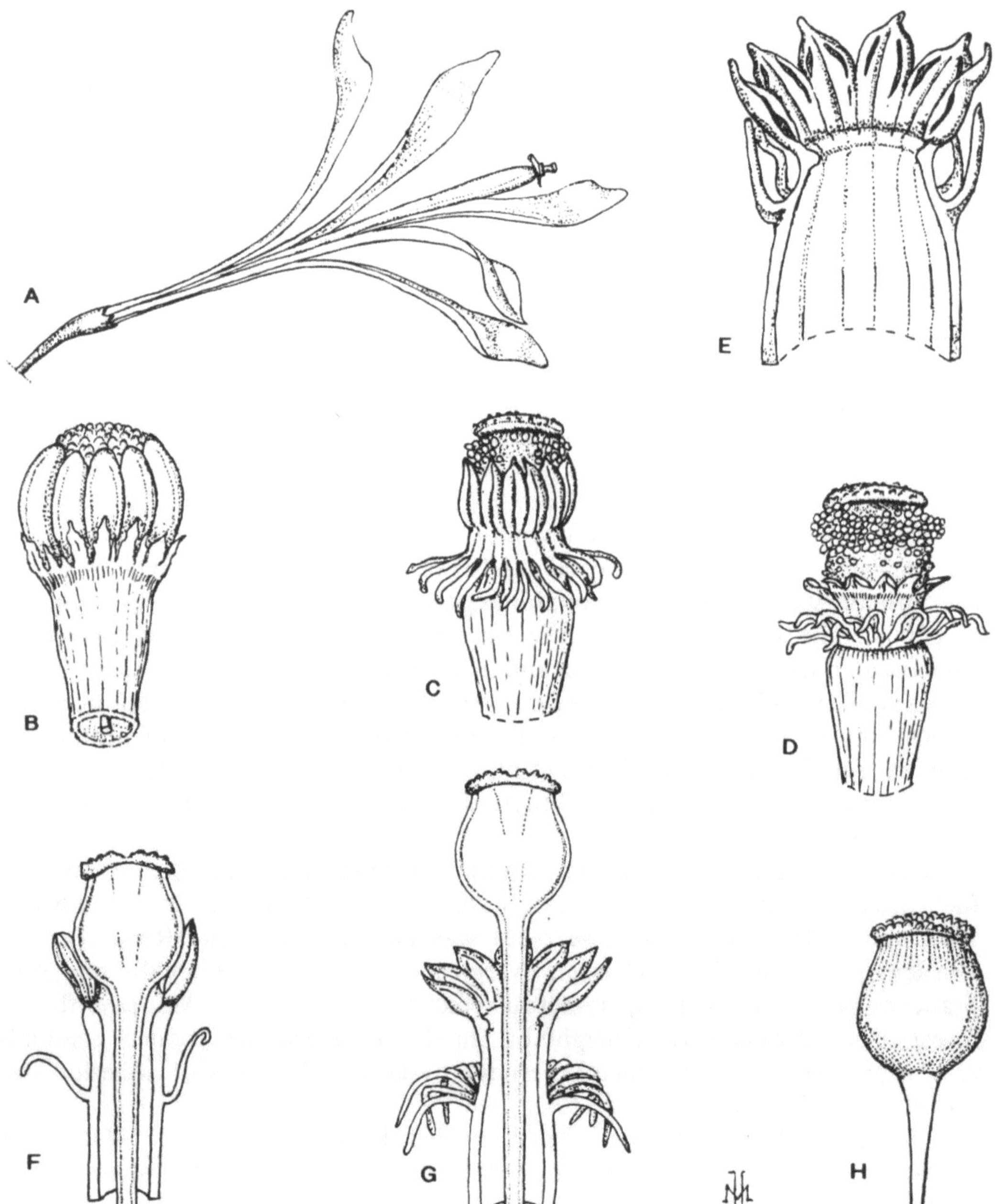

Fig. 26. *Turraea obtusifolia* (*Meliaceae*). *A* side view of flower; staminal column is 31 mm long. *B* androecium and style-head from a bud; anthers about 1 mm long. *C* the same, at time of anther-dehiscence. *D* the same, later, showing pollen deposit on style-head; fringe of staminal appendages is 4 mm in diameter. *E* apex of androecium at beginning of anther-dehiscence, opened out. *F*, *G* sections of androecium and style-head, corresponding to 'C' and 'D' but in 'G' there is greater separation between pollen presenter and anthers. *H* style-head from bud at same stage as 'B'. (M. HICKEY.)

mately 3 cm long or more in order to reach the nectar. The narrow entry to the flower, devoid of proboscis guides, is rather surprising. The lobes of the corona just outside the anthers probably help the insects to orientate visually, but they cannot conduct the proboscis to the entry because of the upstanding anthers; the insects would therefore need to learn by trial and error the position of the entry in relation to the corona.

Although the value of presenting the pollen on the stylar head is not glaringly obvious, it would seem that it must lie mainly in promoting its uptake and proper deposition by pollinators. Two other roles of secondary pollen presentation, namely identical positioning of the pollen-presenting and pollen-receiving organs, and the provision of a more rigid pollen-presenter than is usually provided by stamens, are already carried out in the *Turraea* flower; the floral features that see to this are respectively the narrow staminal tube approximately equalling the style, and the rigidity of the tube.

Sphingophily is not the only pollination syndrome to occur in *Turraea*, for WHITE & STYLES (1963) illustrate species with short and broad staminal tubes and refer to an observation of honey-birds visiting flowers of a species with a long gradually widened staminal tube.

Binucleate pollen is recorded for four genera and trinucleate for one in this family, but *Turraea* is not covered (BREWBAKER 1967).

References for Meliaceae

BREWBAKER, J. L., 1967.

HEYWOOD, V. H., (ed.), 1978.

MARLOTH, R., 1925: The Flora of South Africa **2(1):** 112–114. – Capetown and London: Darter and Wheldon & Wesley.

WHITE, F., 1986: The taxonomy, chorology and reproductive biology of southern African *Meliaceae* and *Pteroxylaceae*. – Bothalia **16**: 143–168.

WHITE, F., 1988: *Turraea obtusifolia*. – Fl. Pl. Africa **50**: t. 1962.

WHITE, F., STYLES, B. T., 1963: *Meliaceae* – In EXELL, A. W., FERNANDES A., WILD, H. (eds.): Flora Zambesiaca **2(1)**: 285–319. – London: Crown Agents.

Rafflesiaceae

Systematic position: *Rosidae, Rafflesiales*
Restriction of occurrence within the family: Tribe *Cytineae*

Distribution within the tribe. Known only in *Cytinus*.
Floral features. Flowers actinomorphic, unisexual, with one syntepalous perianth whorl. Stamens united into a column. Ovary inferior. Style one.
Type of pollen presentation. Pollen is presented on the hairs of the perianth ('Pollenhaufen').

Examples of Secondary Pollen Presentation in Rafflesiaceae

Cytinus L.
C. ruber (FOURR.) KOMAROV. (KUGLER 1977, illus.). **STR p, (a)?. ISS exposed. SEQU na. APPL a, general?. RWD n, p. SYN** melitto. **VIS** mainly large bees (*Hymenoptera-Apoidea*) but also smaller bees and hoverflies (*Diptera-Syrphidae*).

The unisexual flowers of this parasitic plant are borne at ground level in red-bracted clusters of 7–11; they are 20–30 mm long. Both sexes are represented in each inflorescence. The perianth is yellow and in male flowers funnel-shaped, in female campanulate. The tube is divided into four passages by septa connecting the perianth with the staminal or stylar column, as the case may be. A nectary is situated at the base of each passage thus formed. Above the nectary the perianth bears large hairs. In male flowers the sticky pollen is shed into the space between the anthers and the perianth and apparently supported there by the hairs. The flowers appear to be primarily adapted to large bees of the family Anthophoridae which are common in spring in the Mediterranean area where this plant occurs. These bees visit the flowers systematically and pick up pollen from male flowers on their long proboscides and the upper and lower sides of the head. Shorter-tongued insects seek pollen and apparently cause pollination also since they walk over the female flowers while visiting the male.

Discussion of Rafflesiaceae

The pollination system here is not quite a 'mess-and-soil' system, as that is usually understood, because it does not apply pollen to the body of the pollinator generally, at least in the case of those species to which it seems to be adapted. Assuming, as seems likely, that not all the pollen is expelled from the anthers, the effect of the pollen presentation system is similar to that found in *Myristica* (*Myristicaceae*). It seems possible that as there is no directionality in the approach of the insect to the nectar passage it is advantageous to apply pollen all round the proboscis and to both sides of the head. The same consideration could apply to the body of the beetle that pollinates *Myristica*. The function of secondary pollen presentation is not at all clear in this case. Secondary pollen presentation on the 'mess-and-soil' principle, excluded from this review, occurs elsewhere in the family (*Rafflesia*, in tribe *Rafflesieae*–Knuth 1904) and this might have become modified in *Cytinus* to give a less general application of pollen to the pollinator.

Pollen-nuclear number is reported by Brewbaker (1967) for two genera in this family, not including Cytinus: it is two.

References for Rafflesiaceae

Brewbaker, J. L., 1967.
Knuth, P., 1904.
Kugler, H., 1977: Zur Bestäubung mediterraner Frühjahrsblüher. – Flora **166**: 43–64.

Vochysiaceae

Systematic position: *Rosidae, Polygalales*
Restriction of occurrence within the family: Unknown (see following section)

Distribution within the family. The family consists of trees, shrubs and climbers. There is a genus of three species in tropical West Africa, and the rest of the family (5 genera including *Vochysia*, the only one for which secondary pollen presentation is known) is tropical American.
Floral features. Slightly zygomorphic. Sepals 5, connate at base, outer one saccate or spurred at base; petals 2–5, unequal, or only one; stamen 1, staminodes 2–4;

ovary superior or inferior, carpels 1–3, ovules 1–many, style single, with simple stigma.
Type of pollen presentation. The anther lies near the style in the bud and as the flower opens the style elongates, taking pollen with it ('Pseudo-Staubblatt') (observation of Prof. ST. VOGEL, communicated by Dr CHR. WESTERKAMP 1988).

Examples of Secondary Pollen Presentation in Vochysiaceae

Vochysia AUBLET
Vochysia species (P. E. GIBBS, pers. comm. 1990)
STR s. ISS exposed. SEQU sim? APPL ? RWD n. SYN ? VIS –.
In this genus there is a spurred sepal and only one petal which is narrow and rolled round the inner parts; both of these perianth parts are placed on the upper side of the horizontally directed flower. The style is nearly as long as the petal and has a capitate stigma. The stamen has a short filament lying below the style and a long wide anther which embraces the style from below. At maturity the petal partially unrolls and curves back slightly. The anther soon falls away, leaving a large deposit of pollen along the greater part of the style's length. The stylar surface is smooth and some product of the anther appears to be responsible for sticking the pollen to it; this forms a deposit that 'can be wiped cleanly away with the finger'.

Discussion of Vochysiaceae

This family provides case of exposed presentation, with the style acting as pollen presenter as in *Proteaceae*, *Campanulaceae* and *Rubiaceae*.

References for Vochysiaceae

None

Polygalaceae

Systematic position: *Rosidae, Polygalales*
Restriction of occurrence within the family: Tribe *Polygaleae*

Distribution within the tribe. Only certainly known in *Comesperma* and the large genus *Polygala*, which accounts for about half the species in the family. In *Muraltia heisteria* DC. the flower is explosive, with pollen donation and reception taking place during a single pollinator-visit, but insufficient information is given about the method of pollen presentation (SCOTT ELLIOT 1891). *Polyglaeae* is the main tribe of the family.
Floral features. Flowers zygomorphic; members of each whorl usually united to some degree. Sepals usually five with the odd one on the upper side: two upper laterals free and usually enlarged and petaloid, two lower sometimes fused. Petals five or, more usually, three. The lower (median) petal is longitudinally folded into a carina (keel) and embraces the stamens; often it has a crest of filamentous appendages at the tip. Stamens usually eight and with filaments united to form a sheath split along the upper side and more or less united with the petals at base.

Anthers basifixed, usually opening by an apical pore.
Type of pollen presentation. The form of the flower in *Polygala*, particularly as it concerns the keel and the stamen filaments, is comparable with forms found in *Leguminosae*, subfamily *Papilionoideae* (MÜLLER 1881, gives a detailed comparison; see also 'Discussion of Polygalaceae'). Pollen is presented as a deposit on the style ('Pseudo-Staubblatt') or in the keel, whence it is pushed out by the style ('Nudelspritze'?); the stigmatic secretion is frequently employed as an adhesive to stick the pollen to the pollinator. Alternatively, in *Polygala* the style plays a part in an explosive system.

Examples of Secondary Pollen Presentation in Polygalaceae

Polygala L.
P. comosa SCHKUHR and *P. vulgaris* L. (MÜLLER 1883, citing HILDEBRAND for *P. vulgaris*, illus.). **STR s. ISS rel(dose)-poll. SEQU stg. APPL av. RWD n. SYN** melitto. **VIS** bees (*Hymenoptera-Apoidea*), butterflies (*Lepidoptera*). (Fig. 27: *P. comosa*).
The stamen filaments are fused to the inner surface of the keel and nectar is secreted within the base of the latter. The style is stout and somewhat flattened and dilated near the tip which is excavated on the upper side. The proximal edge of the excavation bears a small recurved tongue which is the stigma. Before the

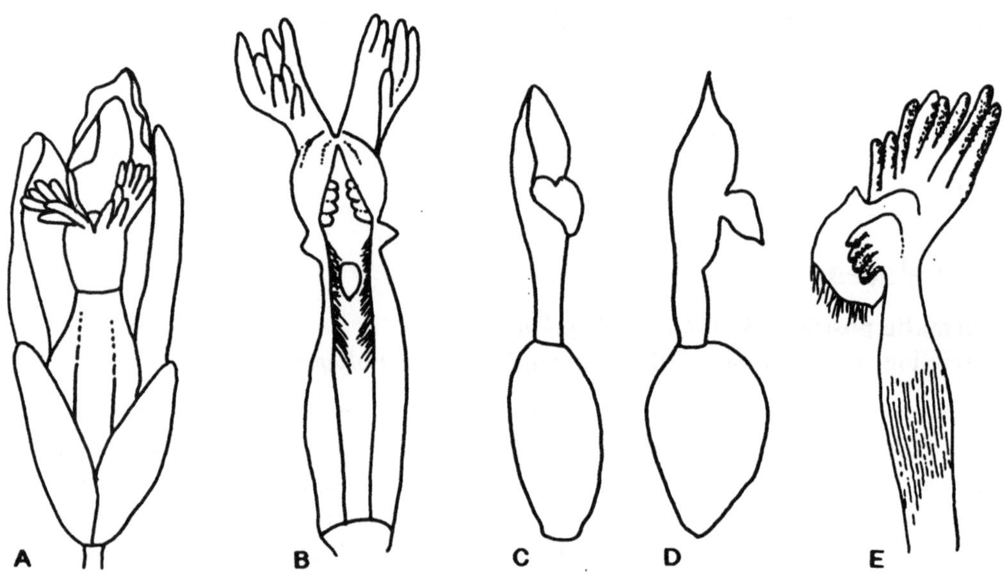

Fig. 27. *Polygala comosa* (*Polygalaceae*). *A* flower from below, showing at the base two protective sepals, at the sides two display-sepals, centrally the lower petal forming the keel and bearing finger-like appendages and, at the apex, two overlapping display-petals. *B* lower petal from above, further enlarged, with stigma and anthers visible within. *C* ovary and style seen obliquely from above, further enlarged, with apical pollen presenting 'spoon' and reflexed tongue-like stigma. *D* the same in profile. *E* inner (adaxial) surface of half of the lower petal just before maturity, with anthers attached. (P. F. YEO, after MÜLLER 1883.)

flower opens the anthers deposit their pollen in the excavation and then shrivel. Insect visitors to the open flower cling to the crest of the lower petal and probe the keel for nectar. As the proboscis enters and withdraws it touches the upwardly projecting stigma and may deposit foreign pollen on it; it becomes gummed by the stigmatic secretion and during withdrawal it picks up the flower's own pollen from the deposit at the tip of the style. The number of pollen grains and ovules in each flower is small. (The keel-petal is pouched near the apex and proximal to the pouch there is a transverse fold on either edge, suggesting that the pouch can be depressed by the insect. These species belong to a group in which polymorphism for flower-colour prevails, the possible colours being blue, pink or white.)

P. chamaebuxus L. (Müller 1881, 1883, both illus.; Knuth 1908, citing Müller, illus.). **STR c, s. ISS rel(dose)-poll. SEQU pln? APPL av. RWD n, p? SYN** melitto. **VIS** various bees (*Hymenoptera-Apoidea*), *Lepidoptera* (stealing).

The flowers of this species are larger and firmer in construction than those of the preceding. The one upper and two lower sepals are sepaloid, firm, and overlapping, functioning as a support for the bases of the other parts and as a protective cover for the nectar. The two lateral sepals are large, white and petaloid, and they diverge widely from the base of the flower. The petals are yellow, the two laterals and the lower one being fused to form the keel. The stamens are related to this as in the preceding species. The ovary is many times shorter than the style, instead of about the same length. The lower part of the style is slender but fused to the stamens, whereas its free distal part is thick and firm and curved upwards. The distal third or quarter of the keel is free from the lateral petals and hinged to allow it to be depressed; when this happens the style tends to emerge. The tip of the style has a truncate and hollowed-out apex from the edge of which emerges a capitate stigma on a comparatively slender stalk. In the bud stage the style-apex is beyond the anthers but their relative position changes before the flower opens and pollen is then deposited in the upper part of the hinged portion of the keel. After this the anthers shrink back again. When visitors depress the keel successive doses of pollen are pushed out by the style apex. Although the petaloid sepals are equivalent to the standard of *Papilionoideae* in the display function they are not able to offer the resistance which the pollinator needs in order to exert downward pressure on the keel; this resistance is provided by the two upper petals with the support of the upper sepal. Müller (1881) envisaged three possible ways for preventing autogamy but did not know which applied in this case; they were delayed receptivity of the stigma (protandry), self-incompatibility and prepotency of foreign pollen. Knuth mentioned the possibility that there might be a stigmatic membrane as in the *Papilionoideae*. If the stigma is receptive at the beginning of anthesis, then the sequence of pollen and stigma presentation is 'stigma first', but if stigmatic receptivity is delayed the opposite will apply. (The flower-scent is described as plum-like, as it is in a number of spring-flowering bee-pollinated plants. The lobes of the crest of the lower petal are few and globose and, being yellow in colour, look remarkably like anthers. The yellow wing petals become dark red at the end of flowering).

P. myrtifolia L. (Müller 1883, citing Delpino; Scott Elliot 1891). **STR c, s. ISS expl depos. SEQU stg or sim. APPL vl. RWD n. SYN** melitto. **VIS** the large bee *Xylocopa violacea* (*Hymenoptera-Apoidea*) (outside native range of plant).

This plant is a South African shrub with conspicuous red-purple flowers, 2cm or more in width. It is slightly asymmetric (as in *Lathyrus–Leguminosae*). Three petals participate in forming the keel. When the flower is mature the style is under tension. A visiting insect depresses the distal pouch of the keel, increasing still further the internal pressure until the sides of the keel part and the style springs free with some force. Details of the form of the style-apex and pollen presentation are not supplied but the anthers dehisce in the bud and it seems that the style-apex and petals must present the pollen.

P. vauthieri CHOD. (BRANTJES 1982, illus.). **STR c, s. ISS rel(dose)-poll. SEQU stg. APPL avl. RWD n. SYN** melitto. **VIS** the large bees *Megachile*–leaf-cutter bee and *Apis*–honeybee (outside its native range), also small bees (stealing) (*Hymenoptera-Apoidea*). (Fig. 28A, B).

The flower of this South American species is blue, with two display sepals exceeding 1 cm in length and breadth and a keel, formed by the lower petal, which is 8 mm long and bent to the left (looking at the flower from the insect's point of view). The keel is deep and the style is bent upwards where it enters the carinal pouch. The stigma is a depression at the tip of the style which secretes a large amount of sticky fluid when touched. Proximal to this the style bears an array of hairs which acts as a pollen basket; some pollen is present in this when the flower first opens but the rest is in the apical folds of the keel. From this reserve the pollen basket is re-loaded after each of several insect visits. The stigmatic secretion is smeared on the insect and helps to attach pollen to it. The stigma can receive foreign pollen only on the first insect visit, because after this a large amount of self-pollen is deposited on it. The pollen is placed on the vectors 8 mm from the

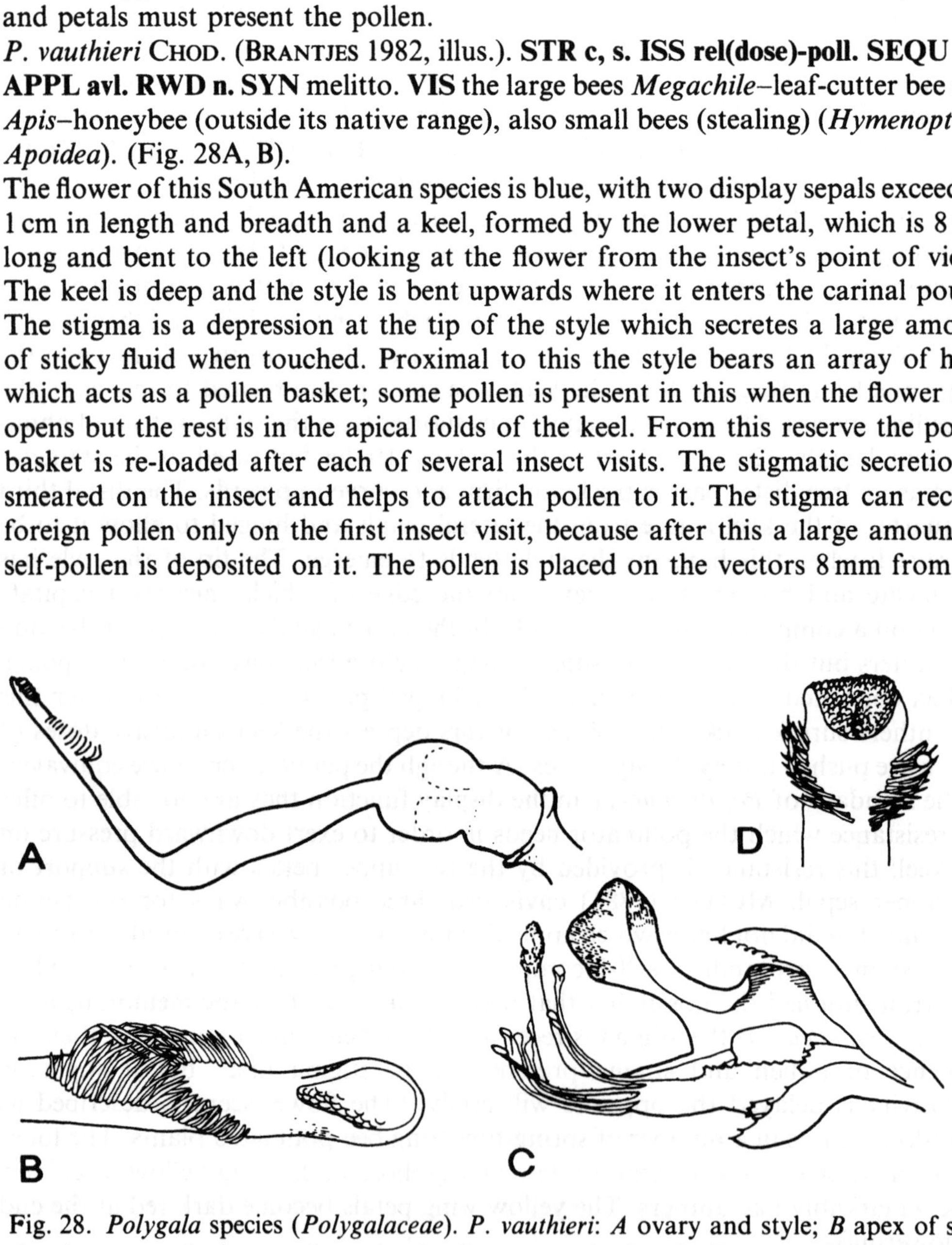

Fig. 28. *Polygala* species (*Polygalaceae*). *P. vauthieri*: *A* ovary and style; *B* apex of style, showing recessed stigma and proximal pollen-presenting brush. *P. monticola* var. *brizoides*: *C* side view after removal of lateral display-sepal, lower petal and one upper petal, showing the partly united stamens; *D* apex of style with papillate stigma and proximal pollen-presenting brush. (BRANTJES 1982.)

tip of the proboscis and on the left side; according to the size of the insect the place of deposition is the cheek or temple (gena or occiput) or both. The principal reward to the pollinator is nectar. The place where the pollen is attached to the insect is difficult for the insect to reach when grooming; however, pollen of *Polygala* was found in the corbiculae of *Apis*.

P. monticola HBK. var. *brizoides* (St. Hil. & Moq.) Steyerm. (Brantjes 1982, illus.). **STR c, s. ISS rel(dose)-poll. SEQU stg. APPL avl. RWD n. SYN** melitto. **VIS** the large of medium-sized bees *Apis*–honeybee (outside its native range), *Exomalopsis*, *Megachile* and *Coelioxys* (*Hymenoptera-Apoidea*); small bees and *Lepidoptera* (stealing). (Fig. 28C, D).

This flower differs from that of *P. vauthieri* in having the keel only 6 mm long, hinged much nearer its base and more strongly bent to the left. Also, when the flower opens the stylar pollen basket is facing left, as a result of torsion through 90°. The operation of the flower is the same as in *P. vauthieri* but the place of deposition of pollen on the pollinator is different; as before it is on the left hand side, but here it is the mandible, coxa or lowest part of the gena. It would appear to be as difficult or more so for the insect to clean it away.

P. bracteolata L. (Scott Elliot 1891, illus.; Brantjes & Van Der Pijl 1980). Full details of this African species are not available, but it is strongly asymmetric in the reverse sense to that of the South American species. The style presents the pollen in a terminal cup (as in *P. vulgaris*) which is apparently twisted towards the right. The stigmatic area is proximal to the terminal cup. The species is said to be bee-pollinated.

Comesperma Labill.

C. virgatum Labill. (Holm 1988, illus.). **STR s. ISS rel(dose)-poll. SEQU pln. APPL v. RWD n. SYN** melitto. **VIS** –.

The flowers of this Australian shrub are arranged in a dense narrowly conical spike and are 10–15 mm wide across the lateral wing-like display sepals. The colour is light violet, with darker violet and yellow areas on the keel petal. The flower has been elegantly described and illustrated by Holm. The description reads remarkably like that of *Polygala comosa*. It differs in the slight asymmetry of the petals and the lateral position of the stigmatic tongue (which, however, is still directed towards the interior of the flower); in addition, the keel petal is distally produced into a lobe on each side so that it clasps the two upper petals: this provides the mechanism for the return of the keel petal to its original position after a pollinator-visit. A further difference is that the transverse folds of the keel petal are so near its apex that there seems to be little scope for movement of the pouched tip. Holm does not mention the stigmatic secretion but as the flower is so like that of *P. comosa* it seems very likely that it has a role in attaching pollen to the vector.

Discussion of Polygalaceae

This family presents on a small scale a situation similar to that of *Leguminosae-Papilionoideae*. The basic construction of the flower is similar enough to result in corresponding floral parts functioning in the same way, and the flowers are melittophilous. Thus similar selective pressures are presumably operating. *Polygala* uses the style to present pollen, it sometimes uses hairs to keep it in place on the style, it issues pollen in measured doses or in an explosion (see also 'Distribution within the tribe') and it employs asymmetry of the flower to place pollen on relatively inaccessible parts of the insect. All these features are found in *Papilionoideae*. The use of adhesives is, however, unknown in that group. The occurrence of both left-handed and right-handed asymmetry, and of two alternative arrangements of the pollen presenter and stigma (presenter terminal or stigma terminal) suggest independent evolution of secondary pollen presentation in different lines within the genus. The roles filled by secondary pollen presentation in the genus are economic use of pollen and its protection from misappropriation, issuance in doses and precise placement on the body of the vector. BRANTJES (1982) found that pollen placement by the two species he studied did not overlap on the same insect, so that precise placement seems to have a role in genetic isolation, although the time of day of anthesis also differed. Precise location of pollen on the insect is also necessary for the pollen to be inaccessible to the insect and it forms part of the broader aim of economy in pollen use and production.

Polygala is one of two genera in the family having trinucleate pollen grains, while two other genera have binucleate pollen (BREWBAKER 1967).

References for Polygalaceae

BRANTJES, N. B. M., 1982.

BRANTJES, N. B. M., VAN DER PIJL, L., 1980: Pollination mechanisms in *Polygalaceae*. – Acta Bot. Neerl. **29**: 56–57

BREWBAKER, J. L., 1967.

HOLM, E., 1988.

KNUTH, P., 1908.

MÜLLER, H., 1881: Alpenblumen, ihre Befruchtung durch Insekten und ihre Anpassungen an dieselben. – Leipzig: Engelmann.

KNUTH, H., 1883.

SCOTT ELLIOT, G. F., 1891.

Apocynaceae

Systematic position: *Asteridae, Gentianales*
Restriction of occurrence within the family: Probably no restriction to subdivisions of the family

Distribution within the family. Secondary pollen presentation occurs sporadically in the two subfamilies currently recognized, namely *Apocynoideae*, in which the thecae of the anthers are basally produced into sterile tails, and *Plumerioideae*, with simple, wholly fertile, thecae. [Two tribes of the *Plumerioideae* are separated off by MELCHIOR (1964) to form a third subfamily, *Cerberoideae*]. The *Apocynoideae* also differ in that their stamens cohere with the stigmatic head (but see discussion).
Floral features. Flowers actinomorphic; perianth comprising calyx and corolla, the latter being sympetalous. Stamens five or occasionally four, attached to the corolla; anthers dithecous, usually with an apically produced connective. Carpels usually two but up to eight, approximately fusiform, free but with the styles united by postgenital fusion, except at their extreme tips, beyond the enlargement of the

style-head, which is elaborate in form and almost isodiametric in transverse section. There is a highly specialized pollination system, characterized by (1) deflecting the insect's proboscis towards the periphery of the corolla tube during penetration, (2) allowing it to be dragged towards the centre during withdrawal and (3) use of adhesives to attach pollen to the insect (SCOTT ELLIOT 1891; CHURCH 1908; SCHICK 1982). The style-head usually has a drum-shaped or conical portion on the outer wall of which is secreted the adhesive. In the conical types (apparently the more primitive) the same secretion functions also as the germination medium (the functional stigma may be co-extensive with the secretory zone or restricted to a small annular portion of it) (SCHICK 1980). In drum-shaped types the underside of the drum is often hollow and acts as the stigma, the outer rim of the hollow forming a scraper, while the lateral surface exudes the adhesive (SCHICK 1980). The scraper can remove pollen from a proboscis that is drawn across it. The anthers are positioned over the upper sterile part of the stigmatic head. Thus during the withdrawal of a proboscis pollen is removed from it, adhesive is added, and the flower's own pollen is carried away on the adhesive. The connection between anthers and style-head in *Apocynoideae* is made by a projection from the connective which is joined to the adhesive region. ROSATTI (pers. comm.) states that pollen may be applied to some part of the insect other than the proboscis in some species; for convenience, only the proboscis is referred to in the descriptions that follow.

Type of pollen presentation. Secondary pollen presentation is not well developed in the family, but the pollen is sometimes deposited from the anthers on to the style-head ('Pollenhaufen') as a concomitant of the elaborate pollination arrangements.

Examples of Secondary Pollen Presentation in Apocynaceae

Subfamily *Apocynoideae*

Apocynum L.

A. androsaemifolium L. (KNUTH 1909, based on LUDWIG; ROSATTI 1989). **STR (a?), s. ISS exposed. SEQU stg. APPL** proboscis. **RWD n. SYN** melitto. **VIS** bees (*Hymenoptera-Apoidea*), hoverflies (*Diptera-Syrphidae*) and houseflies, blowflies etc. (*Diptera-Calypterae*).

The reproductive organs are crowded together in the base of the small (6 mm long and 8 mm wide) cup-shaped flowers. The anthers are connivent over the rhomboidal (not drum-shaped in this case) style-head; the thecae occupy the distal half of the adaxial surface, while two auricles project from the abaxial surface and bend towards each other over the abaxial side of the relatively broad filament. Glandular hairs projecting from the sterile part of the anther and a ring running round the widest part of the style-head interlock and are glued together by their secretions. Although the pollen is described as sticky, it is said (KNUTH) to fall from the anthers on to the upper, non-receptive part of the stylar head and to be picked up from there during the withdrawal of the proboscis of the visitor. Adhesive is applied to it by the projecting ring immediately below the pollen deposit. The proboscis of an insect can be inserted between the auricles of the anthers and a triangular appendage of the corolla standing opposite each anther. The connectives and connective-appendages form hard woody plates, between which the proboscis has to be pulled up; this reduces the circle of pollinators to those that are strong enough to do this. However, smaller insects visit the flowers and frequently become trapped and die there. The flowers have an unpleasant sweetish odour.

A. cannabinum L. (KNUTH 1909, based on LUDWIG; SCHICK 1982).

LUDWIG's study also covered this species which, however, is the same in essentials as *A. androsaemifolium* but it has smaller, more disagreeably scented flowers that attract smaller insects. Nevertheless, most of them get trapped; it must be remembered, however, that these are North American plants that were studied in Europe. SCHICK confirmed LUDWIG's observations; the penetration of pollen tubes showed that the stigma is annular and situated just below the projecting ring of the style-head (SCHICK 1980). SCHICK found that as an insect's proboscis withdrew it pulled out an entire pollen mass (presumably the contents of two thecae belonging to adjacent anthers).

Nerium L.

N. oleander L. (PAGEN 1987, illus.; HERRERA 1991, illus.). **STR (a)?, s. ISS exposed. SEQU st. APPL** proboscis. **RWD** none. **SYN** sphingo? **VIS** the hawkmoths (*Daphnis* and *Macroglossum* (*Lepidoptera*), and the large bee *Xylocopa* (*Hymenoptera-Apoidea*). (Figs. 29, 30).

The flower is generally regarded as adapted to *Lepidoptera* pollination, and there is a report (KNUTH) of visits to the flowers by the Oleander Hawkmoth, *Daphnis nerii*, which seems to have been copied from one author to another. HERRERA heard of a visit by the hummingbird hawkmoth, *Macroglossum stellatarum*, but himself observed only rare visits by *Xylocopa* in south Spain. Dr CHR. WESTERKAMP (pers. comm.) has seen visits by *M. stellatarum* near Mainz, Germany. A receptacular nectary is absent and HERRERA says no nectar is secreted and that the flower is a deceit-flower. [PAGEN does not mention nectar but CHURCH (1908: 206, illus.) says that nectar is produced by the lower inner surface of the corolla-tube, which is wrinkled]. On pollen presentation PAGEN says 'The pollen is shed inside the cone of anthers and collects on the top of the pistil head'; HERRERA agrees, adding that the pollen rests on a drop of mucilage that appears among hairs on the top of the style-head; his drawing shows pollen also in contact with the anthers. Here again there is conflict with CHURCH, who says 'The pollen is viscid, and is shed in definite masses, remaining in position between the lobes of adjacent anthers, which, being in lateral contact themselves, roof in a 'pollenic chamber' around the glandular style-head.' This seems to mean that the pollen does not rest on the style-head. Similarly contradictory reports exist for *Vinca*, described below. (See also 'Floral Features'.)

Stephanostema K. SCHUM.

S. stenocarpum K. SCHUM. (YEO, unpublished 1985). **STR (a), s. ISS exposed. SEQU stg. APPL** proboscis? **RWD n? SYN** micromelitto? **VIS** –. (Fig. 31).

The flower of this East African plant is 8 mm long; the urceolate corolla tube and its spreading lobes are yellow but the tube is continued by a white cup-shaped corona that accounts for about half the length of the flower. The sagittate-based anthers form a conical projection in the throat of the corona. On each stamen-filament there is a dense tuft of hairs projecting inwards and connected with the lower part of the stylar head, with which it makes a firm union. The stylar head is shaped rather like a cotton reel, on top of which is a conical projection. Its

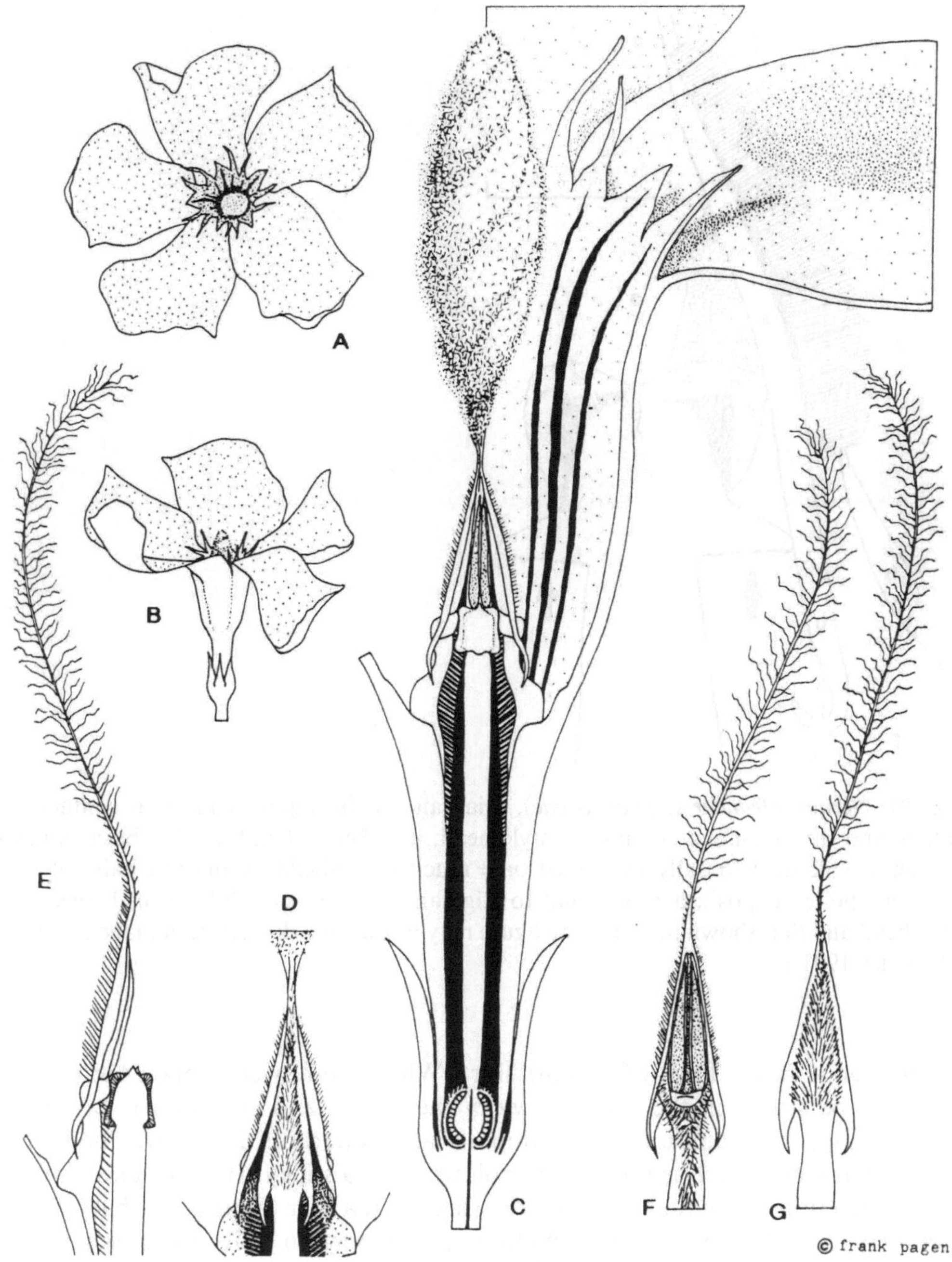

Fig. 29. *Nerium oleander* (*Apocynaceae*). *A* face view of corolla. *B* side view of flower. *C* longisection of flower with two stamens and other parts removed. *D* cone of anthers. *E* relationship of style-head to stamen. *F* adaxial view of stamen. *G* abaxial view of stamen. (PAGEN 1989.)

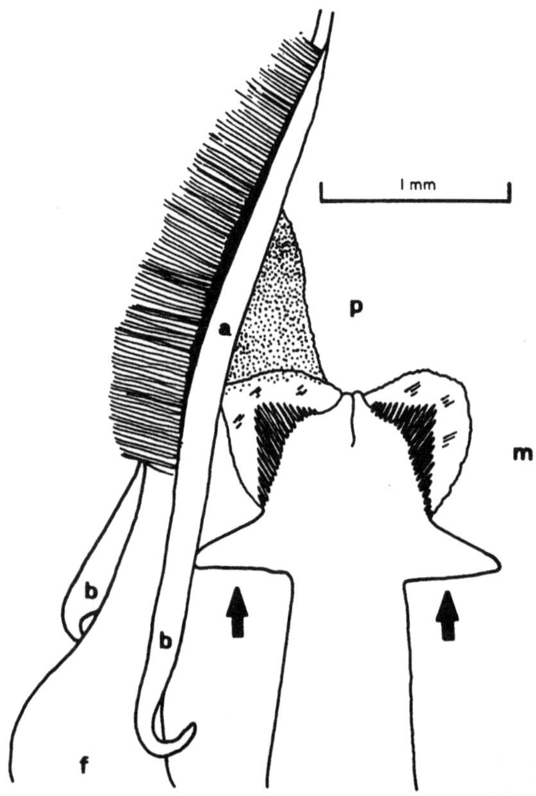

Fig. 30. *Nerium oleander* (*Apocynaceae*). Diagram to show pollen mass in contact with anther and mucilaginous coating of style-head. a = theca of anther, b = basal lobes of thecae, f = filament, basally thickened, m = mucilage embedding massed hairs of style-head, p = pollen deposit; arrows point to stigmatic surface. The differences between this style-head and that shown in preceding figure may be due to a difference in plane of section. (HERRERA 1991.)

lower edge is in the form of a short skirt. When the flower is open the pollen is found to be stored in the space between the thecae and the conical part of the stylar head; a large amount of viscid material is found on the latter. The body of the 'cotton reel', which is greenish in colour, is also sticky. It is not clear how the flower functions, although it must do so in a manner similar to that of other members of the family, and it seems to present a portion of its pollen secondarily, but the secondary presentation is clearly only a minor component of the system. (Observations made in May on a plant in cultivation.)

Subfamily *Plumerioideae*

Amsonia WALT.

A. tabernaemontana WALT. (CHURCH 1908, illus.). **STR s. ISS exposed. SEQU stg. APPL** proboscis. **RWD n. SYN** psycho, melitto. **VIS** butterflies and moths (*Lepidoptera*), and the large bees *Bombus* and *Anthophora* (*Hymenoptera-Apoidea*). The flowers are light blue and have a narrow tube 9 mm long and five narrow spreading corolla-lobes 10–12 mm long. The anthers are without connective-

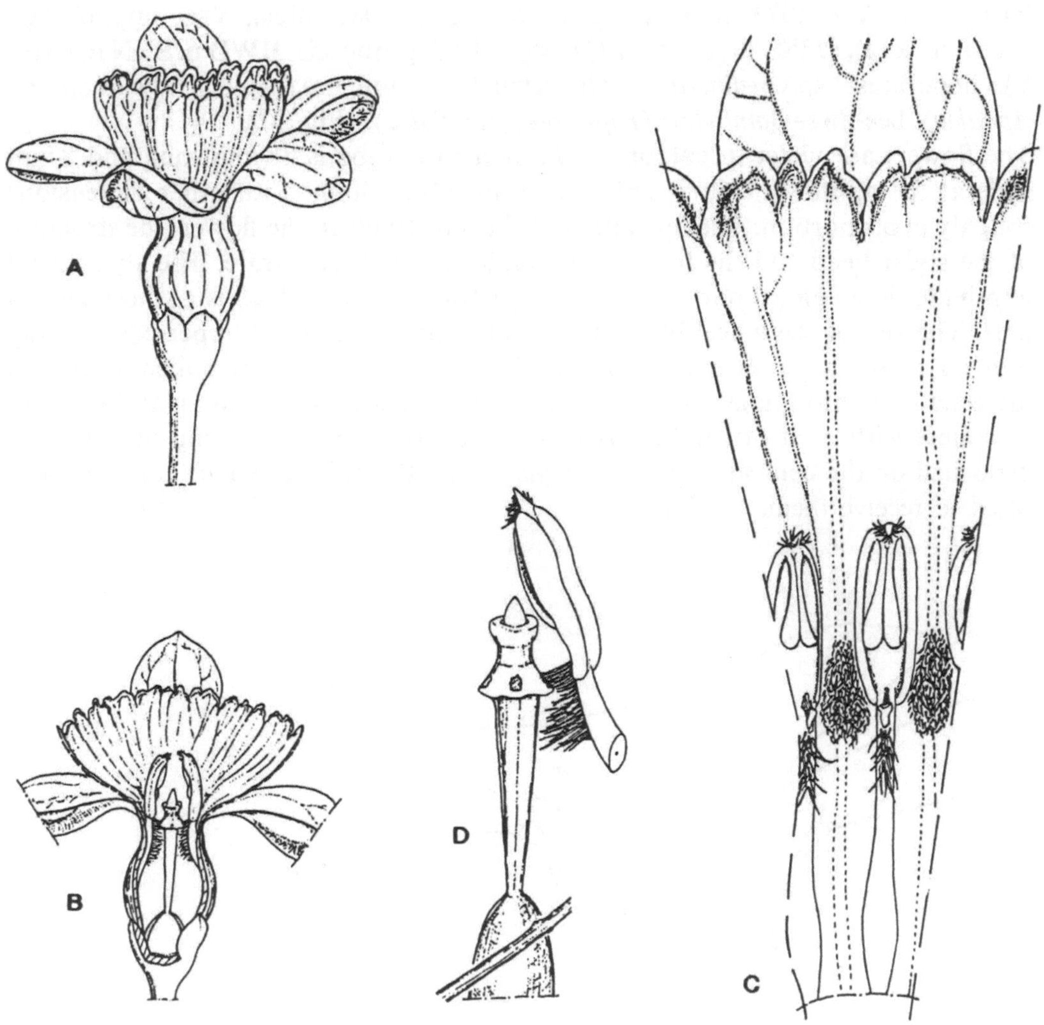

Fig. 31. *Stephanostema stenocarpum* (*Apocynaceae*). *A* side view of flower. *B* longisection of flower. *C* part of flower opened out to show stamens. *D* relationship of style-head to stamen. (BARINK 1984.)

appendages. The style head is comparatively simple, being approximately mushroom-shaped with a sharp edge to the 'cap' acting as a scraper and, on the underside, the stigmatic surface. The top of the 'cap', or part of it, secretes a sticky substance. The floral tube contains dense and elaborate hair-tracts between which there are spaces allowing a proboscis to pass down to the nectar by following the wall of the tube. On withdrawal the usual processes occur (see 'Floral Features'). CHURCH states that the pollen is dry and that some falls onto the stylar cap. Only to the extent that any of this is picked up does this plant show secondary pollen presentation.

Vinca L.

V. major L. and *V. minor* L. (DARWIN 1861; HILDEBRAND & DELPINO in HILDEBRAND 1867; KNUTH 1909, illus., citing several authors; CHURCH 1908, illus.;

PROCTOR & YEO 1973, illus.; SCHICK 1980, illus. 1982, illus.; YEO, unpublished 1985). **STR (a), s? ISS exposed. SEQU stg. APPL** proboscis. **RWD n. SYN** melitto. **VIS** honeybees – *Apis mellifera* – and bees with longer tongues than this (*Hymenoptera-Apoidea*), bee-flies–*Bombylius* (*Diptera-Bombyliidae*). (Fig. 32).

The flowers are violet in colour. In *V. major* the tube is 18 mm long and 8 mm wide at the mouth and in *V. minor* it is 10–11mm long, with other dimensions roughly in proportion. The formation of the hair-tracts in the flower, the structure of the stylar head and the form of the anthers are all elaborate. The stylar head terminates in a conical portion bearing hair-tracts, below which is a wheel-shaped part. The rim of the wheel bears a secretory band and, at its upper edge, a ring of hairs, while the lower edge is produced into a short skirt. The pollen from each theca coheres into a mass, and the masses from thecae of adjoining anthers tend to cohere with each other. In this relationship they are (according to CHURCH) deposited on the conical part of the stylar head, the hairs of which form recesses fitted to receive them.

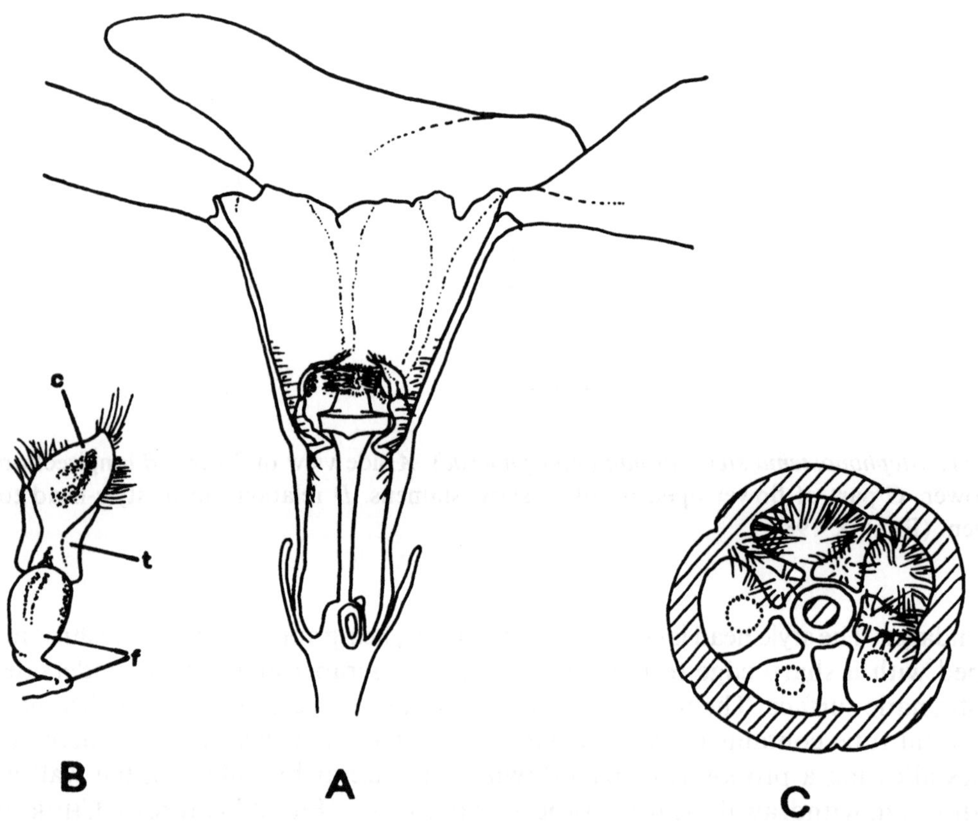

Fig. 32. *Vinca* species (*Apocynaceae*). *A V. minor*, longisection of flower with two stamens visible. *B* stamen, viewed obliquely from abaxial side. *C V. major*, diagram of section of corolla just below stamens, looking upwards at stamen-filaments and style; between the filaments and the abundant hairs (omitted in part of the section) there remain five rather indistinct nectar-passages, three of which are shown by dotted rings. *c* = connective and its appendage, *f* = filament, *t* = theca. (P. F. YEO.)

As in other genera the insect's proboscis, on insertion, appears to be directed towards the wall of the corolla tube, as actually observed by SCHICK (1982) for the honeybee. This guidance is achieved firstly by the horny and slightly channelled backs of the connective appendages of the stamens but the proboscis, having been pushed out towards the corolla wall, must secondly make a slight tangential movement so as to pass between the bases of the filaments; this is probably enforced by the distal part of the filament, which is swollen to equal the size of the anther itself (PROCTOR & YEO, *V. minor*). The insect's head cannot pass beyond the anthers because, in combination with the stylar head, they leave no room for this to happen. HILDEBRAND points out that all the nectar can be obtained by inserting the proboscis once only. On withdrawal it is believed that, as in other genera, the proboscis is dragged towards the axis of the flower. This movement is probably promoted by the downwardly directed hairs in the lower part of the corolla tube and (especially in *V. major*) on the 'knee-bends' of the filaments (YEO, inference from unpublished notes 1964 and 1967; *V. minor*). (The same function is attributed to the coralline hairs of *Nerium oleander* by CHURCH 1908: 207.) The usual processes of pollen transfer then occur (see 'Floral Features'). Pollen scraped from the proboscis can germinate underneath the stylar head (HILDEBRAND). However, further pollen is stripped from the proboscis as it passes over the secretory ring and it can germinate here also (HILDEBRAND, DARWIN) but SCHICK (1980) found in some genera that pollen in this position could penetrate only one or two layers of cells and was then blocked. PAGEN'S observations (1987) on *Nerium* allow for the reconciliation of all previous statements. He found firstly that pollen germinates anywhere on the style-head, the surface of which is made up of the tips of closely packed hairs of graded length, and secondly that although pollen tubes only penetrate the style from the furrow under the head those from grains germinating on the lower part of the drum-shaped part (wheel-shaped part in *Vinca*) have a chance of reaching this furrow. (KNUTH'S statement that adhesive is applied to the proboscis while it is entering the flower is not supported by other observers.)

The most important details in the above account are found in the publications of DARWIN, HILDEBRAND and CHURCH. CHURCH'S claim that secondary pollen presentation occurs is supported by his Fig. 7 (p. 196), showing the double pollen masses deeply lodged in the recesses on the stylar head, as revealed by the removal of part of the corolla and two stamens from a flower of *V. major*. KNUTH'S statement, on the other hand, is as follows: 'The anthers are situated immediately above the stigmatic disk [i.e. stylar head], and dehisce introrsely. Their margins are hairy, so that the pollen can only fall upon the terminal brush of the disk'. The point of this is that the pollen cannot spontaneously reach the stigmatic surfaces, and that there is thus no automatic self-pollination; it seems to be saying what might happen, and does not assert that pollen actually falls or that secondary pollen presentation takes place. Authors who have studied the genus do not claim that the pollen falls to a secondary position.

My own observations do not agree with those of CHURCH. In 1985 I re-examined *V. minor* and investigated *V. major*. When I opened the flowers I found some or all of the (double) pollen masses adhering to the explanate thecal walls and only an occasional one in one of the recesses formed by the hairs on the stylar head. I opened the flowers of *V. major* with progressively increasing care and found that, although the pollen masses projected into these recesses, they came away with the anther when this was removed; when one or two anthers had been removed it was possible to take out the style, and this left all the remaining pollen masses attached to their anthers. Thus, in these flowers, the pollen masses were projecting into the recesses provided for them but were not really touching them or, if they were, they were scarcely adhering to them at all, while continuing to adhere rather firmly to the thecae. One of SCHICK'S (1982) photographs of a dissected flower shows pollen adering to the anthers, as I describe. Although this question does not affect the interpretation of the mechanism of pollen transfer, because it is not in doubt where the pollen masses are in the flower, it does affect the status of these species as secondary pollen presenters. A clue that might explain the apparently conflicting observations is provided by CHURCH. He states (p. 198) that the flowers only attain good condition (in Oxford) in the warmer weather of May and that the pollen is shed as soon as dry air is admitted into the breaking bud or even before in warm weather; he also states that the pollen masses break up on desiccation. Although my work was done at the end of April and the beginning of May, it is possible that the weather was cool and damp enough to prevent normal drying out of the pollen. On the other hand, what I observed could be normal and what CHURCH observed could be abnormal.

It may be noted that the connective appendages, although roofing the chamber where the pollen lies, do not come into contact with the pollen, and so cannot be listed as structures participating in its presentation.

Discussion of Apocynaceae

As stated under 'Type of pollen presentation', secondary pollen presentation in *Apocynaceae* is incidental to the system of using adhesives described under 'Floral features'. The ingredients of such a system are that (1) gumming of the insects must be topographically limited, (2) the pollen must be compactly presented and (3) the gummed organ must be guided so as to make contact with the pollen (an alternative is that the pollen must be actively directed by the plant onto the gummed area of the vector – see *Marantaceae*). The first of these objectives is achieved in *Apocynaceae* by ensuring that only the proboscis enters the flower, and this is done by developments of the corolla, androecium and gynaecium which partly block the floral tube, and by massive hair-tracts that contribute to the same effect. (This seems to restrict the family to insect-pollination.) Other features of the same structures contribute to the third requirement, guidance. The demands of a system of guidance are also probably responsible for the interlocking of the stylar head with the stamens, which is achieved in *Vinca* by the curious knee-bends of the filaments, and in the *Apocynoideae* by physical union between the two types of organ. The second requirement, compact presentation of the pollen, is met by the connivence of the anthers and this in turn apparently easily leads to the deposition of the pollen on the stylar head, so that secondary pollen presentation has arisen in a few instances. As it is not essential it may in some cases be facultative, which could account for conflicting reports (*Nerium*, *Vinca*) as to its occurrence. Species in which it is established that secondary pollen presentation does not occur are listed in Table 2-3.

Although the organization of the flower in this family gives an assurance against self-pollination, geitonogamy is not excluded. However, selfing is at least sometimes excluded by self-incompatibility, which was reported by EAST (1940) to occur in certain strains of *Vinca minor*, in *Nerium oleander* and in the genera *Allamanda*, *Beaumontia* and *Odontodenia*, while ROWLEY (1980) cites a report of 1978 that self-incompatibility has been found in all *Apocynaceae* so far tested.

Table 2-3. Species of Apocynaceae known not to have secondary pollen presentation

Subfamily Apocynoideae
Adenium obesum (FORSK.) ROEMER & SCHULTES (ROWLEY 1980)
Trachelospermum jasminoides (LINDL.) LEM. (YEO, unpublished 1984, 1985)
Subfamily Plumerioideae
Allamanda cathartica L. (SCHICK 1982)
Catharanthus roseus (L.) G. DON (HILDEBRAND 1867)
Plumeria rubra L. (SCHICK 1982)
Thevetia peruviana (PERS.) K. SCHUM. (SCHICK 1982).

Later, however, PAGEN (1987) and HERRERA (1991) found *Nerium oleander* to be self-compatible and SCHICK (1980), although citing two references reporting self-incompatibility in two species of *Apocynum*, thinks that facultative autogamy occurs in genera with a clavate style-head.

CHURCH estimated the number of pollen grains per ovule in *Vinca major* as between 2500 and 3500, observing that this is a relatively low ratio and that the pollination system is efficient. The much lower figure of 32 was recorded by HERRERA (1991) for *Nerium*. My impression is that a low ratio is a general feature of the *Apocynaceae*, as would be expected from the habit of gumming pollen to a small area of the vector's body. The system is also efficient in that it conceals the pollen and prevents its collection by the legitimate pollinators or thieves (CHURCH 1908: 197). Here, a system of reducing expenditure on pollen has evolved that does not employ secondary pollen presentation to that end but has nevertheless given rise to secondary pollen presentation incidentally.

The *Apocynaceae* are less interesting for their secondary pollen presentation than for the fact that they provide the stepping-off point for the evolution of the *Asclepiadaceae* (CHURCH 1908: 209), a family generally agreed to be closely related to and probably derived from *Apocynaceae* or their ancestors. The connivent stamens ensure that entry to and exit from the flower is between the filaments, while the wide connectives cause the thecae of adjacent anthers to be paired off and to present their pollen in a definite relationship to the gap between the stamens. In addition, the stamens are beginning to cohere with the stylar head. In *Asclepiadaceae* the contents of the thecae of adjacent anthers have become pollinia which are attached to a horny retinaculum of stylar origin that lies between them, and the stamens and styles have been fully integrated to form the gynostegium. The adhesive secretion of Apocynaceae also provides a connection with the Asclepiadaceae (see below).

The flower of *Periploca* is functionally the same as that of typical *Asclepiadaceae* but the structural differences are considerable: it is clearly less highly evolved. *Periploca* is therefore sometimes allocated to a family of its own, *Periplocaceae*. The hemispherical stylar head produces elongated detachable retinacula on the radii between the anthers. These retinacula have a gummed expansion at either end. The style lags behind the stamens in length until the flower is ready to open; it then catches up and simply by growth collects the contents of two thecae on the distal expansion of each retinaculum. The proximal gummed dilation acts as a viscidium, adhesive thus being used for two quite different purposes in the same flower (HILDEBRAND 1867; YEO, unpublished 1985).

A morphological curiosity of all these flowers is that the dimerous gynaecium forms a stylar head with a pentamerous organization by postgenital fusion, the two embryonic halves probably developing to different degrees and contributing unequally to the final structure. This is necessitated by the functional relationship of the gynaecium with the pentamerous androecium.

SCHICK (1980) investigated the stylar adhesive of the *Apocynaceae*. On being transferred to a glass slide it appeared milkily opaque and was found to be stiffly fluid, sticky and capable of being drawn out into threads. It quickly became hard and transparent, but the effects of drying out were reversible, even after a year. The adhesive contained a lipophilic component in which six different terpenoids were present. After removal of this component there remained sucrose, glucose and polysaccharides; in addition there was a remnant material that appeared to be cutin or cutin-derived. The secretion is formed under the epidermal cuticle, which becomes free, and it has a bladdery structure formed by cuticles enclosing secretion chambers in which the other components are held. SCHICK found a bladdery structure in the (non-fluid) retinacula of *Periploca* and he cites earlier work that shows a similar constitution of the translators (retinacula) of *Asclepiadaceae* (sensu stricto). Thus there are indications of homology between the *Apocynaceae* and *Asclepiadaceae* here.

References for Apocynaceae

BARINK, M. M., 1984: A revision of *Pleioceras* BAILL., *Stephanostema* K. SCHUM, and *Schizozygia* BAILL. (*Apocynaceae*). – Med. Landbouwhogeschool. **87–3**: 21–53.

CHURCH, A. H., 1908: Types of Floral Mechanism **1.** – Oxford: Clarendon Press.

DARWIN, C., 1861: Fertilisation of Vincas. – Gard. Chron. **1861**: 552.

EAST, E. M., 1940.

HERRERA, J., 1991: The reproductive biology of a riparian Mediterranean shrub, *Nerium oleander* L. (*Apocynaceae*). – Bot. J. Linn. Soc. **106**: 147–172.

HILDEBRAND, F., 1867: FEDERIGO DELPINOS Beobachtungen über die Bestäubungsvorrichtungen bei den Phanerogamen, mit Zusätzen und Illustrationen. – Bot. Zeitung (Berlin) **25**: 265–270, 273–278, 281–286. [Translation with amplification and illustrations of 'Sugli Apparecchi della Fecundazione nelle Piante antocarpee (Fanerogame) etc.', Firenze, 1867].

MELCHIOR, H. (ed.), 1964: A. ENGLER's Syllabus der Pflanzenfamilien, 12th edn. **2.** – Berlin: Borntraeger.

KNUTH, P., 1909.

PAGEN, F. J. J., 1989: Oleander: *Nerium* L. and the oleander cultivars. – Agric. Univ. Wageningen Pap. **87–2**: 1–113 (1987).

PROCTOR, M., YEO, P., 1973: The Pollination of Flowers. – London: Collins.

ROSATTI, T. J., 1989: The genera of suborder *Apocynineae* (*Apocynaceae* and *Asclepiadaceae*) in the southeastern United States. – J. Arnold Arbor. **70**: 307–401. [Excludes *Asclepiadaceae*, for which see op. cit.: 443–514.]

ROWLEY, G. D., 1980: The pollination mechanism of *Adenium* (*Apocynaceae*). – Natl. Cact. Succ. J. **35**: 2–5.

SCHICK, B., 1980.

SCHICK, B., 1982.

SCOTT ELLIOT, G. F., 1891.

Campanulaceae

Systematic position: ***Asteridae, Campanulales***
Restriction of occurrence within the family: None

Distribution within the family. The family is here treated in the strict sense of M. KOVANDA (in HEYWOOD 1978) and divided into three tribes recognized by him at subtribal rank. Secondary pollen presentation occurs in all of these and is indeed so prevalent that it can be described as an attribute of the family (DE CANDOLLE 1830).

The tribal classification is based on ovarian characters: *Campanuleae* has the ovary inferior and dehiscing at the sides (sometimes indehiscent, sometimes a berry); in *Wahlenbergieae* the ovary may be inferior, half-inferior or superior and it opens at the top (a berry in one or two genera). In both tribes, when the carpels are equal in number to the sepals, they lie on the same radii as the latter. In the *Platycodoneae* (YEO 1993), however, the carpels alternate with the sepals; here the ovary may be inferior or semi-inferior and may open either terminally or laterally. Additional tribes of *Campanulaceae* are recognized in Flora of the USSR (FEDOROV 1957). These mainly represent splitting of tribes recognized previously and they thus represent a lower taxonomic level. However, FEDOROV is apparently unimpressed by the ovarian character of *Platycodon*, for he includes the *Platycodoneae* in the *Wahlenbergieae*. The same is done in the recent classification of KOLAKOVSKY (1987), which I have not used because it does not place *Petromarula* and *Physoplexis* and does not mention *Cyananthus*. KOLAKOVSKY raises *Campanuleae* and *Wahlenbergieae*, somewhat remodelled, to the rank of subfamily and describes a new bigeneric subfamily *Canarinoideae* (containing *Canarina*) and an additional subfamily, *Prismatocarpoideae*, accommodating four genera for which I have no information regarding pollen presentation; the genus *Edraianthus* is placed by him in *Campanuloideae*, not *Wahlenbergioideae*. This classification is neutral with respect to the available information on secondary pollen presentation.

Floral features. Flowers actinomorphic; ovary usually inferior. Perianth and androecium usually pentamerous. Petals united for part of their length and corolla very often bell-shaped. Stamens free from the corolla and from each other except in a few cases where there is some degree of concrescence of the anthers. Style single; stigmatic lobes equal in number to the cells in the ovary (from 2 to 8 but most commonly 3). Filaments in most genera with expanded bases inserted at periphery of nectarial disc and bent towards style, frequently forming a cover over the nectar-store; remainder of filament thread-like.

The two parts of the filament have a different course of development. The expanded base is formed early and remains turgid throughout anthesis (KIRCHNER 1897; YEO, unpublished 1964; SHETLER 1979) whereas the slender part develops only after the anther has reached approximately full size (DE CANDOLLE 1830, citing also SPRENGEL and CASSINI), and by the time the flower opens this part has shrivelled, at least distally (KIRCHNER 1897; YEO, unpublished 1964; ERBAR & LEINS 1989).

Type of pollen presentation. Usually stylar ('Pseudo-Staubblatt'), rarely by a piston mechanism ('Nudelspritze'). In view of the prevalence of stylar presentation it is described in full here (see Fig. 33) and in less detail under individual genera. (For a full bibliography see SHETLER's historical review, 1979.)

In the bud the stigmatic lobes are mutually appressed and their outer surfaces are pilose with unicellular hairs. This hair-covering also extends some way down the style; the anthers dehisce in the bud and when they do so they are equal in length to the pilose zone (ERBAR & LEINS 1989) and appressed to it, so that they deposit their pollen among the hairs (the 'collecting hairs' of DE CANDOLLE 1830: they would be better called pollen-carrying hairs). As the flower opens they shrivel and curl away from the style. The timing of the growth of the filamentous part of the stamen (see above) apparently allows the anther to maintain its position relative to the hairy part of the style, for the style is also then growing rapidly in its basal part.

Nevertheless, according to DE CANDOLLE (1830: 22) the stylar brush advances in relation to the anthers, which implies that it actually brushes pollen out of the anthers, as also claimed for *Campanula* by SHETLER (1979: 207) and for *Wahlenbergia* by LLOYD & YATES (1982). However, ERBAR & LEINS (1989) say that the anthers 'open over their whole length and shed pollen grains onto the hairy style', which seems to be correct because it is difficult to reconcile the idea of a sweeping action with the very uniform pollen deposit that is produced on the style. After the shrivelling of the anthers and the slender parts of the filaments, growth of the style continues until the corolla is fully open.

The stylar pollen-carrying hairs of *Campanula* disappear during the first (male) stage of the flower. SHETLER (1979) reviewed the literature on this phenomenon and contributed his own observations. Each hair has a swollen base sunk into the epidermis of the style; the narrow part disappears into this by invagination. Anatomical details are given by JOST (1918).

The invagination happens at the same time as the 'disappearance' of the pollen, and both processes seemed to SHETLER to be the result of insect activity, since neither took place fully and regularly in the absence of insect visits and the invagination of the hairs was much more rapid during periods of high insect activity. The mechanism of retraction has not been identified but JOST's observations suggest an active process involving some change in the hair-cells; JOST's timings show complete retraction in the space of 40 minutes and the accomplishment of most of the process in about the first ten minutes. Pollen removal took place from the tip downwards. It is clear that before retraction the hairs resist the removal of pollen. Thus KIRCHNER (1897) stated that the retraction released the pollen layer by layer from the outside of the deposit inwards towards the stylar surface, and from the apex downwards, leading to its removal progressively by insects (being unsure of KIRCHNER's exact meaning, I have taken up the interpretation given by ERBAR & LEINS 1989: 48). Similarly, HILDEBRAND (1870) found in *Trachelium* that the pollen was difficult to remove before the invagination of the hairs and easy afterwards. LLOYD & YATES (1982) consider the retraction of the hairs in *Wahlenbergia* to be spontaneous and affirm that its gradual progress has the effect of extending the liberation of the pollen over a sequence of pollinator-visits. ERBAR & LEINS (1989) agree that the invagination of the hairs is spontaneous and proceeds from the top of the hair-zone downwards. They state that it finishes with the complete disappearance of the hair into its hole [though there are, in addition, much smaller, non-invaginating, hairs which might be mistaken for the tips of invaginated hairs, reported to be visible above the surface by SHETLER

(1979)]. Thus the method of issue of pollen is measured dosing, and I have shown this (in a few cases with a query) in the coded summaries for most species covered below.

Striking departures from the above-described structures and processes occur in the *Phyteuma* group, and in *Phyteuma* itself the style acts as a piston and the corolla as the cylinder.

Examples of Secondary Pollen Presentation in Campanulaceae

Tribe *Campanuleae*
Campanula L.

De Candolle's monograph (1830) clearly depicts the floral details of *C. rotundifolia* and *C. rapunculoides* L. in plate 1, and of several others species in other plates where the whole plant is also portrayed. Knuth illustrates only *C. pusilla* Haenke (correctly *C. cochleariifolia* Lam.). He quotes a useful synopsis of the range of floral characters in *Campanula* from Kirchner (1897). Shetler's (1979) review paper deals with the genus but I have cited it and quoted it under *C. rotundifolia*. The conclusion of his survey of the breeding system is that most species of *Campanula* do not self-pollinate and that artificial selfing fails in most cases. Four species studied by Erdelská were found not to be completely dichogamous. Erdelská found a zone of distinctive small hairs (processes of epidermal or subepidermal cells) between the zone of pollen presenting hairs and the zone of stigmatic papillae. She suggested that their function was to provide protection against selfing.

There is great variety in the posture and aggregation of the flowers and a wide range of sizes and shapes of corolla in this large genus. Although the indications are that melittophily is characteristic, this variation suggests adaptation to different species (see *C. persicifolia* below). However, the records of pollinators given by Knuth provide little evidence to support this supposition, and closer study of species is needed.

C. rotundifolia L. (De Candolle 1830, illus. [plate 1D]; Knuth, citing also work of Sprengel, Müller, Macleod and Kirchner; Proctor & Yeo 1973, illus.; Shetler 1979; Erbar & Leins 1989, illus.). **STR s. ISS rel(dose). SEQU pln. APPL v. RWD n, p. SYN** melitto. **VIS** a wide range of bees (*Hymenoptera-Apoidea*), especially *Melitta haemorrhoidalis* (syn. *Cilissa haemorrhoidalis*) and *Megachilidae* (mainly *Chelostoma* and *Megachile*). (Fig. 33).

The flowers are nodding, mostly 2–2.5 cm long and held on thin stalks in a loose inflorescence well above the foliage. The colour is a normal 'Campanula-blue'. Insects enter the bell of the flower, clinging to the style in order to probe beneath the staminal nectar covers, since these are most easily parted and lifted by their tips, which rest against the style. The posterior ventral part of the body thus rests on the distal pollen-laden part of the style or on the recurved stigmatic lobes, depending on the age of the flower. The *Megachilidae* have their pollen scopae in the form of a brush occupying the entire underside of the abdomen and this inevitably brings pollen from other flowers in contact with the stigmas. They actively gather the pollen with their legs. Although *Melitta haemorrhoidalis* is well known to have a strong association with *C. rotundifolia*, it adds honey to the

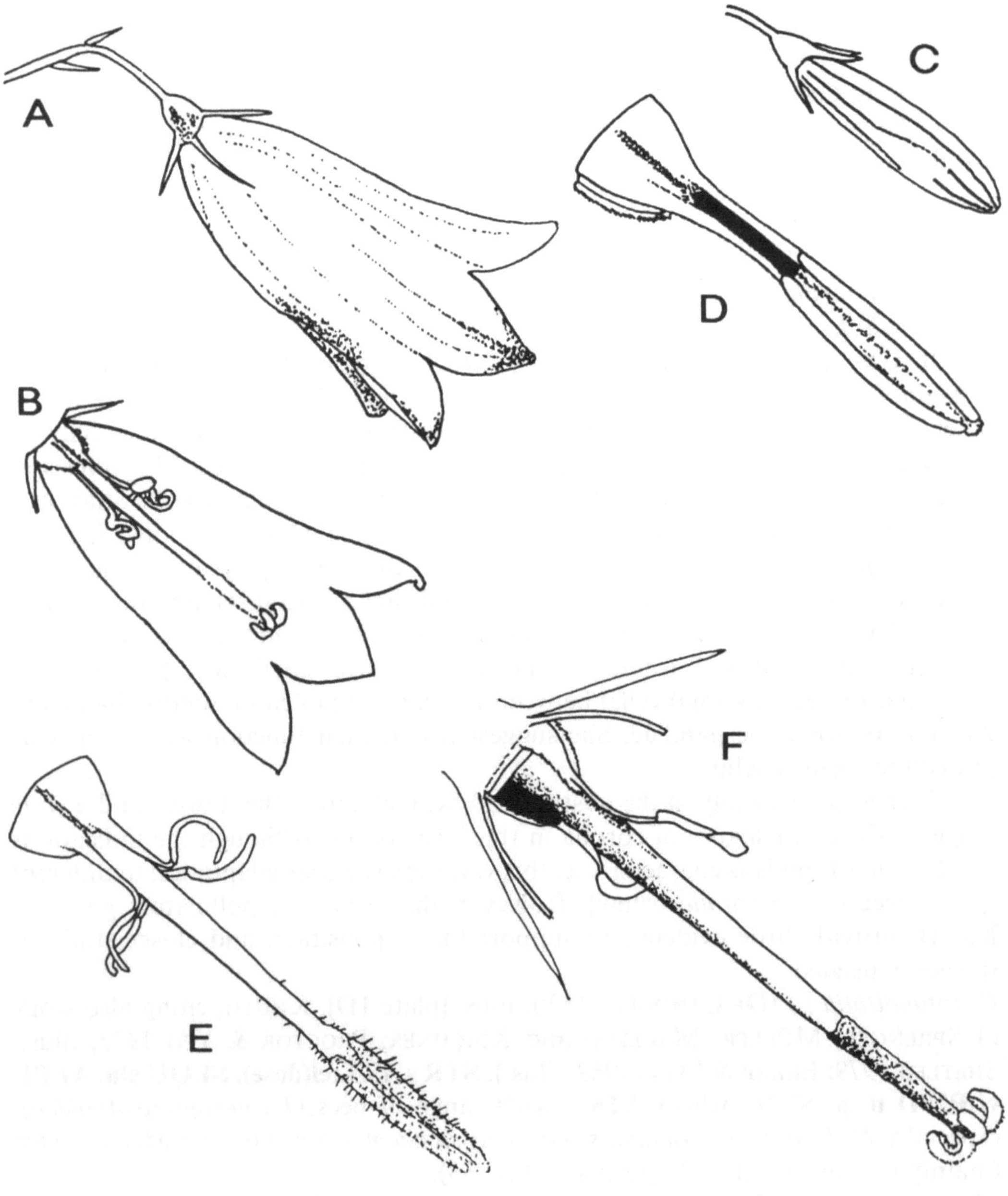

Fig. 33. *Campanula rotundifolia* (*Campanulaceae*). *A* side view of flower in natural posture; corolla 20 mm long. *B* section of flower in female stage. *C* bud with corolla 13.5 mm long. *D* stamens and style (dark) from 'C', 8.5 mm long, therefore not filling space in corolla; the loaded pollen-presenting brush is visible between the two anthers shown. *E* style and stamens from flower in male stage; the hairs of the pollen brush protrude through the pollen deposit. *F* the same in the female stage with a stamen removed to show the (dark) nectarial chamber; the style-branches are separated and the hairs of the pollen brush have been retracted, leaving a rough surface. (P. F. YEO.)

pollen-load in the scopae on its hind legs, so effective pollination of *Campanula* by this bee must depend on grains that escape grooming. Many of the wide range of other bees that are recorded as visitors to the flowers of this species can probably cause pollination, but the insects of other orders that have been seen in the flowers may not and they are unlikely to be able to obtain the nectar. Although the pollen is fully exposed the method of pollen issue is shown as 'rel(dose)' because the hairs on the style hold the pollen until they shrink (see 'Type of pollen presentation' and authors cited by LLOYD & YATES 1982).

C. persicifolia L. (JANSON 1983). At the study-site in Sweden there were two common and effective pollinators, *Chelostoma campanularum* (*Apoidea, Megachilidae*) and *Thricops hirsutula* (*Diptera, Muscidae*). Both are small insects, ill-adapted to pollinating the large bowl-shaped flowers of this species, but they were abundant and their movements included contacting the pollen presenter and stigmas. A rare but effective pollinator was *Bombus lucorum* (*Apoidea, Apidae*) which touched the style and stigmas dorsally. Janzon concluded that the flower was adapted to some pollinator that was not present in the study-area.

C. macrostyla BOISS. & HELDR. This species has an erect bowl-shaped flower 5–6.5 cm wide, about 2.5 cm deep and hairy inside. The interior colouring is reticulated on a pale ground proximally. The style-length is more than twice the depth of the corolla and the style-head is fusiform and up to 2 cm long. In a colony observed in Turkey in the late afternoon (YEO, unpublished 1989), little pollen remained on the pollen-presenter. Partially separated stigmatic branches were seen only in a withering flower (a condition shown also by J. D. HOOKER 1878). However, in some flowers the edges of the stigmatic branches were recurved in a manner recalling those of some *Asteraceae* and were presumably receptive. The staminal nectar cover is normal for the genus. DAMBOLDT (1978) states that the floral features are an adaptation to pollination by flies.

Symphyandra A. DC.

In this genus the flower is a large nodding bell as in *Campanula* but it may be yellow in colour. The anthers are permanently united by their edges to form a tube round the style.

S. cretica A. DC., *S. pendula* (BIEB.) A. DC., *S. armena* (STEVEN) A. DC. (DE CANDOLLE 1830, illus.; KIRCHNER 1897). **STR s. ISS rel(dose)? SEQU pln. APPL v. RWD n, p. SYN** melitto. **VIS** –.

KIRCHNER states that the stylar hairs take up the pollen as the style grows through the anther-tube. In *S. cretica* the style is finally very much longer than the anthers. The cited authors do not describe the nature of the union of the anthers. The genus ranges from the east Mediterranean region to central Asia.

Adenophora FISCHER

This genus differs from *Campanula* in having a cup-shaped or cylindric structure on the upper surface of the ovary which secretes and stores the nectar. The expanded bases of the filaments are enlarged further to encase it.

A. liliifolia (L.) A. DC. and other species. (DE CANDOLLE 1830, illus.; KIRCHNER 1897). **STR s. ISS rel(dose). SEQU pln. APPL v. RWD n, p. SYN** melitto? **VIS** –. (Fig. 34A, B).

Pollen presentation arrangements are as in *Campanula*. *A. liliifolia* is the only central European species of a mainly Asiatic genus; it has a flower-scent like that of *Narcissus*.

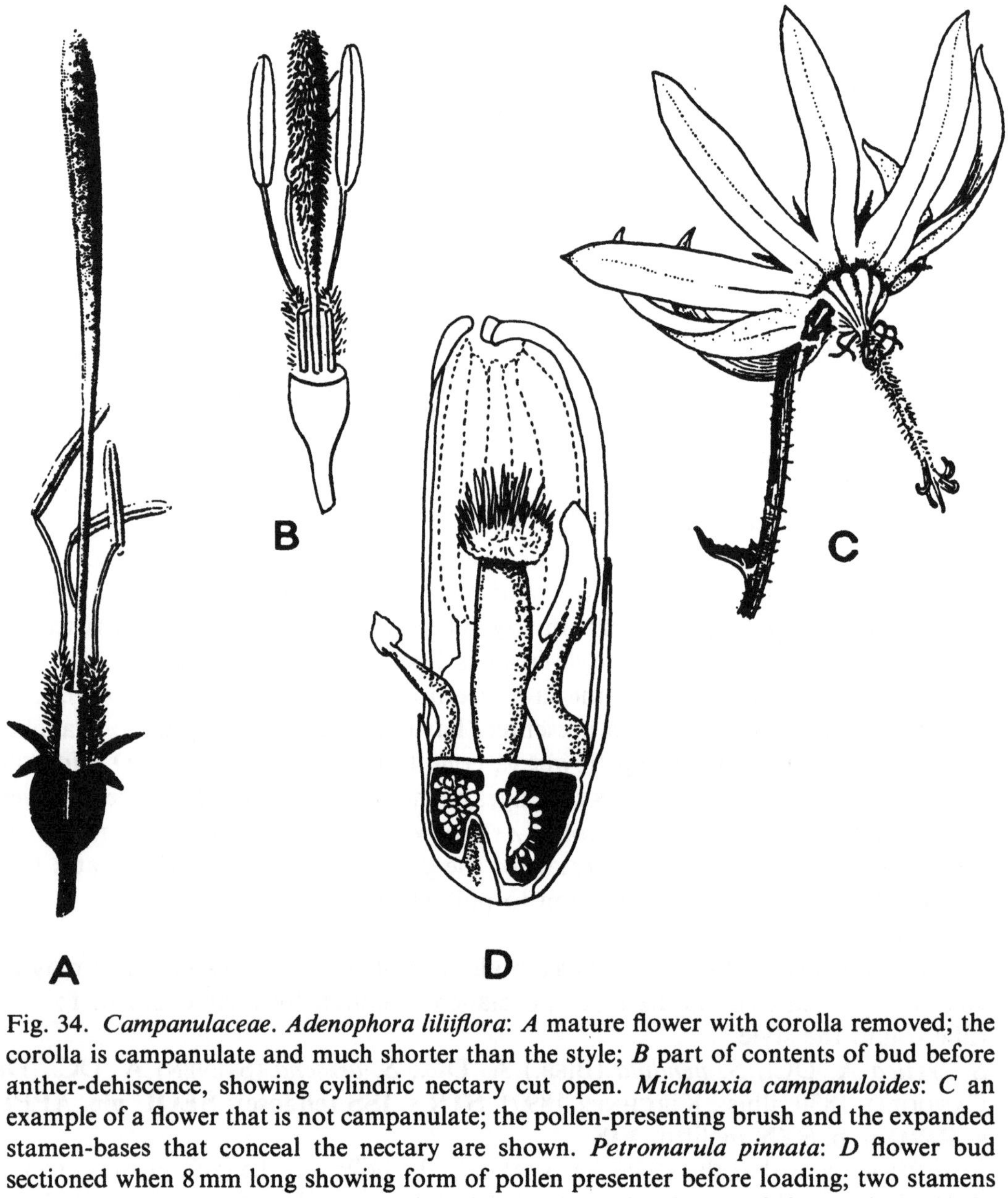

Fig. 34. *Campanulaceae. Adenophora liliiflora*: *A* mature flower with corolla removed; the corolla is campanulate and much shorter than the style; *B* part of contents of bud before anther-dehiscence, showing cylindric nectary cut open. *Michauxia campanuloides*: *C* an example of a flower that is not campanulate; the pollen-presenting brush and the expanded stamen-bases that conceal the nectary are shown. *Petromarula pinnata*: *D* flower bud sectioned when 8 mm long showing form of pollen presenter before loading; two stamens have been cut and displaced while dotted lines show the thecae of three others. (A, B, A. De Candolle & Heyland in De Candolle 1830; C, L. H. Bailey 1953; D, P. F. Yeo.)

Legousia Durande (*Specularia* A. DC.)

This is a genus of annual plants with short-lived flowers opening in sun. The corolla has a short tube and the lobes spread widely.

L. speculum-veneris (L.) Chaix (*S. speculum* auctt.). (Knuth, quoting Kerner and Kirchner; Kirchner 1897). **STR c?, s. ISS rel(dose). SEQU pln. APPL v. RWD n?, p. SYN** micromelitto. **VIS** the small bees *Halictus* and *Lasioglossum* (*Hymenoptera-Apoidea*).

The flower in this species is about 2 cm wide. The filament-bases are hardly enlarged and nectar-secretion seems to be slight. Most of the pollen is presented as in

Campanula except that the anthers do not dehisce until the flower is opening. On the first day pollen is presented and on the second the stigmas are receptive. The small bees that visit the flowers apparently straddle the corolla and touch the style and stigmas with their undersides. The closed corolla has five inwardly directed folds and during opening some of the pollen is deposited on these. When it closes at night and during rainy or dull weather, pollen from the folds may be transferred to the flower's own stigmas and effect fertilization. The existence of arrangements for self-pollination is more likely in this annual species than in the perennials that form the greater part of the *Campanulaceae*. Since pollen is deposited on the corolla it would seem possible that some cross-pollination could take place by this alternative secondary route. KIRCHNER also described the smaller-flowered *L. hybrida* (L.) DELARBRE which is similar to *L. speculum-veneris*.

Ostrowskia REGEL

O. magnifica REGEL. (J. D. HOOKER 1896, illus.).

According to HOOKER's taxonomic account this tall tuberous-rooted herb from Central Asia varies from five to nine in the number of parts in each floral whorl (the plate shows seven or eight); its capsule opens by twice as many pores as the number of stigmas. The bluish lilac bell-shaped corolla is 6 cm long and 10 cm wide. The style is swollen in the stigmatic region, which has pollen-carrying hairs as in genera already described; secondary pollen presentation apparently takes place in the usual way. The expanded filament-bases are relatively small but may help to hold the nectar even if they do not conceal it.

Trachelium L.

T. caeruleum L. (DE CANDOLLE 1830; KNUTH, quoting KIRCHNER and DELPINO; KIRCHNER 1897, citing DELPINO). **STR s. ISS rel(dose). SEQU pln. APPL v,** proboscis? **RWD n. SYN** psycho?. **VIS** day-*Lepidoptera* (*Pieris sp.*) and the bee *Halictus* sens. lat. (*Hymenoptera-Apoidea*).

This herbaceous plant bears its numerous small, slender, blue flowers in a dense corymb. The flowers are flimsily constructed and nectar can be seen through the translucent wall of the corolla-tube; the stamen-bases are not widened and do not have a role in concealing nectar. Before the flower opens the anthers completely fill the width of the corolla. The style at first does not reach the anthers but later it grows more rapidly and as it has a thickened tip it is obstructed by the anthers and is forced to bend. At about this time the anthers dehisce. The pressure from the style finally forces the corolla to open and the style then passes quickly between the anthers and sweeps up their pollen. The pollen is held firmly among the presenting hairs until they shrink (HILDEBRAND 1870: 624–5). These hairs are of the usual unicellular kind with a swollen base sunk into the epidermis; they are found only on the outsides of the three stigmatic branches which are very small. Presentation of pollen and later of stigmas (which has a longer duration) takes place outside the corolla. The narrow flowers make them suitable for the insertion of the proboscis of a lepidopteran, and the blue colour, although strongly associated with bee-pollination, is also appropriate for the attraction of members of the butterfly family *Pieridae* (PROCTOR & YEO 1973: 128). In its pollen presenting arrangements *Trachelium* is distinguished from the preceding genera in that the style is not placed between the anthers at the time of their dehiscence and is loaded later by sweeping up the pollen as it passes between them.

Michauxia AITON (*Mindium* ADANSON).

The floral parts in this genus, including the carpels, are in eights. The corolla-lobes are widely spreading or reflexed and, as in the remaining genera of the tribe, they separate from the base first. The whole or almost the whole of the stamen filament remains turgid during anthesis.

M. campanuloides AITON. (CURTIS, 1793, illus.). **STR s. ISS rel(dose). SEQU pln. APPL a? RWD n, p? SYN** sphingo? **VIS** –. (Fig. 34C).

This species from the East Mediterranean Region is a biennial growing to nearly 2 m in height. The mainly white corolla is cut nearly to the base into strap-like lobes up to 4.5 cm long that are reflexed from the base; thus the flower, though large, has a diameter less than twice the length of the corolla-lobes. The style is stout and the stigmatic lobes linear. The expanded stamen-bases are large and firm and they form a dark-coloured flask-like vessel around the style. The account by CURTIS does not deal with floral biology, but the illustration shows the deposit of pollen on the style and fine hairs on the latter below this deposit. The form of the flowers and their position in a large loose panicle suggest pollination by hawkmoths or possibly by *Xylocopa* bees.

M. laevigata VENT. (W. J. HOOKER 1832, illus.)

This species, from Transcaucasia, N. Iran, Iraq and E. Turkey, is a perennial, growing to a height of over 3 m. Its flowers are slightly smaller than those of *M. campanuloides*, with greenish white, widely spreading, ribbon-like lobes, and might be pollinated by moths smaller than those suggested for that species.

M. tchihatchewii FISCHER & MEYER. (J. D. HOOKER 1900, illus.).

This southern Turkish species is described as 'a majestic biennial' and it has several differences from the two preceding. The corolla has a bowl-shaped base and broad recurved lobes and the style has a rather large ovoid head. The inflorescence is columnar and all the flowers open at about the same time; in the plant illustrated all the flowers are in the male stage with the stigmatic branches still coherent. It seems likely that this species is bee-pollinated.

Petromarula HEDWIG f.

The flowers of this monotypic genus resemble those of *Michauxia* (particularly *M. campanuloides*) in the form of the corolla and the stamen-filaments, and in the thickness of the style. Differences are given below.

P. pinnata (L.) A. DC. (*Phyteuma pinnatum* L.). (DE CANDOLLE 1830; KIRCHNER 1897; YEO, unpublished photographs 1983 and unpublished observations 1990). **STR s. ISS exposed. SEQU pln. APPL av? RWD n. SYN** melitto. **VIS** –. (Fig. 34D).

This Cretan endemic has been known since 1594, and has been cultivated in Europe intermittently since then. I photographed it on walls in Chania harbour in 1983, unfortunately a year before I decided to prepare the present work, but have since had it flowering in cultivation. It is a rather robust monocarpic chasmophyte with a branched ascending inflorescence of narrow thyrsiform units.

The flowers are pentamerous and the corolla is cut virtually to the base into pale violet-blue linear lobes, 10–12 mm in length. The anthers are linear and they shrivel after giving up their brick-pink pollen. The flask-shaped vessel formed by the filaments encloses the style for almost half its length. The colouring of this vessel is intensified from the pale base towards the apex and especially along the crests, edged with interlocking papillae, that are formed where neighbouring filaments meet, so that these are deep violet. The flowers face more or less horizontally and the stout style is descending but slightly turned up at the tip, which is broadly capitate. The hair-covering of the style-head is quite dense and is graded from small papillae underneath to large conical unicellular hairs up to 1mm long bordering the three lobes at the centre of the upper surface. In the earlier stages of the bud the base of the style-head is well above the bases of the anthers. Just before the anthers dehisce, however, it is below them, only its tip and the greater part of its hairs being between the anthers as these begin to dehisce. The hairs on the sides of the style-head diverge by 60–70° from the stylar axis and sweep the pollen from the anthers as the style grows up between them, helped by the fact that the tips of the anthers are incurved. In the newly open flower the pollen forms a smooth cake on top of the style, apparently moulded by the spatulate tips of the anthers. It completely conceals at least the larger hairs; in unvisited flowers this cake becomes disrupted as the stigmas are exposed and the hairs can then be seen. As a rule a small amount of pollen remains in these anther-tips, the edges of which finally fold inwards. Even when the female stage is reached it is difficult to brush all the pollen from the style-head, suggesting that pollen is available over a sequence of insect visits.

The stylar furrows underneath the pollen mass open after a few days, apparently as a result of growth within the furrow; the outer borders of the furrows then become cushion-like and slightly overhang the pollen-presenting surface which is pushed away from the centre of the style-head.

The colour and size of the flowers and the robust construction of the nectar-vessel suggest that pollination is by *Xylocopa* bees, and that pollen would be placed on the undersides of their heads; they would be unlikely to be able to gather much of this, and their reward would be mainly in the form of nectar.

Asyneuma GRISEB. & SCHENK (*Phyteuma* sect. *Podanthum* DOSS)

Although this genus has often been treated as part of *Phyteuma*, DAMBOLDT (1978) suggests that 'it is more closely related to different groups of the genus *Campanula*'.

A. canescens (WALDST. & KIT.) GRISEB. & SCHENK. (KIRCHNER 1897; KNUTH, citing KIRCHNER). **STR s. ISS rel(dose). SEQU pln. APPL probably a, v. RWD n, p? SYN** melitto. **VIS** *Hymenoptera.*

This plant has narrow violet-blue flowers in clusters of 2 or 3 in a dense narrow spike-like arrangement. The crowded flowers are slender, with almost no corolla-tube and linear corolla-lobes separating from the base to become widely spreading. Although the flower is narrow, the filament-bases are widened. Pollen is deposited on to a pilose zone of the style in the advanced bud stage as in *Campanula*. Once the corolla-lobes have separated, the three-lobed style is well exserted.

Phyteuma L., sensu stricto (*Phyteuma* sect. *Hedranthum* G. DON)

The flowers of *Phyteuma*, narrow as in *Asyneuma*, are borne in dense heads; I deal first with the species in which the heads are cylindric or conical and second with those in which the heads are at least as broad as long.

P. michelii ALL., *spicatum* L., *nigrum* F. W. SCHMIDT and *P. ovatum* HONCK. (*P. halleri* ALL.). (KIRCHNER 1897; MÜLLER, illus.; KNUTH, citing earlier authors, illus.). **STR c, s. ISS growth. SEQU pln. APPL av. RWD n, p? SYN** melitto, psycho? **VIS** varied but especially bees (*Hymenoptera-Apoidea*) and *Lepidoptera.* (Fig. 35).

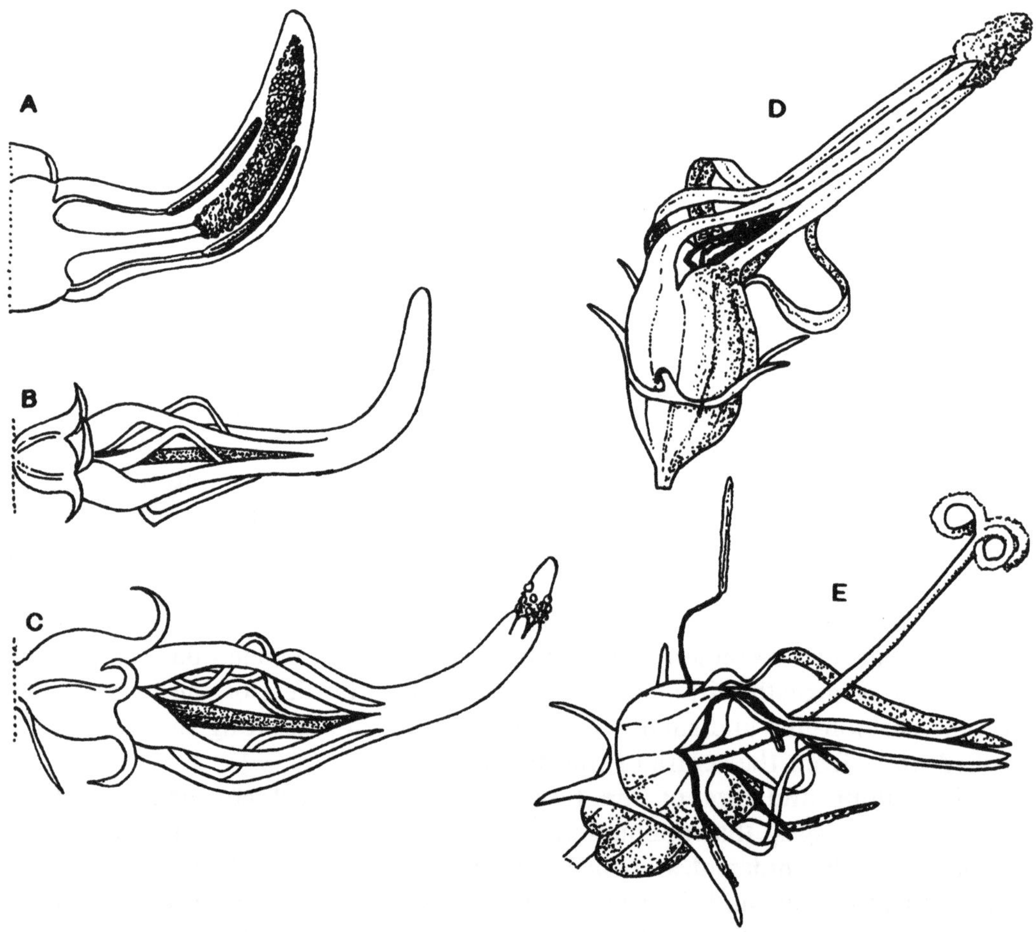

Fig. 35. *Phyteuma* species with spicate inflorescences (*Campanulaceae*). *P. spicatum*: *A* young bud with corolla opened to show two stamens and deposit of pollen on style; *B* flower in late bud stage with corolla lobes separating from base, revealing style and stamens; *C* flower in early stage of anthesis with pollen-laden style emerging from tubular apex of corolla. *Phyteuma michelii*: *D* flower at same stage as 'C', with pollen being pushed out of corolla by style (the stamens have shrivelled); *E* female stage with stigmas exposed and corolla lobes fully separated. (A–C, M. Hickey, after Warming 1895; D, E, P. F. Yeo after Müller 1883.)

The flowers of these species open from the base of the head upwards. The short inflated corolla-tube is continued into five linear lobes which separate first at their bases. Before the flower opens the anthers deposit their pollen onto a stylar brush similar to that of *Campanula*. This pollen mass is tightly enclosed by the still coherent corolla-lobes which form a cylindrical tube in this region; the stamens shrivel so that the anthers are pulled downwards and are removed from the tube. As the bases of the corolla-lobes separate they become bowed outwards, so shortening the corolla. This process, together with the continuing growth of the style, raises the pollen upwards and sweeps pollen accumulated above the style out of the tip of the tube. Thus the corolla stores and actually presents the pollen, while the style acts as a piston. As soon as the tip of the style itself reaches the apex of the corolla the two or three stigmatic lobes roll back and the corolla-lobes fall apart. (These details, given by KIRCHNER, do not amount to a full explanation of what happens, inasmuch as it is not made clear how the pollen gets above the style-head, so as to be pushed out of the corolla by it.) The nectar (which is somewhat shielded by the hairiness of the stamen-bases) is accessible between the bases of the petals. The flowers of the species named are coloured in shades of greyish blue, except in those of *P. spicatum* which is greenish white; this species is exceptional in the weak vanilla-like scent of the flowers. KIRCHNER says the less well-adapted flies do not find the nectar whereas the better-adapted, together with the bees and the butterflies, do; bees worked the inflorescences from the base up. He found *P. spicatum* to be regularly visited by flies and bumblebees (*Bombus* spp.).

P. orbiculare L., *scheuchzeri* ALL., *hemisphaericum* L., *humile* GAUDIN, etc. (KIRCHNER 1897; KNUTH, citing MÜLLER and KIRCHNER; CAROLIN 1960; YEO, unpublished 1974). **STR c, s. ISS growth. SEQU pln. APPL av. RWD n, p? SYN** melitto, psycho? **VIS** varied, bees (*Hymenoptera-Apoidea*) and, especially, butterflies (*Lepidoptera*).

Apart from the shape of the inflorescence and the fact that they are perhaps more favoured by *Lepidoptera*, the structure and behaviour of these species is the same as in the preceding group. CAROLIN'S statement (supported by BRANTJES 1983: 221) that the activity of insects in probing the base of the corolla can shorten it and so promote the expulsion of pollen needs confirmation. If correct, it would alter the entry above under 'ISS'.

Physoplexis (ENDL.) SCHUR (*Phyteuma* sect. *Synotoma* G. DON)

P. comosa (L.) SCHUR. (KIRCHNER 1897; KNUTH, citing KIRCHNER). **STR s. ISS rel(dose), growth? SEQU pln. APPL a, av? RWD n. SYN** psycho. **VIS** –.

The flowers are arranged in a flat capitulum; they are few in number compared with those of *Phyteuma* but much larger, with greatly inflated corolla-bases. The plant is dwarfed and the flowers are held near the ground. As in *Phyteuma*, pollen is deposited from the anthers on to a pilose region of the style before anthesis and slits appear at the base of the corolla as anthesis begins. At the same time, however, the narrow rod-like style protrudes from the corolla by several millimetres, carrying pollen instead of pushing it out as in *Phyteuma*; the style continues to grow and finally splits distally into two slender recurved stigmatic branches, yet the corolla-lobes never part at their tips. Another difference from *Phyteuma* is that the anthers are not withdrawn from the narrow part of the corolla. The main difference from *Asyneuma* is that the pollen on the style is brought forth by the emergence of the style while the corolla remains unchanged, whereas in *Asyneuma* it is uncovered by the separation of the corolla lobes. Although the corolla plays a part by encasing the stylar pollen-deposit for a time, it is not directly involved in its presentation. The capitulum is strikingly coloured, with the inflated part of the corolla being white, merely tinged with lilac, and the narrow part, together with the exserted part of the style and the pollen it bears, being blackish red. The corolla is 16mm long, and the style comes to exceed it by a further 16mm. The slits in the corolla are very narrow, and KIRCHNER deduced that pollination is by *Lepidoptera*. The cluster of spiky flowers and the great separation between the nectar and the place of pollen exchange seem to support this.

Tribe *Wahlenbergieae*
Wahlenbergia ROTH

The corollas of this genus are usually bell-shaped. DE CANDOLLE (1830) states that the style is pilose, as is shown in his illustrations of *W. denticulata* A. DC., *W. lobelioides* (L. f.) A. DC. and *W. capensis* (L.) A. DC., but not *W. foliosa* A. DC. and *W. procumbens* (THUNB.) A. DC. Hairs are clearly shown in an illustration of *W. tuberosa* J. D. HOOK. (J. D. HOOKER 1875) but not in that of the Tasmanian endemic, *W. saxicola* A. DC. (J. D. HOOKER 1882). Contractile hairs of the type found in *Campanula* are found in the second species described below.

W. gracilis (FORST. f.) A. DC. (HAVILAND 1885). **STR s. ISS rel(dose)? SEQU pln. APPL a? RWD n. SYN** melitto? **VIS** 'insects'. (Fig. 36: *W. consimilis*, see this paragraph).

HAVILAND's work on this Australasian species was done in New South Wales; he said that it was variable, with flowers 8–20 mm long, but in the present century

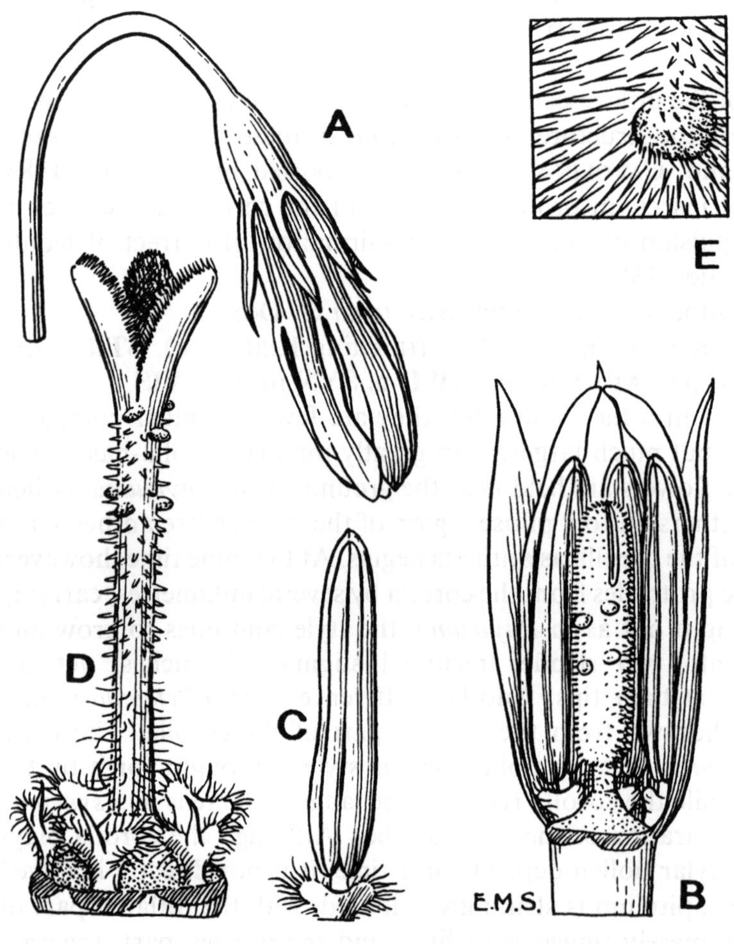

Fig. 36. *Wahlenbergia consimilis* (*Campanulaceae*). *A* flower bud. *B* dissection of flower bud at time of pollen release, a few grains being shown on the pollen presenter. *C* stamen, adaxial view. *D* nectarial disc, the auriculate filaments and the style; the style shows, below, hairs and, above (in the pollen presenting zone), 'papillae' and 'rounded protruberances.' *E* 'papillae' and one 'protuberance', which bears smaller 'papillae', at a higher magnification. (E. M. STONES in MELVILLE 1959.)

it has been split up by some authors, and it is not known which form or forms HAVILAND studied. According to him the form and behaviour of the stamens is as in *Campanula*, except that the shrivelled anthers are shed. There is no mention of hairs on the style; instead, when the anthers dehisce several large glands on the adjacent part of the style secrete a viscid fluid to which the already released pollen adheres. Development then proceeds as in *Campanula*. When the stigmatic lobes diverge there is rarely any pollen left on the style. The receptive stigmatic surfaces are moist. HAVILAND several times mentions insects without saying he has actually seen them. He says that the expanded stigmas form 'a convenient stage upon which any insect may alight'; however, as these are situated in the mouth of the bell formed by the corolla I infer that pollen is applied to and received from the anterior part of the pollinator. In *W. consimilis* LOTHIAN, one of the larger-flowered segregates of *W. gracilis*, the pollen presenter is covered with short stout pointed hairs ('papillae'). In addition, near the top of the papillate zone are 'three rows of three rounded protruberances', which themselves bear papillae, though these are smaller than those on the rest of the style (MELVILLE 1959). There can be little doubt that the protruberances are the glands observed by HAVILAND; as their presence coincides here with that of pollen-carrying hairs it seems that a further investigation of their function is needed. The dilations at the bases of the filaments in *W. consimilis* are in the form of auricles too small to cover in the nectary.

W. albomarginata W. J. HOOK. (LLOYD & YATES 1982, illus.). **STR s. ISS rel(dose). SEQU pln. APPL a? RWD n. SYN** melitto. **VIS** the hoverfly *Austrosyrphus* (two species) (*Diptera-Syrphidae*) and several species of solitary bees (*Hymenoptera-Apoidea*: *Halictidae* and *Colletidae*).

In this species there are retractile hairs on the style that gradually release the pollen from the top downwards. They are depicted in scanning electron micrographs showing that the upper half of the hair sinks into the lower half, while the exposed lower half sinks into the enlarged base that is from the beginning immersed in the stylar tissue, as described for *Campanula* by BRONGNIART in 1839. This process of contraction takes place before the removal of the pollen (LLOYD, pers. comm.). The male phase of the flower was found to last 3.8 days on average. By the time the stigmas are receptive all the flower's own pollen has normally disappeared. The stigmas were receptive for 2 days on average. Plants are self-compatible.

Codonopsis WALLICH

This is a genus of erect or twining herbaceous perennials from eastern Asia.

C. clematidea (FISCH. & MEY.) C. B. CLARKE. (ANON. 1986, illus.). **STR s. ISS rel(dose)? SEQU pln. APPLIC d** or **v. RWD n. SYN** melitto. **VIS** –.

Pollen from the anthers of the nodding bell-shaped flowers of this species is deposited before anthesis on the obovoid style-head formed from the arched immature stigmatic branches, which are shortly pilose on their outer surfaces. The pollen deposit is moulded so that at first glance it looks like the anthers themselves. As the flower opens nectar begins to be secreted by the black and yellow nectary at the base of the (otherwise pale blue) flower. The diagrams of the longitudinal section of the flower show nectar accumulating both inside and outside the stamen-bases. Insects become dusted with pollen in approaching this nectar. The stigmatic lobes only begin to unfold when their outer surfaces are more or less freed of pollen by nectar-taking insects in concert with the shrinkage of the hairs.

C. pilosula (FRANCH.) NANNF.
YEO (1972) observed the green and brown flowers of this species in cultivation in England where they were freely visited by two species of *Vespula* sens. lat. (social wasps). The insects became dusted dorsally with pollen while taking nectar that had accumulated in grooves at the base of the corolla.

Canarina L.

This genus cannot be assigned to a tribe on the basis of fruit-dehiscence because the fruit is a berry; its position was discussed by HEDBERG (1961), who opted for placing it in *Wahlenbergiae*.

C. canariensis (L.) VATKE. (DE CANDOLLE 1830, illus.; YEO 1972, illus.; VOGEL & al. 1984; OLESEN 1985, illus.). **STR s. ISS rel(dose). SEQU pln. APPL a. RWD n. SYN** ornitho. **VIS** the birds *Sylvia* and *Phylloscopus* (*Aves-Sylviidae*).
This climbing species has 6-merous brownish orange bell-shaped flowers nearly 5 cm wide and about the equally long. The ornithophil features were described by YEO; OLESEN stated that the plant has secondary pollen presentation as in *Campanula* and DE CANDOLLE showed the pollen-collecting hairs on the style-head and the appearance of the spread stigmas. VOGEL and OLESEN observed feeding visits by birds; both authors supposed that the plant evolved in the presence of nectar-feeding birds more specialized than the *Sylviidae* (warblers) but now absent from the Canary Isles.

Jasione L.

J. montana L. (MÜLLER 1883, illus.; ERDELSKÁ 1983; PARNELL 1985). **STR s. ISS rel(dose). SEQU pln. APPLIC a, v. RWD n, p. SYN** entomo. **VIS** many species from a wide range of orders: social and solitary bees and solitary wasps (*Hymenoptera*); flies in the families *Syrphidae*, *Conopidae*, *Empididae*, *Muscidae* etc. (*Diptera*); butterflies and moths, including *Sphingidae* (*Lepidoptera*); some beetles (*Coleoptera*) (Fig. 37).
The flowers are small, normally blue, and arranged in a capitulum that may contain well over 100 but averages 70 (MÜLLER) to 40 (PARNELL) flowers, perhaps depending on the population. The flowers are slender, with linear sepals and free linear petals. In bud the anthers deposit their pollen on a stylar brush as in *Campanula* (illustrated by SEM in *J. perennis* by ERBAR & LEINS 1989). The anthers are lightly coherent at the base. There is free access to the nectar between the bowed, ribbon-like filaments. As the flowers open the style grows to exceed the petals. The male stage of the flower lasts up to a week; then the two small stigmatic lobes diverge sufficiently to expose the receptive surface. ERDELSKÁ found the male stage lasting several days in a greenhouse, but out of doors in sunny weather the pollen cannot cling to the pollen presenter for more than two hours, and the female stage follows 3–5 hours after this, so that the individual floret is fully dichogamous and completes anthesis within the day. The sequence of opening is from the outside of the head inwards. Plants have a very strong self-incompatibility mechanism. Capsules can produce several seeds each but PARNELL (1982) obtained an average of only 1.18 by artificial outcrossing.

Edraianthus A. DC.

In this genus of small European and Caucasian mountain plants, differing from *Campanula* in the irregular splitting of the capsule, the solitary or clustered bell-shaped blue flowers are always erect (KIRCHNER 1897).

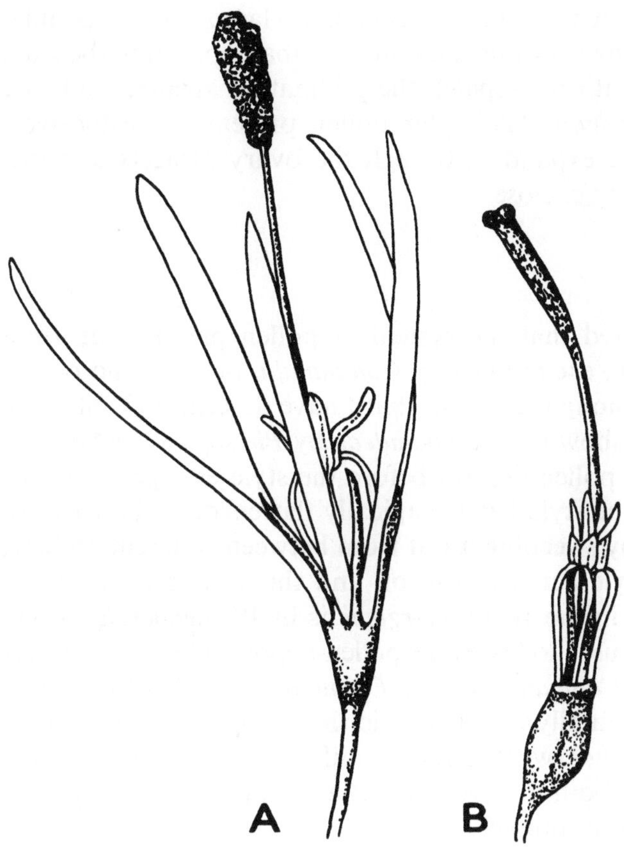

Fig. 37. *Campanulaceae*. *A* flower of *Jasione montana* at male stage. *B* the same at female stage, with sepals and petals removed. (P. F. YEO, after MÜLLER 1883.)

Cyananthus BENTH.

This genus is remarkable in the family for its fully superior ovary. Like *Jasione* it has slightly coherent anthers. The floral organization is distinctive. The corolla is salverform (hypocrateriform) rather than campanulate and the throat is narrow and sometimes partly filled by hairs. The style is long and very slender, with a small expansion at the top, and the anthers are correspondingly small. The ovary is conical and at anthesis the anthers, being coherent, are found forming a garland round its apex (the filaments always seem to be the same length as the ovary). Otherwise, floral behaviour is much as in the majority of the family, with pollen being deposited before anthesis on a stylar pollen presenter which (at least sometimes) is pilose. The five stigmatic lobes only diverge to a small extent. This information has been compiled from plates in Curtis's Botanical Magazine ('Bot. Mag.') (BALLARD 1939, 1940; COWAN 1943; MARQUAND 1935).

Tribe *Platycodoneae*

Platycodon A. DC.

This monotypic genus is the only one of its tribe.

P. grandiflorus (JACQ.) A. DC. (DE CANDOLLE 1830, illus.). **STR s. ISS rel(dose). SEQU pln. APPLIC v? RWD n, p? SYN** melitto. **VIS** –.

This perennial, native to northern Asia, China and Japan, has large blue campanula-like flowers with a wide bell, facing upwards. Despite its tribal separation (because the cells of the ovary alternate with the sepals), the pollination arrangements are remarkably like those of *Campanula*. The stylar pollen presenter is extensively pilose and the filament-bases are expanded, though the ovary projects upwards and partly fills the space that they enclose.

Discussion of Campanulaceae

In introducing this family I stated that the system of pollen presentation on a central free-standing stylar brush, exemplified by *Campanula*, prevails and occurs in the three recognized tribes. Among the genera that I have described (Table 2-4), departures from this system are shown in *Campanuleae* by *Physoplexis*, where the corolla closely invests the stylar pollen deposit before the style emerges from the corolla, by *Petromarula*, where the style-head is apically flattened, by *Trachelium*, where the style picks up pollen by sweeping it out from between coherent anthers, and by *Phyteuma*, where the style acts as a piston and the corolla as a pollen-magazine. A possible departure in the *Wahlenbergieae* is in *Wahlenbergia*, where in some cases an adhesive may be used to fasten the pollen-deposit. Coherent anthers in *Jasione* and *Symphyandra* in *Campanuleae* and *Cyananthus* in *Wahlenbergieae* do not affect presentation. As a family characteristic the *Campanula* system thus probably arose once and then underwent a few modifications. Throughout the family there is strong protandry, both in development and in presentation; selfing of the individual flower is therefore unlikely.

Flower form in the family, crudely classified into four shapes, is distributed as shown in Table 2-4. It is evident that the bell-shaped flower is prevalent, and the fact that the greater part of the large genus *Campanula* has it adds to its prevalence. This is relevant to the function of secondary pollen presentation in the family. Two key features of the flower are that the sexual organs are central in a bell-shaped corolla, serving as a pathway to the nectar for insect pollinators, and that the anthers are introrse. If the anthers were pressed against the outside of the corolla they would dust the dorsal side of an insect clinging to the style, while its ventral side would contact the stigmas; on the other hand, if the anthers were moved to a central position round the style, they would be facing the wrong way because they are introrse. Deposition of pollen on a specially modified zone of the style solves the problem. Secondary pollen-presentation here is therefore concerned with correct positioning of the pollen. Only in the modified system of *Phyteuma* does there appear to be an element of pollen protection. However, in most, and perhaps all, species that hold the secondarily presented pollen in a brush of contractile hairs there is a pollen-dosing system, as mentioned in the introduction to the family. Lloyd & Yates (1981) confirmed this by trying to shift the pollen from the stylar hairs with a brush and found that it could only be done with difficulty until the hairs had contracted, while in *Petromarula* Yeo found difficulty in dislodging part of the pollen.

Despite the heavy prevalence of one system of pollen presentation and one pollination syndrome, small numbers of species, either in the main genera or in satellite genera, show not only some deviations in pollen presentation but also a

Table 2-4. Flower form of genera of Campanulaceae treated in this review

Tribe	Flower form			
Genus	Bell	Tube	Salver	Star
Campanuleae				
Campanula	+	+		+
Symphyandra	+			
Adenophora	+			
Legousia				+
Ostrowskia	+			
Trachelium		+		
Michauxia	+			+
Petromarula				+
Asyneuma		+		
Phyteuma		+		
Physoplexis		+		
Wahlenbergieae				
Wahlenbergia	+			
Codonopsis	+			
Canarina	+			
Jasione		+		
Edraianthus	+			
Cyananthus			+	
Platycodoneae				
Platycodon	+			

wide range of pollination syndromes, including the specialized ones of sphingophily (*Michauxia*), ornithophily (*Canarina*) and possibly some form of myiophily (*Campanula macrostyla*). In addition, there are undoubtedly specializations within the syndrome of melittophily to particular groups of bees and (in the case of *Codonopsis pilosula*) to vespids. In the latter pollination is nototribic, though still dependent on the central placing of both pollen and stigmas.

It is interesting to note that in *Asyneuma* and *Physoplexis* the pollen presentation differs from that in *Phyteuma*, in which these two genera are sometimes included.

De Candolle (1830), in examining transections of the styles of many *Campanulaceae*, found that the pollen-carrying hairs of the style were arranged in longitudinal bands equal in number to the thecae of the anthers (and twice the number of the anthers), regardless of the number of stigmatic branches. In the cases where there are 5 stamens (10 thecae) and 3 stigmatic branches he found that one of the latter appeared to carry four rows of hairs and the others three each. Close examination showed him that the hairs actually arise uniformly over the stylar surface and are pressed apart by the thecae, and displaced either into the gap between two anthers or the groove between two thecae of the same anther. There is thus no structural adaptation of the style to the different numeric condition of the androecium, as there is in *Apocynaceae*.

When observing some nearly finished flowers of *Phyteuma* in 1985 I noted that some pollen was wasted by retention in the anthers after the flower had opened. In *Petromarula* there is a constant but small wastage of pollen that is not cleared from the anthers.

The closely related family *Lobeliaceae* differs from *Campanulaceae* in having the odd petal on the side of the flower away from the parent axis (MELCHIOR 1964), in having zygomorphic flowers, and in always having coherent anthers that form a pollen magazine, from which pollen is expelled by growth of the style. The family *Asteraceae*, which shows more than one type of secondary pollen presentation, seems to be the next most closely related family to the *Campanulaceae*. The relationships of *Campanulaceae* are further discussed under families *Lobeliaceae* and *Asteraceae*.

The stigmas of *Campanulaceae* are dry (HESLOP-HARRISON & SHIVANNA 1977) and the pollen is binucleate in six out of seven genera examined (trinucleate in the remaining one – BREWBAKER 1967). SHETLER (1979) concluded from his survey of the breeding system of *Campanula* that most species do not self-pollinate and that artificial selfing fails in most cases. Self-incompatibility mechanisms seem to be prevalent, but within species there can be individual variation in the degree of self-sterility. *Wahlenbergia* can be self-compatible (*W. albomarginata*, described here) or fully self-incompatible (*W. androsacea* A. DC. in tropical Africa–THULIN 1975). Self-incompatibility also occurs in *Jasione montana*, in which it would have value in preventing selfing by transfer of pollen from flower to flower within the capitulum.

References for Campanulaceae

ANON., 1986: Wie die Glockenwinde Selbstbestäubung einschränkt. – Mitt. Bot. Gart. **35(6)**: 2pp. (reprinted in Gärtn.-Bot. Briefe **87**: 34–35, 1986).

BAILEY, L. H., 1953: The Garden of Bellflowers. – New York: Macmillan.

BALLARD, F., 1939: *Cyananthus macrocalyx*. – Bot. Mag. **162**: t. 9562.

BALLARD, F., 1940: *Cyananthus microphyllus*. – Bot. Mag. **162**: t. 9598.

BRANTJES, N. B. M., 1983.

BREWBAKER, J. L., 1967.

COWAN, J. M., 1943: *Cyananthus sherriffii*. – Bot. Mag. **164**: t. 9655.

CURTIS, W., 1793: *Michauxia campanuloides*. Rough leav'd Michauxia. – Bot. Mag. **7**: t. 219.

DAMBOLDT, J., 1978: *Asyneuma, Campanula*. – In DAVIS, P. H., (ed.): Flora of Turkey **6**: 65–81 and 2–64. – Edinburgh: University Press.

DE CANDOLLE, A., 1830: Monographie des Campanulées. – Paris: Desray.

EAST, E. M., 1940.

ERBAR, C., LEINS, P., 1989: On the early floral development and the mechanisms of secondary pollen presentation in *Campanula, Jasione* and *Lobelia*. – Bot. Jahrb. **111**: 29–55.

ERDELSKÁ, O., 1983: Dichogamy and pistil hairs in the *Campanulaceae*. – Preslia **55**: 269–271.

FEDOROV, A. A., 1957: Family *Campanulaceae*. – In SHISHKIN, B. K., BOBROV, E. G., (volume eds.): Flora of the USSR. **24**: 126–447. – Leningrad: Botanical Institute of the Academy of Sciences of the USSR.

HAVILAND, E., 1885: Occasional notes on plants indigenous in the immediate neighbourhood of Sydney (No. 9). – Proc. Linn. Soc. New South Wales **9**: 1171–1174.

HEDBERG, O., 1961: Monograph of the genus *Canarina* L. – Svensk Bot. Tidskr. **55**: 17–62.
HESLOP-HARRISON, Y., SHIVANNA, K. R., 1977.
HEYWOOD, V. H. (ed.), 1978.
HILDEBRAND, F., 1870: FEDERIGO DELPINO's weitere Beobachtungen Über die Dichogamie im Pflanzenreich mit Zusätzen und Illustrationen. – Bot. Zeitung (Berlin) **28**: 585–594, 601–609, 617–625 (*Campanulaceae* here), 633–641, 649–659, 665–675.
HOOKER, J. D., 1875: *Wahlenbergia tuberosa*. – Bot. Mag. **101**: t. 6155.
HOOKER, J. D., 1878: *Campanula macrostyla*. – Bot. Mag. **104**: t. 6494.
HOOKER, J. D., 1882: *Wahlenbergia saxicola*. – Bot. Mag. **108**: t. 6613.
HOOKER, J. D., 1896: *Ostrowskia magnifica*. – Bot. Mag. **122**: t. 7472.
HOOKER, J. D., 1900: *Michauxia tchihatchefii*. – Bot. Mag. **126**: t. 7742.
HOOKER, W. J., 1832: *Michauxia laevigata*. Smooth Michauxia. – Bot. Mag. **59**: t. 3128.
JANSON, L.-Å., 1983: Pollination studies of *Campanula persicifolia* (*Campanulaceae*) in Sweden. – Grana **22**: 153–165.
JOST, L., 1918: Die Griffelhaare der Campanulablüte. – Flora **111**: 478–489.
KIRCHNER, O., 1897: Die Blüteneinrichtungen der Campanulaceen. – Jahresh. Vereins Vaterl. Naturk. Württemburg **53**: 193–228.
KNUTH, P., 1909.
KCLAKOVSKY, A. A., 1987: Sistema semeistva *Campanulaceae* starogo sveta. (System of the *Campanulaceae* family from the Old World.) – Bot. Zhurn. **72**: 1572–1579 (Russian with English summary).
LLOYD, D. G., YATES, J. M. A., 1982.
MARQUAND, C. V. B., 1935: *Cyananthus longiflorus*. – Bot. Mag. **158**: t. 9387.
MELCHIOR, H. (ed.), 1964: A. ENGLER's Syllabus der Pflanzenfamilien, 12th edn. **2**. – Berlin: Borntraeger.
MELVILLE, R., 1959: *Wahlenbergia consimilis*. – Bot. Mag. **172**: n.s., t. 343.
OLESEN, J. M., 1985: The Macaronesian bird-flower element and it relation to bird and bee opportunists. – Bot. J. Linn. Soc. **91**: 395–414.
PARNELL, J. A. N., 1982: Some observations on the breeding biology of *Jasione montana*. – J. Life Sci. Royal Dublin Soc. **4**: 1–7.
PARNELL, J. A. N., 1985: *Jasione montana*, in Biological Flora of the British Isles, no. 157. – J. Ecol. **73**: 341–358.
PROCTOR, M., YEO, P., 1973.
SHETLER, S. G., 1979: Pollen-collecting hairs of *Campanula* (*Campanulaceae*), I: historical review. – Taxon **28**: 205–215.
THULIN, M., 1975: The genus *Wahlenbergia* s. lat. (*Campanulaceae*) in tropical Africa and Madagascar. – Symb. Bot. Upsal. **21**: 1–223.
VOGEL, S., WESTERKAMP, CH., THIEL, B., GESSNER, K., 1984: Ornithophilie auf den Canarischen Inseln. – Pl. Syst. Evol. **146**: 225–248.
YEO, P. F., 1972.
YEO, P. F., 1993: *Platycodoneae*, a new tribe in *Campanulaceae*. – Taxon **42**: 109.

Cyphiaceae

Systematic position: *Asteridae, Campanulales*
Restriction of occurrence within the family: None

Distribution within the family. Occurs in all species studied in this unigeneric family.
Floral features. Calyx 5-partite. Petals free, merely cohering by their claws above the base, narrow, three forming an upper lip, two forming the lower. Stamens free or connate (or perhaps always coherent) into a hairy tube. Ovary half-inferior

with several to many seeds; style one, enlarged at the apex which is asymmetric with a pore on one side leading into a stigmatic cavity.
Type of pollen presentation. Pollen presentation is from a box formed by the anthers and the apical surface of the style-head ('Nudelspritze'?). (Information from HILDEBRAND 1870, MARLOTH 1932 and DYER 1975).

Examples of Secondary Pollen Presentation in Cyphiaceae

Cyphia BERGIUS

Different accounts of this genus do not fully agree as to structure, while information on pollen presentation is scanty. Species on which some information is available are named and dealt with collectively. The genus is South African and the plants herbaceous, sometimes twining.
C. assimilis SONDER, *C. bulbosa* (L.) BERGIUS, *C. phyteuma* (L.) WILLD., *C. volubilis* (THUNB.) WILLD., *C. zeyheriana* PRESL. (HILDEBRAND 1870; LINDLEY 1822, illus.; MARLOTH 1932, illus.). **STR a, s. ISS growth? SEQU pln. APPL v. RWD n. SYN** sphingo (*C. volubilis*), melitto? (*C. phyteuma*). **VIS** hawkmoths (*Lepidoptera-Sphingidae*).
The sexual organs reach as far as the mouth of the functionally tubular part of the corolla; according to HILDEBRAND they lie on the lower side and touch the insects ventrally. HILDEBRAND states that nectar is secreted in the upper half of the corolla. The pollen is released from the anthers before the flower opens and remains in a single large mass in a box formed by the anthers and the top of the expanded style-head, which forms the floor of the box. This part of the style is usually covered with hairs, but HILDEBRAND, while admitting that further investigation is needed, states that the style does not grow between the anthers and that the hair-tuft has no sweeping function. (If this were true it would be difficult to see how the pollen is released or how the stigma could later be exposed for the receipt of pollen.) The lateral opening on the style-head may be at the tip of a beak-like process or be surrounded by a rim; it is also usually surrounded by additional hairs. The stylar cavity is filled with a viscid fluid and according to MARLOTH its mouth regularly becomes surrounded or filled by pollen grains which germinate here or within the chamber into which it may be drawn by shrinkage of the body of fluid.

Discussion of Cyphiaceae

Apart from being organized for sternotribic pollination, the flowers of *Cyphiaceae* seem to function like those of *Lobeliaceae*, though apparently lacking specialized arrangements for controlling pollen delivery. There seems to be some degree of pollen protection, and there is near-identity in the place of presentation and receipt of pollen. At present there is no evidence on the regulation of the issue of pollen.

References for Cyphiaceae

DYER, R. A., 1975: The Genera of Southern African Flowering Plants **1**. – Pretoria: Republic of South Africa Department of Agricultural Technical Services.
HILDEBRAND, F., 1870: F. DELPINO's weitere Beobachtungen über die Dichogamie im

Pflanzenreich mit Zusätzen und Illustrationen. – Bot. Zeitung (Berlin) **28**: 585–594, 601–609, 617–625, 633–641 (*Cyphiaceae* here), 649–659, 665–675.
LINDLEY, J., 1822: *Cyphia Phyteuma.* Rampion-flowered Cyphia. – Bot. Reg. **8**: t. 625.
MARLOTH, R., 1932: The Flora of South Africa **3(2)**. – Capetown: Darter; London: Wheldon & Wesley.

Lobeliaceae

Systematic position: *Asteridae, Campanulales*
Restriction of occurrence within the family: None

Distribution within the tribes. Two tribes, *Lobelieae* and *Delisseeae*, are recognized (WIMMER 1943; M. KOVANDA in HEYWOOD 1978). Secondary pollen presentation apparently occurs throughout both, except in isolated instances of autogamy, but descriptive accounts of pollen presentation are not available for any member of the *Delisseeae.*

Floral features. Calyx lobes five, narrow. Corolla zygomorphic, of five united petals, usually with a slit along the upper side. Stamens five; anthers coherent to form a cylinder into which pollen is shed before anthesis. Ovary inferior, though with a conical apex within the corolla in tribe *Lobelieae* (fully superior in rare exceptions). Style simple, with two or three stigmatic lobes. The disposition of the flower described here results from resupination. The odd petal (middle lobe of lower lip) is the one that is initially nearest to the axis. (Resupination is caused not by the twisting of the pedicel but by its curvature, so that the flower is flipped over and appears to come from the side of the stem opposite to that on which the pedicel is attached.)

The bases of the stamen-filaments may be expanded and incurved over the ovary, forming a nectarial chamber, as in *Campanulaceae.* However, the narrow part of the filament does not wither early, as it is functional throughout the male stage of the flower. The basal part cannot form a flap that can be raised by a pollinator, because that would displace the distal part and the anther-cylinder; instead, the lower ends of the filamentous parts of some stamens are bowed outwards, to leave gaps giving access to the nectar reservoir. The distal parts of the filaments are usually coherent, and perhaps also sometimes the lower ends of the narrow parts of those that are not bowed outwards. The tip of the androecium is curved towards the lower lip of the flower. Frequently the two corolla-lobes on either side of the slit are erect and form an upper lip, albeit cleft; occasionally the distal part of the corolla is flattened out so that there is no upper lip, but a five-lobed lower lip. Both these forms of the corolla are matched in *Goodeniaceae.* Sometimes the whole corolla, while being more or less tube-like, is decurved distally, leaving the distal part of the staminal column and style, which are less curved, protruding from the slit; the appearance of the flower then recalls that of many *Proteaceae*, though there only the style protrudes.

Type of pollen presentation. When the pollen is shed the tip of the style is positioned inside the base of the anther cylinder, and during anthesis it passes through the cylinder and pushes out the pollen with the aid of a brush of hairs surrounding it just below its head ('Nudelspritze'). Finally the style-head itself emerges and the two or three stigmatic lobes that compose it expand, as in *Campanulaceae.* The system varies in that in some genera the pollen is pushed out simply by growth of the style, while in others its release is triggered by the pollinator. Although the pollen is discharged directly from the thecae of the anthers it is also in contact with the style-head, which plays an essential role in presentation, and the Lobeliaceae may therefore be regarded as showing secondary pollen presentation.

Examples of Secondary Pollen Presentation in Lobeliaceae

Tribe *Lobelieae*

Lobelia L.

L. laxiflora HBK. (*Siphocampylus bicolor* G. DON). (MÜLLER, quoting HILDEBRAND; HILDEBRAND 1866, illus.; WAGNER 1917, illus.). **STR a, s. ISS growth. SEQU pln. APPL ad. RWD n. SYN** ornitho. **VIS** –.

This species from the American tropics has flowers with a scarlet tube and yellow throat. WAGNER recognized that pollination was most likely to be by birds. The brush-tipped style-head pushes pollen out of the anther-cylinder by growth. The two stigmatic lobes part only when the style has grown several millimetres beyond the anther-cylinder, which has a bearded mouth. HILDEBRAND suggested that in the female stage of the flower the sweeping hairs of the style might provide additional protection against self-pollination. WAGNER was mainly concerned with a malfunction of the system in cultivated plants.

L. erinus L. (KNUTH, quoting HILDEBRAND, DELPINO and T. H. FARRER, illus.; STUMPF 1987; ERBAR & LEINS 1989, illus.). **STR a, s. ISS growth. SEQU pln. APPL ad. RWD n. SYN** melitto. **VIS** bees (*Hymenoptera-Apoidea*), hoverflies (*Diptera-Syrphidae*) (these records from Europe).

The small, typically blue, flowers of this horticulturally popular South African annual herb appear to be adapted to bees. *Lepidoptera* are likely to be able to take the nectar without causing pollination, while hoverflies that were seen on it were most likely gleaning pollen. Pollen is swept out of the anther-cylinder by growth. STUMPF says that the two anthers on the lower side of the flower each bear a tuft of small bristles and one slightly larger one which, however, probably does not possess the triggering function of the much larger hairs in *Isotoma*. The development of the flower from its earliest stages is beautifully illustrated by SEMgraphs by ERBAR & LEINS; this study showed that during part of the floral development the style-apex lies well above the base of the anthers but shortly before the flower is due to open the filaments elongate to bring the bases of the anthers level with the style-apex; pollen is then shed into the anther-cylinder ready to be pushed out by further growth of the style.

L. cardinalis L. (DEVLIN & STEPHENSON 1984; STUMPF 1987, illus.). **STR a, s. ISS rel-(dose). SEQU pln. APPL ad. RWD n. SYN** ornitho. **VIS** hummingbirds (*Aves-Trochilidae*). (Fig. 38).

The flower of this North American species is about 2.5 cm long and 3 cm wide, with a uniformly scarlet corolla which is split to the base along the upper side. The staminal column emerges from the slit well before the corolla lobes separate; its filaments are fused into a glabrous cylindrical tube at the apex but are free and hairy towards the base. In this lower region the lower filaments are swollen and the (interrupted) tube is dorsiventrally compressed. The throat region of the corolla has folded commissures, and the lateral folds grip the sides of the widened filament

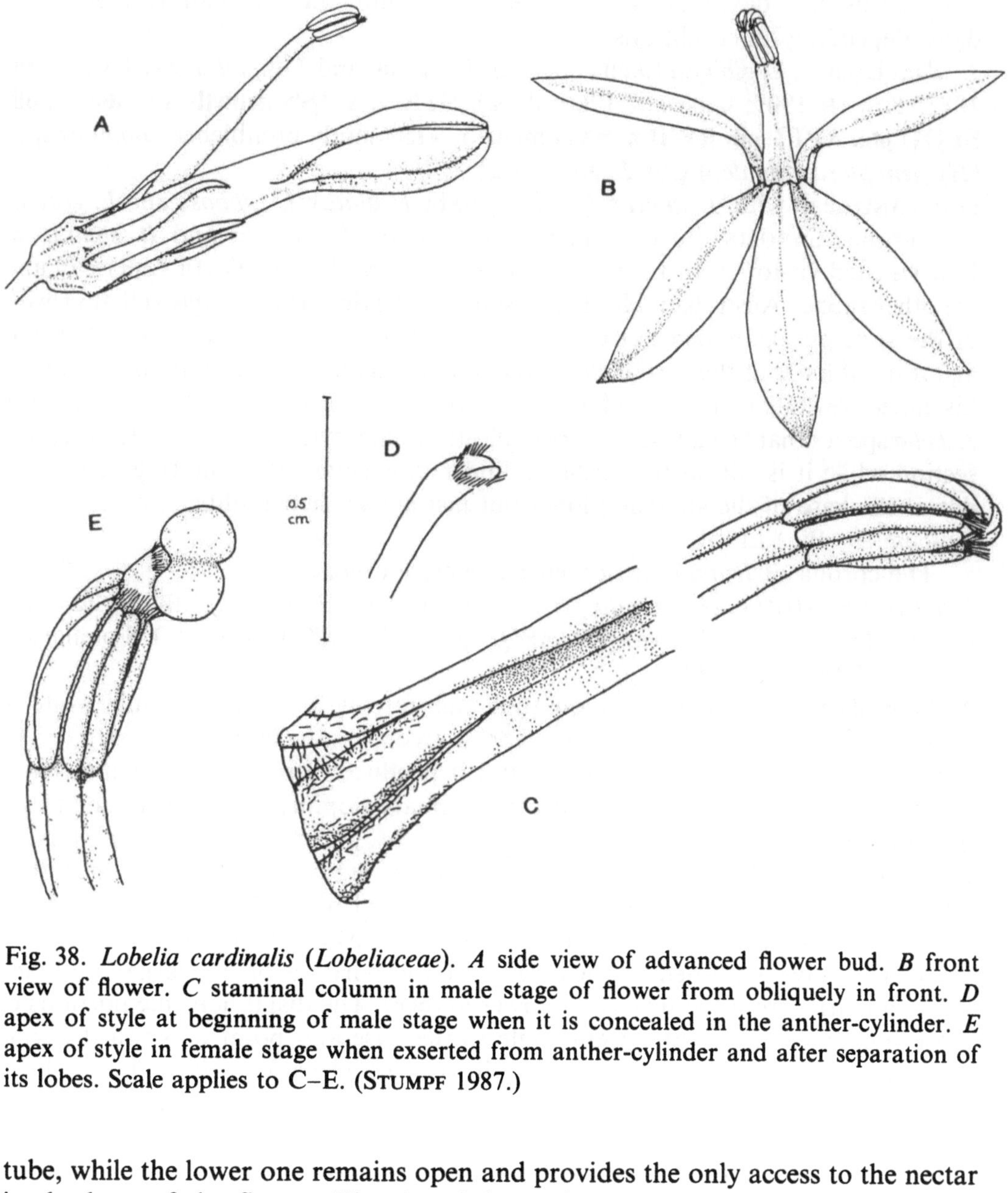

Fig. 38. *Lobelia cardinalis* (*Lobeliaceae*). *A* side view of advanced flower bud. *B* front view of flower. *C* staminal column in male stage of flower from obliquely in front. *D* apex of style at beginning of male stage when it is concealed in the anther-cylinder. *E* apex of style in female stage when exserted from anther-cylinder and after separation of its lobes. Scale applies to C–E. (STUMPF 1987.)

tube, while the lower one remains open and provides the only access to the nectar in the base of the flower. The tip of the anther-cylinder is covered with small papillae curved towards the apex. The two shorter anthers on the lower side of the cylinder bear an apical tuft of about 100 short stiff hairs that close the orifice. DEVLIN & STEPHENSON found that the flowers are self-fertile. By rubbing something against the brush on the anther-cylinder they could cause a dose of pollen to be released. In some experiments this was done until no more pollen emerged; in flowers treated in this way further pollen was available next day, up to a maximum of three or four days in all. Flowers protected from pollination had a male phase lasting 9.6 days on average. Intermediate durations of the male phase occurred when an intermediate amount of pollen was removed daily, or when all was removed on alternate days. The female stage was one to three days long if pollination took place on the first day, and longer if pollen was withheld, up to an average of 3.4 days, depending on conditions.

L. alata LABILL., *L. gibbosa* LABILL., *L. gracilis* ANDR. and *L. siphilitica* L. (HAVILAND 1883; KNUTH 1909; CAROLIN 1960, illus.). **STR a, s. ISS growth, rel-(dose)-poll. SEQU pln. APPL ad. RWD n. SYN** melitto. **VIS** mainly bumblebees *Bombus* spp. (*Hymenoptera-Apoidea*) (for *L. siphilitica*).

HAVILAND dealt with '*L. anceps*' (assumed to be *L. alata*), *L. gibbosa* and *L. gracilis* as a group; CAROLIN did the same with *L. alata*, *L. gracilis* and *L. siphilitica*; KNUTH cited visitor-records for *L. siphilitica*. *L. siphilitica* is North American and the other species Australian. HAVILAND states (as he does for *Monopsis*) that growth of the style pushes pollen out of the anther-cylinder but CAROLIN says that it is agitation of hairs at the tips of the anthers that causes pollen to fall out (however, his collective statement covered also *Isotoma petraea* and may not be true for the *Lobelia* species that he named). CAROLIN illustrates the style of *L. alata* in longitudinal section while it is still in the anther-cylinder; this shows that the ring of hairs is not at the base of the stigmatic lobes but just above their middle.

Isotoma (R. BR.) LINDL.

The corolla of *Isotoma* is not slit but merely notched.

I. petraea F. MUELL. (*Laurentia petraea* (F. MUELL.) E. WIMM.). (BRANTJES 1983, illus.). **STR a, s. ISS rel(dose)-poll. SEQU pln. APPL ad. RWD n. SYN** phalaeno, sphingo. **VIS** –. (Fig. 39).

The corolla has a long and narrow tube and narrow lobes. It is mainly white in colour but the base of the lower lip bears a greenish area edged with purple. The flower opens initially in the evening and it produces scent only at night. These are features of the pollination syndrome of phalaenophily. There is a male phase of anthesis of three or four nights, followed by a female phase. The stamen-filaments are partly fused to the corolla and joined to each other immediately below the anthers. The anthers form a hard closed box into which the pollen is released. At the extreme tip of the box is a downward-facing hole closed by a valve. The valve is formed by the enlarged basal portions of two large non-living hairs that arise from the tips of the lower anthers. The distal hair-like portions of these trichomes are set at a right-angle to the base and point downwards, lying close together as a single prong; when this is pressed backwards, as by the head of an insect, the leverage on the basal portions makes them bend at a point half way along their length, thus slightly opening the valve and allowing pollen to emerge. On release of this pressure, the valve is pressed back to its original position by a group of very short stiff non-living hairs lying against its outward-facing surface. During the male stage of anthesis the stigmatic lobes are pressed together and bear in their basal half an annular brush of blunt hairs (papillae) that reach obliquely forward to the wall of the anther-box. Together with the style-head these form a piston which, as the style grows, presses the pollen forward in the box. Every time the valve is opened, therefore, some pollen escapes. If, however, the valve is not opened by pressure from a visitor the pollen grains are forced backwards between the hairs of the stylar brush, and the pressure in the box never becomes sufficient to open the valve from inside. Thus, although growth of the style is an essential part of the mechanism it is not on its own responsible for the discharge of pollen. The female phase of the flower is initiated when the style breaks out of

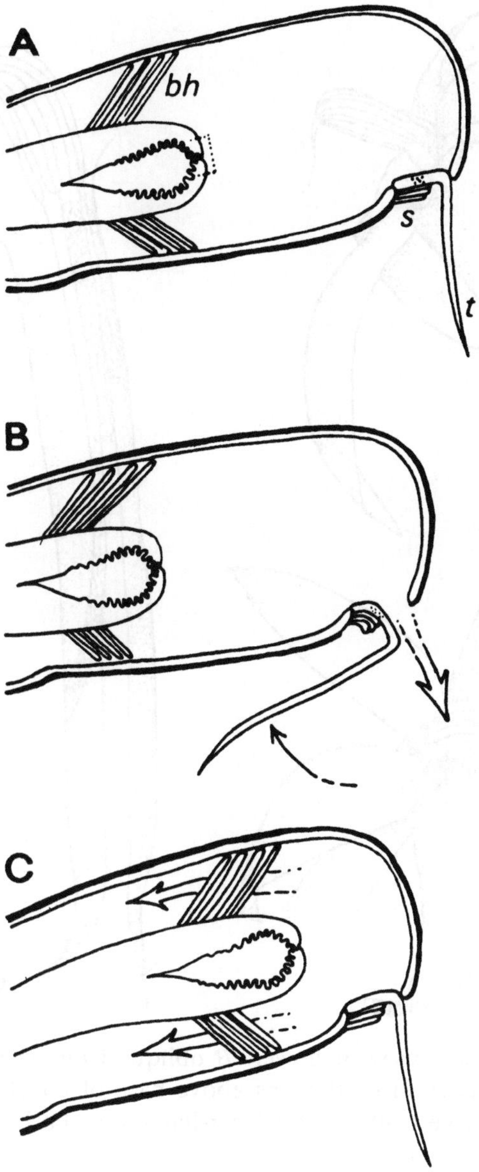

Fig. 39. *Isotoma petraea* (*Lobeliaceae*), longisections through the anther-cylinder; the connivent style-branches enclose a cavity. *A* at beginning of anthesis; *bh* = brush, *s* = spring-hairs, *t* = trigger-hairs. *B* the same, but with the trigger-hairs pushed back allowing pollen grains to come out (arrow); the dots mark the hinge. *C* later during anthesis; if the valve has not been opened, some of the pollen passes between the hairs of the brush (arrows). (BRANTJES 1983.)

the anther box and its lobes diverge; as this happens some residual pollen is allowed to fall out.

I. axillaris LINDL. (MELVILLE 1960, illus.; STUMPF 1987, illus.). **STR a, s. ISS rel-(dose)-poll. SEQU pln. APPL ad. RWD n. SYN** psycho. **VIS** –. (Fig. 40).

The flower of *I. axillaris* is structurally like that of *I. petraea* and it has a similarly long corolla (4 cm) but the limb has shorter lobes and is blue within. MELVILLE

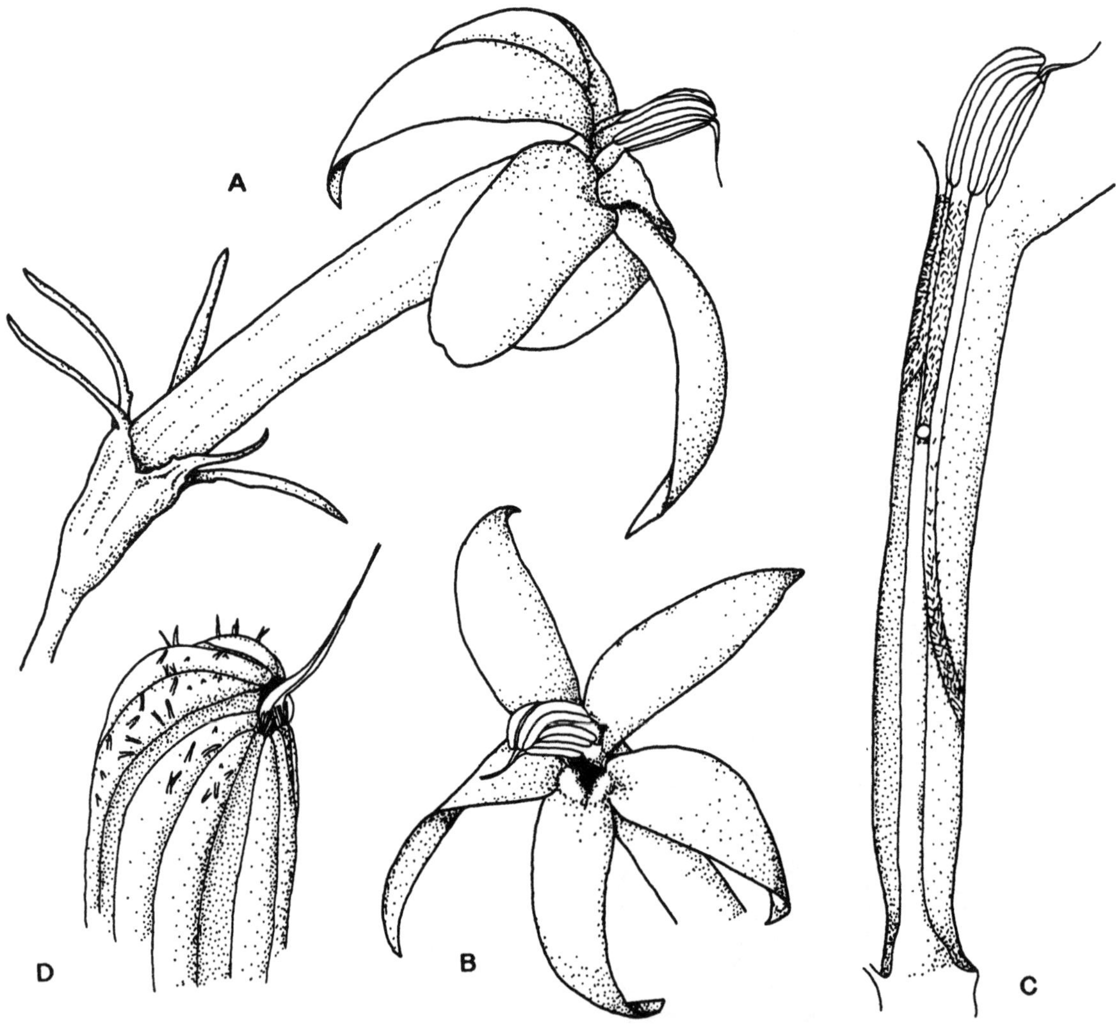

Fig. 40. *Isotoma axillaris* (*Lobeliaceae*). *A* side view of flower. *B* oblique front view of flower. *C* longisection of flower showing insertion of stamens above base of corolla. *D* apex of anther-cylinder, showing the two large contiguous trigger-hairs and the bristles that press them back after displacement. (Stumpf 1987.)

gave less functional detail than Brantjes gave for *I. petraea* but both species are Australian and he states that all Australian species of *Isotoma* have 'trigger-hairs' on the anthers. He shows that the same two kinds of hair are present at the apex of the anther-cylinder as in *I. petraea*, which suggests that the mechanism is exactly the same. Wimmer (1953) placed these two species next to each other in his monograph and remarked that they are very similar. Stumpf describes the union of the two large trigger-hairs as due to the interlocking of papillate swellings and cuticular flanges that arise from the cell walls. She also gives details of the extent of union of the stamen filaments with each other and with the corolla, and states that the bases of the two upper corolla lobes form a collar grasping the base of the anther-cylinder.

I. fluviatilis (R. Br.) F. Muell. ex Benth. (*Laurentia fluviatilis* (R. Br.) Wimm.). This species was covered by Carolin (1960) in his short statement about *Lobeliaceae* quoted under Lobelia; thus it should function similarly to *I. petraea* and, like *I. axillaris*, it does in fact have the same two types of hair on the anthers (Wimmer 1953).

I. hypocrateriformis (R. Br.) Druce. (Holm 1988).

This West Australian species has a shorter corolla tube and much larger limb than the species described so far. The construction and mechanism are very similar to those already described but the two trigger-hairs do not seem to have a basal expansion; however, as the anthers are thin-walled, deflection of the trigger-hairs is sufficient to cause the orifice to open. The pollen is issued in small portions. There are rugosities at the mouth of the corolla which are thought to serve as proboscis guides and also small visual guide-marks.

Hippobroma G. Don

H. longiflora (L.) G. Don (*Isotoma longiflora* (L.) Presl; *Laurentia longiflora* (L.) Endl.). (Melville 1960).

Melville states that this plant has only undifferentiated hairs at the mouth of the anther-tube, and he thinks that this may justify its exclusion from the genus *Isotoma*. It is not clear whether the hairs in any way regulate the issue of pollen. Brantjes (1983) appears to suggest that they do, but not in such a refined way as in *I. petraea*. Stumpf (pers. comm. 1989) states that the opening of the anther-cylinder is very large, and that all anthers have apical hairs.

Downingia Torr.

D. elegans Torr.

Stumpf (1987 and pers. comm. 1990) found that the anther-cylinder of this species bears at its tip two large flattened trigger-hairs as well as a group of thinner hairs, as in *Isotoma*. The large trigger-hairs are attached to the two lower anthers, as in *Isotoma*, but are free from one another and divergent. The plant bears small light blue short-tubed flowers with yellow marks in the throat. It is presumed to be pollinated by bees and is evidently self-fertile.

Monopsis Salisb.

M. debilis (L. f.) Presl (*Lobelia debilis* L. f.; *M. simplex* sensu Wimm., non (L.) Wimm.). (Haviland 1883). **STR a, s. ISS growth. SEQU pln. APPL a. RWD n. SYN** melitto. **VIS** –.

Monopsis is unique in the *Lobeliaceae* in having linear stigmatic lobes; these have the usual ring of hairs at the base and this is presumably positioned at the base of the anther-cylinder as the anthers begin to shed pollen. *M. debilis* is a native of South Africa but is naturalized in Australia, and it was included in Haviland's observations made near Sydney. However, Haviland considered this and three species of *Lobelia* to be so similar that they could be dealt with by a combined description; in fact he only observed a single individual of *M. debilis*. The common description says that the stylar hairs sweep pollen out of the anther-cylinder and that it may fall away; also that practically no pollen clings to the outside of the part of the style that emerges from the anther-cylinder. The corolla of *M. debilis* is almost actinomorphic and coloured purplish blue; its diameter is apparently 6–10 mm (Phillipson 1986). From the colour of the flower I suppose the species to be melittophilous and although the corolla lobes are nearly equal the androecium is set to one side in the usual way, and I therefore presume that application of pollen to the visitor is anterior but not necessarily dorsal.

Discussion of Lobeliaceae

Secondary pollen presentation is an inherent family characteristic of the *Lobeliaceae*. Union of the anthers, which is incipient in some *Campanulaceae* (most notably in *Trachelium*), is here complete. In *Campanulaceae* the deposition of pollen all round the style is perhaps responsible for restricting that family to actinomorphy. The quite different form of pollen presentation in *Lobeliaceae* represents a different way of utilizing introrse anther-dehiscence, and lends itself readily to zygomorphy, which is the norm in the family. It clearly has a pollen-saving role (since the pollen is well hidden in the staminal box), it has the effect of presenting pollen and stigma in the same part of the flower and it enables provision to be made for pollen-dosing triggered by the pollinator. Descriptions suggest that only a small proportion of the pollen production is delivered at one time, and there may be cases where pollen is placed where the pollinator finds it difficult to groom off. Certainly, the flowers seem well adapted to providing nectar as a reward and withholding pollen. Regarding the slit corolla, see the discussion of *Goodeniaceae*. One gets a general impression that melittophily is the prevalent syndrome in Australasia, but ornithophily is strongly developed in the Americas and the Hawaiian Islands, and our examples also cover two versions of the lepidopterophil syndrome. STUMPF (1987: 19, *q.v.* for references) emphasizes the wide range of pollinators of this family, which includes also bats.

As with the other families of *Campanulales* and *Asterales* that have secondary pollen presentation, the relative height of the anthers and style-head changes during development (see *Lobelia erinus*) so that the style-head can fulfill its role in presenting pollen.

There is one aspect of the mechanics of the pollen presentation in the family that has not been taken into account in the writings that I have seen. BRANTJES makes it clear that pollen in the anther-cylinder is under some pressure, which is what causes it to emerge when the valve is opened. Similarly, DEVLIN & STEPHENSON found that after a number of doses of pollen had been taken no more emerged, but that the supply was resumed next day. Therefore, there must be elasticity in the system somewhere; the possibilities are that the pollen grains could be compressed to a more angular shape, or that the anther-cylinder or the filaments could be stretched.

The *Campanulaceae*, despite their difference from *Lobeliaceae* in floral symmetry (odd petal away from the parent axis, actinomorhpy) and pollen presentation, have generally been regarded as closely related to them, and the two groups are often treated as subfamilies of *Campanulaceae*. They agree in a very strong overall similarity in vegetative characters, the usually inferior ovaries and the presence of latex ducts. Both families can have either binucleate or trinucleate pollen, though the *Campanulaceae* are usually binucleate (BREWBAKER 1967). They differ in their stigmatic surfaces – dry in *Campanulaceae* and wet in *Lobeliaceae* (HESLOP-HARRISON & SHIVANNA 1977). EAST (1940) was able to cite one definite case of self-incompatibility in the *Lobeliaceae* (*Lobelia fulgens*) but the site of the incompatibility reaction was unknown; he claimed incomplete evidence for self-incompatibility in the closely related *L. cardinalis* and species of *Isotoma*, but as shown above *L. cardinalis* has been found to be self-compatible, as has *Downingia elegans*. The two families, curiously enough, have no similarity in their pollen grain morphology (DUNBAR 1975b). ZAVADA (1984) has claimed that differences in pollen grain structure are correlated with the type of self-incompatibility system (gametophytic or sporophytic). He attributes this to the role of the exine in acting as a repository for tapetally derived recognition substances which operate in sporophytically controlled self-incompatibility systems. If it turns out that in *Lobeliaceae* sporophytic self-incompatibility is the rule this might account for the difference, because it appears that *Campanulaceae* are frequently self-compatible. However, ZAVADA'S conclusions have been refuted by GIBBS & FERGUSON (1987) on the grounds of factual errors, faulty sampling and an unjustified supposition that large pollen wall cavities and large pores are necessary for the functioning of sporophytic incompatibility systems. Nevertheless, DUNBAR'S photographs show gaps between the reticulations of the exine of most *Lobeliaceae* (DUNBAR 1975b) but not usually between those of *Campanulaceae* (DUNBAR 1975a). The relationships of *Lobeliaceae* are further discussed under family *Asteraceae*.

References for Lobeliaceae

BRANTJES, N. B. M., 1983.

BREWBAKER, J. L., 1967.

CAROLIN, R. C., 1960.

DEVLIN, B., STEPHENSON, A. G., 1984: Factors that influence the duration of the staminate and pistillate phases of *Lobelia cardinalis* flowers. – Bot. Gaz. **145**: 323–328.

DUNBAR, A., 1975a: On pollen of *Campanulaceae* and related families with special reference to the surface ultrastructure. I. *Campanulaceae* Subfam. *Campanuloideae*. – Bot. Not. **128**: 73–101.

DUNBAR, A., 1975b: idem. II. *Campanulaceae* Subfam. *Cyphioidae* and Subfam. *Lobelioidae*; *Goodeniaceae*; *Sphenocleaceae*. – Bot. Not. **128**: 102–118.

EAST, E. M., 1940.

ERBAR, C., LEINS, P., 1989: On the early floral development and the mechanisms of secondary pollen presentation in *Campanula*, *Jasione* and *Lobelia*. – Bot. Jahrb. **111**: 29–55.

GIBBS, P. E., FERGUSON, I. K., 1987: Correlations between pollen exine sculpturing and angiosperm self-incompatibility systems – a reply. – Pl. Syst. Evol. **157**: 143–159.

HAVILAND, E., 1883: Occasional notes on plants indigenous in the immediate neighbourhood of Sydney (No. 3). – Proc. Linn. Soc. New South Wales **8**: 182–186.

HESLOP-HARRISON, Y., SHIVANNA, K. R., 1977.

HEYWOOD, V. H. (ed.), 1978.

HILDEBRAND, F., 1866: Über die Vorrichtungen an einigen Blüthen zur Befruchtung durch Insektenhülfe. – Bot. Zeitung (Berlin) **24**: 73–78.

KNUTH, P., 1909.

MELVILLE, R., 1960: Contributions to the flora of Australia: VI. The pollination mechanism of *Isotoma axillaris* LINDL. and the generic status of *Isotoma* LINDL. – Kew Bull. **14**: 277–279.

MÜLLER, H., 1883.

PHILLIPSON, P. B., 1986: Taxonomy of *Monopsis* (*Lobeliaceae*): *M. simplex*, *M. debilis* and a new species. – Bot. J. Linn. Soc. **93**: 329–341.

STUMPF, S., 1987, unpublished.

WAGNER, A., 1917: Über eine unzweckmässige Einrichtung im Blütenbaue von *Lobelia laxiflora*. – Ber. Deutsch. Bot. Ges. **35**: 360–369.

WIMMER, F. E., 1943: *Campanulaceae-Lobelioideae*, I. Teil. – In ENGLER, A., & al., (eds.): Das Pflanzenreich **IV.276b** (**106**. Heft). – Berlin: Akademie Verlag.

WIMMER, F. E., 1953: *Campanulaceae-Lobelioideae*, II. Teil. – In ENGLER, A., & al., (eds.), Das Pflanzenreich **IV.276b** (**107**. Heft). – Berlin: Akademie Verlag.

ZAVADA, M. S., 1984: The relation between pollen exine sculpturing and self-incompatibility mechanisms. – Pl. Syst. Evol. **147**: 63–78.

Goodeniaceae

Systematic position: *Asteridae*, *Campanulales*
Restriction of occurrence within the family: None

Distribution within the family. Secondary pollen presentation appears to be characteristic of the whole family. The genus *Brunonia* is sometimes placed in a family by itself (*Brunoniaceae*).

Floral features. Ovary usually inferior to half-inferior, producing one to many seeds. Corolla sympetalous, actinomorphic in *Brunonia*, otherwise zygomorphic; in zygomorphic genera 5-lobed and slit along the upper side, and with 'tube' saccate to calcarate at base (STUMPF 1987); lobes usually arranged to form two lips but sometimes spread out as a single 5-lobed lower lip, often conspicuously flanged (Fig. 41A). Stamens five, free from the corolla or nearly so; filaments free from one another; anthers either free or joined. There are 1–5 nectarial swellings at the base of the style (STUMPF, pers. comm.). The style lies along the upper side of the flower. At its apex it develops a purse-like structure, the indusium or pollen-cup, enclosing and projecting beyond the stigmatic region. The pollen-cup may be formed as a single envelope or be made up of two or four lobes.

Type of pollen presentation. The anthers mature before the flower opens and their pollen is scooped into the pollen-cup (Fig. 41B) which, in completing its growth, closes its mouth; at the same time the distal part of the style bends downwards bringing the cup into a nodding position (not *Brunonia*).

After the flower has opened the columnar tip of the style grows up within the pollen-cup and gradually forces pollen out ('Nudelspritze'). In some genera lobes on the upper petals (auricles) form a casing round the pollen-cup. Insect visitors rub the lips of the pollen-cup and acquire pollen from it. Finally, most of the pollen is lost and the tip of the style reaches a position where it can be touched by pollinators (Fig. 41C). Often a furrow lies across its surface, and the areas on either side of this, or sometimes only its rims, become receptive as stigmatic surfaces.

Floral details of several species are illustrated in KRAUSE (1912). The form and anatomical details of the pollen-cup and the form of the stigma have been

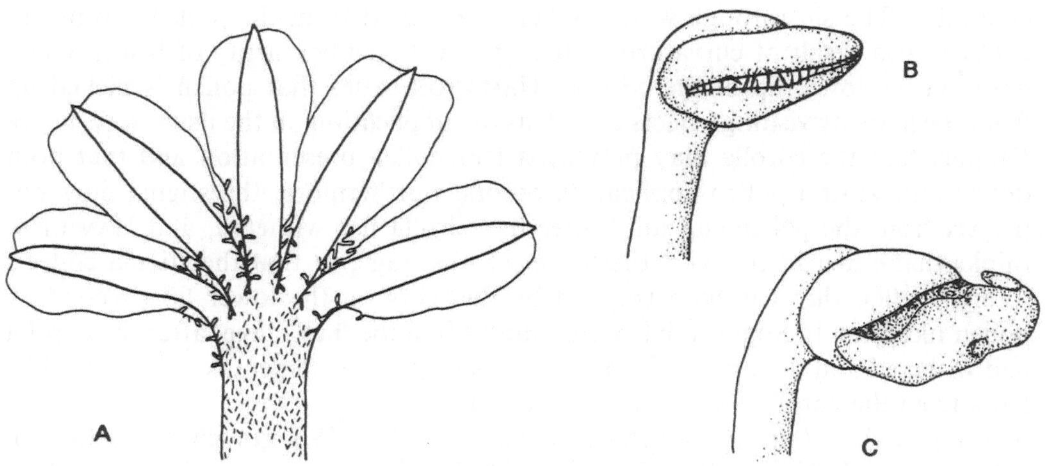

Fig. 41. *Scaevola crassifolia* (*Goodeniaceae*). *A* corolla limb spread out, adaxial view. *B* stylar pollen cup at pollen presenting stage, and *C* at female stage, with the large stigma emerging from the cup. (STUMPF 1987.)

described by CAROLIN (1960). In the development of the flower the tip of the style, bearing the pollen-cup, is at first above the tips of the anthers; the anthers then grow rapidly to overtop the pollen-cup, after which the style grows up between them again, allowing the loading of the pollen-cup with pollen. Stability of the anthers during loading can be provided by pressure from the corolla, coherence of connivent apical appendages of the anthers or lateral fusion of the anthers into a tube (STUMPF 1987). HAMILTON (1894) stated that the sequence of stages just described occurred in all the species of the family that he had examined; it has been illustrated by scanning electron micrographs by LEINS & ERBAR (1989) for *Selliera* and by ERBAR & LEINS (1988) for *Brunonia*. The presence of nectar is reported in some members of the family. *Lechenaultia* departs significantly from the above-described condition.

Examples of Secondary Pollen Presentation in Goodeniaceae

Scaevola L.

This is a widespread genus of 80–100 species, many of them shore-plants. *S. hispida* CAV. and *S. hookeri* F. MUELL. (HAMILTON 1894, illus. – *S. hookeri*). **STR s. ISS rel, growth. SEQU pln. APPL d. RWD?. SYN** melitto?. **VIS** 'insects' (*S. hispida*). For illustration, compare Fig. 41 (*S. crassifolia*).

The lower corolla-lip of these species bears lines of soft hairs or crests; the pollen-cup of *S. hispida* has marginal hairs that are lacking in *S. hookeri*. The style apex projects well from the corolla and pollination is nototribic. Neither scent nor nectar could be detected in the flowers but at least those of *S. hispida* received frequent insect visits (see discussion of *Goodeniaceae*).

S. suaveolens R. BR. (HAMILTON 1894, illus.). **STR c?, s. ISS rel,growth. SEQU pln. APPL d, v. RWD?. SYN** melitto. **VIS** moths (*Lepidoptera*).

This east Australian species shows certain peculiar features. The leaf that subtends the flower is said to provide a supplementary alighting platform and also to move round above the flower and turn sideways after the flower has withered so as to conceal it. The slit through which pollen is extruded from the pollen-cup makes contact, as a result of curvature of the style, with a thick array of brush-tipped hairs on the lower lip of the corolla. HAMILTON states that pollen is picked up from the hairs by visiting insects as well as, by implication, in the usual way, hence I report that the corolla may play a part in pollen presentation and that both dorsal and ventral pollen application occurs. Furthermore, the stigma does not project from the pollen-cup until after the corolla has withered, and HAMILTON thinks that either insects visit the flower at this stage, or that the stigma collects foreign pollen that has been trapped by the hairs on the lower lip ('secondary pollen reception'!). Stigmas that were examined in the final stage, after the corolla had withered and the bract had turned sideways, bore fragments of the brush-like hairs from the corolla and scales from moths.

S. plumieri (L.) VAHL. (KUGLER 1973, illus.). **STR s. ISS growth** and **rel(dose)-poll, growth. SEQU pln. APPL d. RWD n. SYN** melitto. **VIS** bees of the genera *Apis*, *Halictus*, *Prosopis* [*i.e. Hylaeus*] and family *Andrenidae* (*Hymenoptera-Apoidea*); solitary wasps of the genera *Bembex* and *Cerceris* (*Hymenoptera-Sphecoidea*) and *Eumenes* (*Hymenoptera-Vespoidea*); the beetle *Spermophagus* (gleaning pollen) (*Coleoptera-Bruchidae*).

This shrubby coastal species has white flowers that attain a length of 30 mm. All the corolla-lobes belong to the lower lip; the channelled base of the corolla is filled with retrorse hairs. The loading of the pollen-cup with pollen follows the sequence described by HAMILTON (see 'Floral Features' above). At the beginning of anthesis the style becomes sickle-shaped and a little pollen is forced out of the pollen-cup to remain between the bristles that border its two edges. When a pollinator of the right size probes for nectar it will usually touch the hair-fringe of the pollen-cup and the pollen held within the fringe falls out and sticks to its back. Details were observed by pushing dead *Apis adansonii* bees into the flower; only the pollen that was already held between the bristles was deposited on the insect. Additional pressure from the insect did not bring more pollen out of the pollen-cup; it merely deflected the elastic style. KUGLER could not confirm a statement by SCOTT ELLIOT (1891) for *S. thunbergii* ECKL. & ZEYH. that the upper edge of the orifice of the pollen-cup was longer than the lower and that this allowed pollen to emerge when pressure was applied. The fully developed stigma reaches just to the mouth of the pollen-cup and so is itself screened by the fringe of bristles which now have the role of sweeping pollen off incoming insects, whence it becomes transferred to the stigma itself. All the Hymenopteran visitors observed in this study on the coast of Kenya could carry out pollination, but the Andrenid bees were too small to function normally; however, they clung to the style during visits and it was therefore thought that they might cause some pollination. The gleaning activities of the beetles reveal that some pollen gets left in the anthers. Some pollen remains in the pollen-cup (KUGLER; KRAUSE 1912) and some may adhere .to the corolline hairs (KRAUSE 1912); in both these places it might cause self-pollination, should the plants be self-fertile.

In *S. thunbergii* SCOTT ELLIOT (1891) imitated the action of an insect by pushing a finger into the flower. This pushed back the pollen-cup, causing its mouth to close up and the style to become more curved. On withdrawal a ribbon-like mass of pollen was left on the finger.

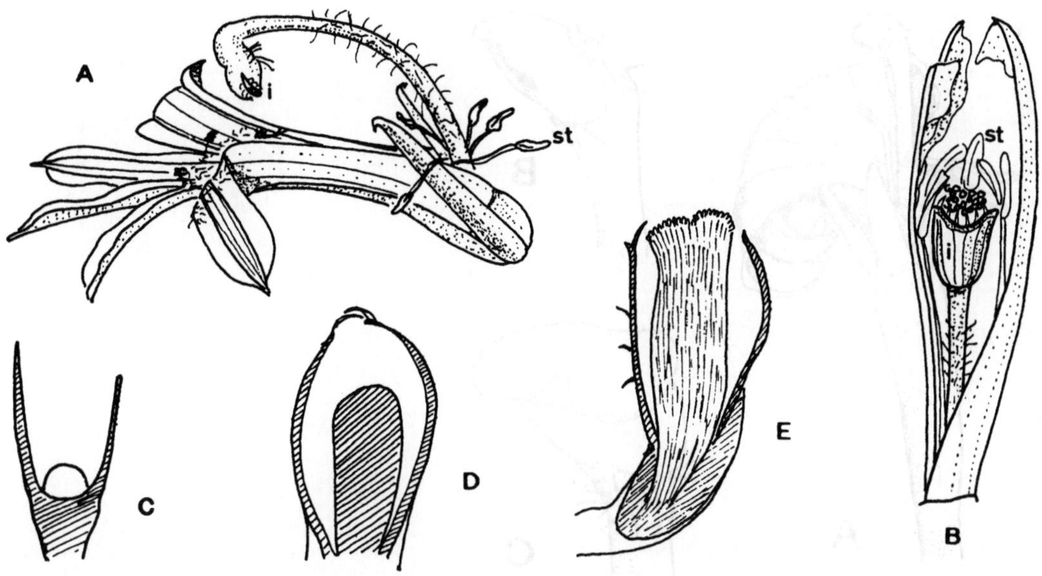

Fig. 42. *Scaevola thesioides* var. *filifolia* (*Goodeniaceae*). *A* side view of flower. *B* longisection of flower bud at stage of transfer of pollen from anthers to pollen cup. *C–E* stages in development of style-head (pollen cup plus stigma). *i* = pollen cup, *st* = stamens. (HOLM 1988.)

S. thesioides BENTH. var. *filifolia* E. PRITZ. (HOLM 1988, illus.). **STR s. ISS rel?, growth. SEQU pln. APPL d. RWD? SYN** melitto. **VIS** winged ants (*Hymenoptera-Formicoidea*), *Diptera*. (Fig. 42).

The plant is a shrub, 50 cm tall, covered with slender wand-like inflorescences with the flowers clustered at the tips. The flowers are about 2 cm wide, one-lipped, blue or whitish. HOLM could find no nectar, but as there are specialized hairs at the throat of the corolla he suggested that these might be providing some alternative reward. (HOLM made the same suggestion for the second species that he described: *S. glandulifera* DC.) Although it looked as if the visiting insects ought to be pollinators, HOLM could find no pollen on them. The way the style is hooked suggests that visitors may receive pollen on withdrawal from the flower, as described for the preceding species.

Selliera CAV.

S. radicans CAV. (HAMILTON 1894, illus.; LEINS & ERBAR 1988, illus.). **STR s. ISS rel, growth or rel(dose), growth. SEQU pln. APPL d. RWD n. SYN** melitto?. **VIS** –. (Fig. 43).

The plant is a low creeper on which the flowers are borne erect. They have nectar but no perceptible scent. The corolla is greenish (HAMILTON; elsewhere described as whitish internally and red or purplish externally) and is split to the base along the back. The five lobes are on the abaxial side of the flower and are more or less recurved. The upper part of the style is bent towards the lip and as the pollen-cup is purse-like it functions as an upper lip; its edges bear bristles forming a grille through which pollen is extruded. LEINS & ERBAR consider that these have a role in 'portioning' the release of pollen. It is not clear that the pollinator has any role in triggering pollen-release. When the pollen-cup is in a position to be loaded it

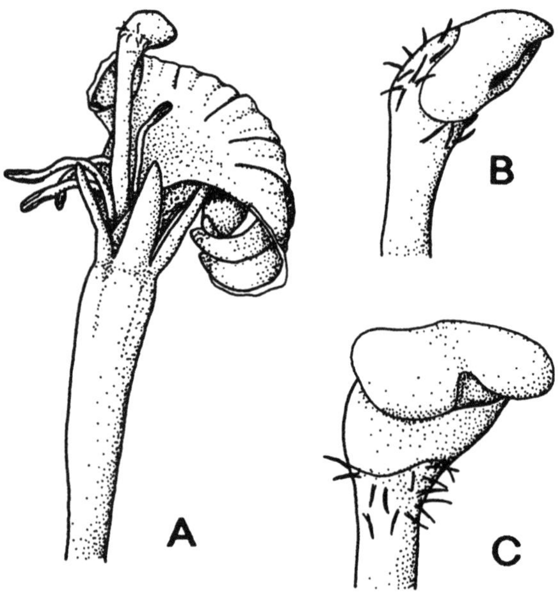

Fig. 43. *Selliera radicans* (*Goodeniaceae*). *A* flower viewed obliquely from behind. *B* pollen cup at male stage of flower, and *C* at female stage, with the large stigma emerging. (STUMPF 1987.)

is wide-open and five-angled; the pollen is then shed from the anthers in masses in which the grains are held together by pollenkitt (adhesive coating derived from the anther). Only when the pollen-cup has grown up between the anthers and gathered up the pollen does it become purse-shaped. LEINS & ERBAR thought that during the loading of the pollen-cup its marginal bristles might help to scoop the pollen into it, but HAMILTON expressed the contrary view. STUMPF also doubts that the bristles have this function (STUMPF, pers. comm. 1990). Indeed, such bristles were absent from the material used for her Diplomarbeit.

Dampiera R. BR.

D. purpurea R. BR. (*D. brownii* F. MUELL.). (HAMILTON 1894, illus.). **STR c, s. ISS rel(dose) and growth. SEQU pln. APPL d or v. RWD n. SYN** melitto. **VIS** –.

HAMILTON furnished much detail about this flower but it is not easy to understand the principles on which it functions. It is one of four genera in which the anthers cohere (KRAUSE 1912; CAROLIN, in MORLEY & TOELKEN 1983) but this does not seem to affect what happens during anthesis, although HILDEBRAND (1870) pointed out that it apparently compensates, during the loading of the pollen-cup, for the absence of hairs on the margin of the latter. The flowers are scented and the corolla tube contains a tasteless liquid. The corolla has auricles (see 'Floral Features') and the pouch-like parts of these are intensely purple-coloured (so recalling the colour often seen on the apex of the keel of *Leguminosae-Papilionoideae*). An insect visiting the flower has to force the auricles apart in order to contact the pollen-cup or stigma. The pollen is extruded from the pollen-cup in a 'worm-like string' which dries to a powder as it emerges and falls into the auricular chamber, and from here falls 'in a small shower' on an insect which parts the auricles. When the flower was withering HAMILTON found pollen not only on the stigmatic surface but all over both the inside and the outside of the pollen-cup, and he thought this was the flower's own pollen; the auricles were also still full of pollen. This suggests that these flowers were receiving inadequate pollinator service. The stigma never became completely free of the pollen-cup, and it was assumed that deposition of foreign pollen requires that an insect push back the lower rim of the pollen-cup. A large proportion of the flowers was resupinate, a condition which would preclude fall of previously extruded pollen onto the visitor; it was assumed that in this case pollen from the pollen-cup was smeared directly onto the insect's underside.

HAMILTON stated that *Dampiera stricta* (SM.) R. BR. (illus.) is similar to the preceding and has the auricles full of powdery pollen, but that it is not resupinate. BROUGH (1927) stated that the stigma very gradually forces the microspores (pollen) out through the pore at the apex of the pollen-cup; he also oberved that papillate receptive cells in the apical region of the stigma do not appear until the stigma has fully occupied the indusial cavity. He suggested that self-pollination is possible in the absence of insect visitors. In *D. lanceolata* DC. the growth of the stigma within the pollen-cup was so slight that HAMILTON supposed cross-pollination to be impossible. HAMILTON studied further species in the dried state only. The auricles that he illustrated range from simple to highly complicated arrangements of folds and trichomes.

D. helmsii KRAUSE. (STUMPF 1987).

One edge of each of the upper corolla lobes forms an intensely coloured auricle. The upper and lower lips are so far apart that it seems unlikely that insects reaching for the nectar would part the auricles; however, the flanges of the upper corolla lobes that do not form auricles are specially shaped so that they would apparently guide the head of a withdrawing insect into the space between the auricles, where it would pick up pollen. The lips of the pollen-cup are arranged vertically and wide apart which, it is thought, would facilitate the application of pollen to the insect.

D. coronata LINDL., *D. hederacea* R. BR., *D. sericantha* BENTH. and *D. teres* LINDL. (HOLM 1978). **STR s. ISS rel(dose)-trigg. SEQU pln. APPL d. RWD n? SYN** melitto? **VIS** –.

HOLM stated that when an insect touches the 'nose' of the pollen-cup of these West Australian species the pollen comes out, probably all at once, and, being sticky, it fixes itself to the insect. The insect has to use force to part the auricles and it needs to do this in order to reach the nectar.

D. cuneata R. BR. (HOLM 1988, illus.). **STR s. ISS rel(dose)-trigg. SEQU pln. APPL d. RWD n?. SYN** melitto. **VIS** –. (Fig. 44).

Deep blue, yellow-centred, flowers are borne singly or in small groups on stems 10–15 cm tall. The corolla tube is thick-walled with two lateral slits as well as the adaxial one. The limb is two-lipped, the lobes of the lower lip having claws with strongly undulate margins. The medial wings of the upper lobes interlock by means of a prong on one and a socket on the other; proximal to this arrangement they each bear a pouch-like auricle. The auricles are also unequal so that one fits inside the other; together they envelop the pollen-cup. The style is not arched but is bent immediately below the pollen-cup so that the latter opens downwards. The opening of the pollen-cup is circular, smaller in diameter than the cup itself and guarded by two scales forming a beak. Experiments with a tool showed that a strong insect

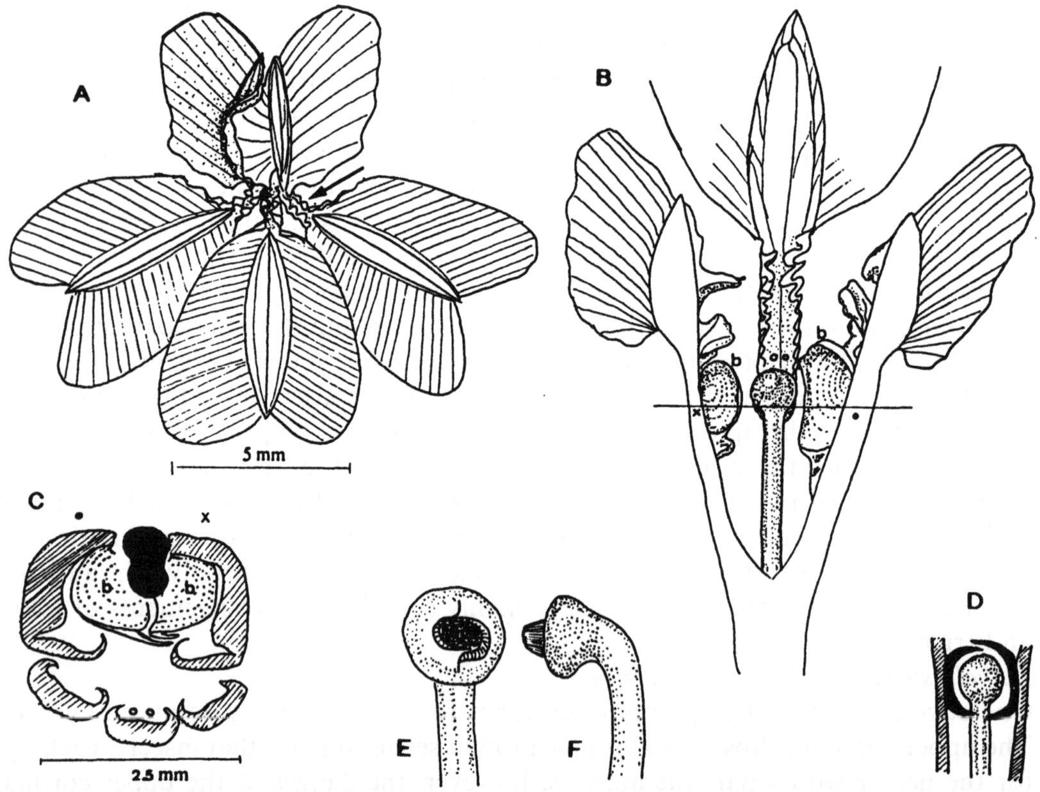

Fig. 44. *Dampiera cuneata* (*Goodeniaceae*). *A* front view of flower; arrow points to wavy-edged claws of corolla lobes. *B* flower throat partly dissected, and with the two lower lateral petals removed. *C* transection at level of ruled line in 'B'; pollen cup shown black. *D* diagram of pollen cup and auricles. *E* and *F* lateral and face views of the style-head. *b* = auricle. (HOLM 1988.)

would be required to part the upper corolla-lobes; when this is done the pollen-cup can be touched, which causes the release of part of the sticky pollen. Elasticity of the style appeared to be important in ensuring contact between the visitor and the mouth of the pollen-cup. The stigma lies at the bottom of the pollen-cup and no flowers in which it grew up to the mouth of the latter were found, so the method of. pollen reception remains unknown. HOLM (1988) found essentially the same structure in *D. cauloptera* DC. When he touched the beak of the pollen-cup all of the pollen came out in two flowers and only a portion in another.

In *D. teres* LINDL. HOLM (1988) found a more normal pollen-cup with wide hairy lips in which the stigma grew out to the mouth, pushing the pollen before it.

Goodenia SM.

G. grandiflora SIMS. (HILDEBRAND 1870, illus.). **STR s. ISS rel(dose)-poll** and **growth. SEQU pln. APPL ad. RWD n. SYN** melitto. **VIS** –.

Three stages in the development of the flower are recognized. In the first the anthers standing around the pollen-cup open; the pollen-cup is ringed by hairs which, during the ensuing growth of the style, brush the pollen into the cavity of the pollen-cup. In the second the flower opens, and the pollen-cup, now closed, is presented for contact with insects by curvature of the style; the hairs of the pollen-cup are considered to have two functions at this stage: prevention of spontaneous loss of pollen, and catching on the insect's body so as disturb a portion of the pollen and allow it to fall onto the insect. There is thus an arrangement to release the pollen a little at a time with each insect-visit. The third stage is marked by the beginning of the growth of the stigma; at first this pushes pollen towards the mouth of the pollen-cup, so that new pollen is continually ready to be swept away. Finally the stigma emerges from the pollen-cup and two broad stigmatic lobes separate and curve back over the edges of the pollen-cup. By this time all the flower's own pollen has gone, and pollination is dependent on the arrival of pollen from younger flowers.

G. cycloptera R. Br. (F. E. Haviland 1915, illus.). **STR s. ISS rel(dose)-trigg** and **growth. SEQU pln. APPL d. RWD n?. SYN** melitto? **VIS** –.

The tip of the style is firmly held in the auricles but the subapical part is arched so as to project from the upper side of the corolla. At anthesis the pollen-cup opens; pollen is then dusted onto the backs of insects that force apart the auricles. Haviland makes no mention of loose pollen in the auricles. The stigma develops rapidly, pushing forward the pollen; it becomes viscid only when it has emerged from the pollen-cup. Haviland bagged some clusters of flowers in the field and all of them dropped off without setting seed.

G. pinnatifida Schlecht. (Stumpf 1987, illus.). **STR s, c. ISS growth. SEQU pln. APPL d. RWD n. SYN** melitto. **VIS** –.

Stumpf considers that the firmness and strong curvature of the style in this species would prevent an entering insect from raising the pollen-cup and setting free the pollen. Pollen is continuously extruded as a result of growth of the style and collects on the auricles and on a hair-tuft just below the lower lip of the pollen-cup. As the insect withdraws it would be expected to collect pollen from the auricles on its back. Pollen that falls on to the hair tuft can be contacted by the stigmas when mature; good fruit-set in the greenhouse suggested that self-pollination by this means is effective.

Some observations on other species of *Goodenia* may be added here. Hamilton (1885) studied *G. hederacea* Sm. and described the filling of the pollen-cup, suggesting that the hairs on its lips help to collect the pollen from the anthers while the style is growing but the stamens are not, as described also for *Selliera* (see above). As with *G. cycloptera* the subapical part of the style arches out of the corolla. Despite the existence of a structure and development that is normal for the family, Hamilton concluded that once loaded the pollen-cup never re-opened and that the flower was habitually self-fertilising. E. Haviland (1884, 1885) showed that he was aware of the normal basic mechanisms of pollen presentation in the family by referring to the summary of earlier work presented by Müller (1883); nevertheless, he was convinced by a 'very careful study of the genus' that insects carry away the pollen directly from the anthers. In his detailed report on *G. ovata* Sm. he stated that the anthers when mature (before the opening of the flower) emerge from the slit in the corolla and that in the bud or partly open flower the pollen-cup was open and empty; although at the same stages he found unbroken anthers he could apparently see no alternative to the above-mentioned conclusion. As there are now known to exist bees specializing in the pollen of *Goodeniaceae* (references in ARMSTRONG 1979: 479) it seems possible that HAVILAND was witnessing the effects of these insects. It seems likely that they are pollen robbers, as the pollen presenting mechanism of the family would seem to be directed towards limiting pollen removal to small amounts on each visit. An alternative possibility is that some species of the genus have changed their pollination method as a result of selection pressure by robber bees, as otherwise it is difficult to see why the vulnerable anthers should be exposed outside the corolla. (STUMPF – pers. comm. 1990 – found pollen, though not much, in the pollen-cup of open flowers of *G. ovata*).

Lechenaultia R. BR.

A combined account, based on information from various authors, has been pieced together here.

Lechenaultia spp. (CAROLIN 1960, illus.)

L. formosa R. BR. (F. W. BURBIDGE 1871, illus.).

L. formosa R. BR., *L. biloba* LINDL. (DARWIN 1871, illus.).

L. biloba LINDL. (SARGENT 1918).

L. linarioides DC., *L. biloba* LINDL., *L. tubiflora* R. BR. (HOLM 1978, illus., 1988, illus.).

L. linarioides DC. (STUMPF 1987, illus.).

STR s. ISS rel-trigg. SEQU? APPL v or d. RWD n. SYN melitto, ornitho (according to species). **VIS** –. (Fig. 45: *L. Tubiflora*).

The corolla usually has auricles; sometimes the flowers are resupinate and there is then a distinct superficial resemblance to the flowers of *Leguminosae-Papilionoideae*. The pollen-cup is zygomorphic and bilabiate, the morphological lower lip being nearly flat or concave and the upper convex. The pollen-cup is shaped and positioned so that the inserted part of the pollinating visitor first touches a belt of large fleshy hairs (SARGENT) on the summit of the convex lip (STUMPF), to which no function has been ascribed. This contact was ignored by HOLM but from his account the visitor (imitated by a tool) next touches a secretory zone on the outside of the same lip; its movement is thus from the 'back' of the cup to the 'front'. Next it pushes the thin lower lip and when it does so the pollen (often all of it – HOLM) comes out and is affixed to the area that has been gummed by the secretion of the upper lip (HOLM). When the pollen has been removed a band of papillae or hairs on the edge of the upper lip is exposed (various authors); this stains red with safranin (STUMPF) and appears to be the place of receipt of foreign pollen. HOLM (1988) thought the stigma was inside the pollen-cup though he could not locate it. There are no reports of stigmas growing up from the base of the cup as in other genera. It is not yet clear whether the pollen-fixing secretion is stigmatic fluid or a separate secretion from an adjacent area. STUMPF's description of hairs on the pollen cup of *L. linarioides* corresponds with what is illustrated for *L. tubiflora* by HOLM. The hairs on the upper edge of the mouth might be the source of the adhesive secretion.

There is thus a remarkable transformation compared with other members of the family: the stigmatic surface has moved to some part of the stylar pollen-cup and an adhesive is used to fasten pollen to the visiting insect. It has not been explained what causes the emission of pollen but the convexity of the back of the floor (CAROLIN 1960) of the pollen-cup (or perhaps a change of contour) may provide the mechanism by creating pressure, though STUMPF's drawings provide

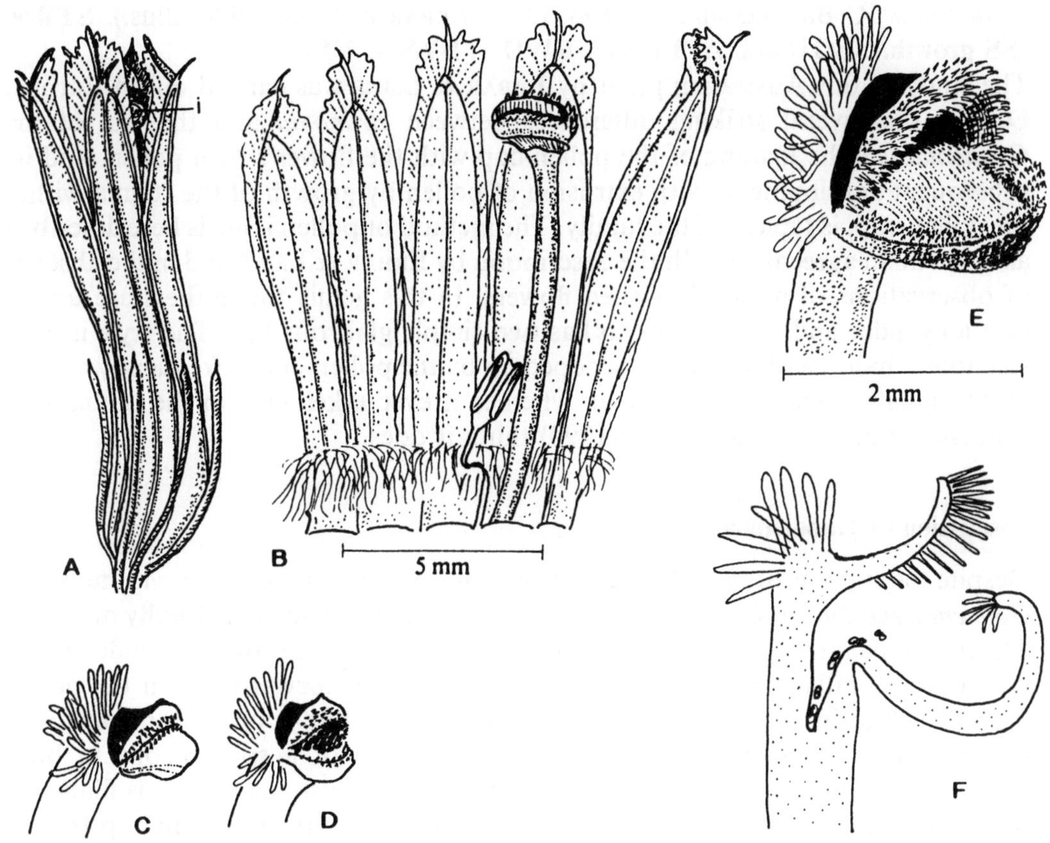

Fig. 45. *Lechenaultia tubiflora* (*Goodeniaceae*). *A* side view of flower. *B* flower opened out, one stamen remaining. *C* pollen cup with mouth closed, and *D* with mouth open, revealing pollen mass. *E* pollen cup further enlarged with mouth open, empty of pollen. *F* longisection of almost empty pollen cup. Black area is zone of exposure of adhesive. *i* = pollen cup. (HOLM 1988.)

no support for this. Furthermore, observers have not described the deposition of foreign pollen. One would expect this to take place before the application of fresh adhesive to the visitor and its contact with the flower's own pollen, as in other flowers employing adhesive (*Polygalaceae*, *Apocynaceae*, *Marantaceae*). (Another analogy with the *Apocynaceae* is the transposition of the stigmatic surface.)

Of the species covered here, *L. biloba* shows the syndrome of melittophily, while *L. formosa*, *L. linarioides* and *L. tubiflora* show that of ornithophily. The last species is the most extreme, with an erect flower on a short shoot arising from a trailing stem and a tubular lipless and nearly actinomorphic corolla.

SARGENT found that *L. biloba* forms large clonal patches and by field observation and experiment obtained indications that individuals are self-incompatible.

Brunonia R. BR.

This is the genus that, as mentioned in the introduction to *Goodeniaceae*, is sometimes treated as constituting a monotypic family, *Brunoniaceae*. It has the flowers in capitula, a reduced and actinomorphic corolla, sepals in the form of a pappus and a superior one-seeded ovary.

B. australis R. BR. (HAMILTON 1894, illus.; ERBAR & LEINS 1988, illus.). **STR s. ISS growth. SEQU pln. APPL av?. RWD –. SYN –. VIS –.**
The study of this Australian plant by ERBAR & LEINS was carried out in Europe. Despite the several striking differences between *Brunonia* and the rest of the *Goodeniaceae*, the loading of the pollen-cup with the flower's own pollen, and its subsequent development, with extrusion of pollen by growth of the stigma within the pollen-cup, is typical of the family. The method of pollen issue is here described as 'growth' in agreement with the description by ERBAR & LEINS and in the absence of observations of naturally visited flowers. In this genus the anthers are united into a cylinder and the pollen-cup has two simple glabrous lips. The style is conspicuously hairy, and pollen not scooped up by the pollen-cup is frequently caught by these hairs, but it is not known whether it ever takes part in the pollination process, though LEINS & ERBAR (1990) think it does.

Discussion of Goodeniaceae

Despite the superficial resemblance between flowers of some *Lobeliaceae* and some *Goodeniaceae* the structure of the pollen-presenting mechanism is totally different. The functional arrangements, however, are (to the extent that they are understood) remarkably similar. In both families pollen is pressed forward within the pollen magazine, and in many cases the mouth of the magazine is opened in some way by contact with the pollinator. The prevalence also of a split corolla in both families suggests that allowance has to be made for the style (style plus filaments in *Lobeliaceae*) to increase its curvature in response to pressure at the tip from a pollinator (see accounts of *Scaevola* and *Dampiera cuneata*). KUGLER (1973) points out that *Goodeniaceae* and *Campanulaceae* resemble each other in that the anthers are [usually] free and deposit pollen on the style, whereas in *Lobeliaceae* and *Asteraceae* the anthers cohere and release their pollen into the anther-cylinder.

The system in *Goodeniaceae* is well adapted for the presentation and reception of pollen in the same part of the flower and to protect pollen at all times except during the visit of a potential pollinator. It also allows regulation of the quantity released, which seems in some cases to be a portion and in others the whole. The mechanism involves highly developed alterations to the style, and has in one case (*Lechenaultia*) progressed to the use of adhesive and the re-positioning of the stigmatic surface. Many species are insect-pollinated and the floral reward is nectar. It is not clear to what extent insect visitors can or do appropriate the pollen that is presented to them, but it could be that the bees that are specialized for using pollen of *Goodeniaceae* are robbers, who have gained access to a food source closed to other insects. Where auricles are present a further obstacle has to be overcome by the pollinator. They make the flower functionally similar to that of *Leguminosae-Papilionoideae* and strengthen the impression that the family is mainly specialized towards bees; in some cases they seem to act as a secondary pollen magazine. Resupination is most prevalent in *Lechenaultia* where both the melittophil and the ornithophil syndrome are seen. Although enough has been published to make it possible to give an introductory account of pollen-presentation in this family few of the accounts are anything like complete, and a great deal of clarification and amplification remain to be done. There seems to be marked variation within genera in the pollination arrangements. An observation by SCOTT ELLIOT (1891) and the suggestion of STUMPF (1987) that in some cases pollen is applied to the pollinator during withdrawal from the flower need to be followed up.

The growth sequence in which the style is at first longer than the stamens, is then overtaken by them but is finally the longer organ shows agreement with other members of the *Campanulales/Asterales* alliance (ERBAR & LEINS 1988; LEINS & ERBAR 1989, 1990).

There is one instance of apparent self-incompatibility (*Lechenaultia*). Pollen grains are binucleate (three genera; BREWBAKER 1967) and the stigma is dry (*Scaevola*; HESLOP-HARRISON & SHIVANNA 1977).

References for Goodeniaceae

ARMSTRONG, J. A., 1979: Biotic pollination mechanisms in the Australian flora – a review. – New Zealand J. Bot. **17**: 467–508.

BREWBAKER, J. L., 1967.

BURBIDGE, F. W., 1871: Fertilisation of *Leschenaultia formosa*. – Gard. Chron. **1871**: 1103.

BROUGH, P., 1927: Studies in the *Goodeniaceae*. I. The life-history of *Dampiera stricta* (R. BR.). – Proc. Linn. Soc. New South Wales **52**: 471–498.

CAROLIN, R. C., 1960.

DARWIN, C., 1871: Fertilisation of *Leschenaultia*. – Gard. Chron. **1871**: 1166.

ERBAR, C., LEINS, P., 1988: Studien zur Blütenentwicklung and Pollenpräsentation bei *Brunonia australis* Smith (*Brunoniaceae*). – Bot. Jahrb. **110**: 263–282.

HAMILTON, A. G., 1885: On the fertilization of *Goodenia hederacea* SM. – Proc. Linn. Soc. New South Wales **10**: 157–161 (1885).

HAMILTON, A. G., 1894: Notes on the methods of fertilisation of the *Goodeniaceae*. Part. I. – Proc. Linn. Soc. New South Wales, ser. 2, **9**: 201–212.

HAMILTON, A. G., 1895: Notes on the methods of fertilisation of the *Goodeniaceae*. Part. II. – Proc. Linn. Soc. New South Wales, ser. 2, **10**: 361–373.

HAVILAND, E., 1884: Occasional notes on plants indigenous in the immediate neighbourhood of Sydney, no. 7. – Proc. Linn. Soc. New South Wales **9**: 449–452.

HAVILAND, E., 1885: Some remarks on the fertilisation of the genus *Goodenia*. – Proc. Linn. Soc. New South Wales **10**: 237–240.

HAVILAND, F. E., 1915: The pollination of *Goodenia cycloptera* (N. O. *Goodeniaceae*). – Proc. Linn. Soc. New South Wales **39**: 851–854.

HESLOP-HARRISON, Y., SHIVANNA, K. R., 1977.

HILDEBRAND, F., 1870: F. DELPINO's weitere Beobachtungen über die Dichogamie im Pflanzenreich mit Zusätzen und Illustrationen. – Bot. Zeitung (Berlin) **28**: 585–594, 601–609, 617–625, 633–641 (*Goodeniaceae* here), 649–659, 665–675.

HOLM, E., 1978: Some unusual pollination mechanisms in western Australian wildflowers. – W. Austral. Naturalist **14**: 60–62.

HOLM, E., 1988.

KRAUSE, K., 1912: *Goodeniaceae* and *Brunoniaceae*. – In ENGLER, A., (ed.): Das Pflanzenreich **IV.277 & 277a**. – Leipzig: Engelmann.

KUGLER, H., 1973: Zur Bestäubung von *Scaevola plumieri* (L.) VAHL und *Ipomoea pes-caprae* SWEET, zwei tropischen Strandpflanzen. – Flora **162**: 381–391.

LEINS, P., ERBAR, C., 1989: Zur Blütenentwicklung and sekundären Pollenpräsentation bei *Selliera radicans* CAV. (*Goodeniaceae*). – Flora **182**: 43–56.

LEINS, P., ERBAR, C., 1990: On the mechanisms of secondary pollen presentation in the *Campanulales-Asterales*-complex. – Botanica Acta **103**: 87–92.

MORLEY, B. D., TOELKEN, H. R., 1983: Flowering Plants in Australia. – Adelaide: Rigby.
MÜLLER, H., 1883.
SARGENT, O. H., 1918: Fragments of the flower biology of Westralian plants. – Ann. Bot. (Oxford) **32**: 215–231.
SCOTT ELLIOT, G. F., 1891: Notes on the fertilisation of South African and Madagascar flowering plants. – Ann. Bot. (Oxford) **5**: 333–405.
STUMPF, S., 1987, unpublished.

Rubiaceae

Systematic position: *Asteridae*, *Rubiales*
Restriction of occurrence within the family: Subfamilies *Cinchonoideae* (syn. *Coffeoideae*, *Ixoroideae*) and *Rubioideae*

Distribution within the subfamilies. The classification and nomenclature of the *Rubiaceae* are still in a state of flux (VERDCOURT 1976; ROBBRECHT 1980, 1988; BRIDSON & ROBBRECHT 1985; ROBBRECHT & PUFF 1986; BRIDSON & VERDCOURT 1988). The last of these references has here been followed as far as possible. Subfamilies *Cinchonoideae* and *Rubioideae* are the two large subfamilies of the *Rubiaceae* (the third, *Antirheoideae* or *Guettardoideae*, comprises only one tribe). BREMEKAMP (1934) emphasized the widespread occurrence of secondary pollen presentation in this large, mainly tropical, family and defined his *Ixoroideae* (approximately equal to *Cinchonoideae*) by the possession of arrangements for secondary pollen presentation. Accordingly all the tribes that he recognized in the subfamily are accounted for among the 11 tribes dealt with here (Table 2–5). In *Rubioideae* secondary pollen presentation is known only in a single monotypic temperate genus.
Floral features. Calyx, corolla and androecium tetramerous or pentamerous and usually actinomorphic. Anthers usually dorsifixed, sometimes basifixed. Ovary

Table 2-5. Tribes of Rubiaceae treated here

Subfamily *Rubioideae*
Tribe
Rubieae
Subfamily *Cinchonoideae* (*Coffeoideae*) (Tribes arranged alphabetically)
Tribe
Acrantherae
Aulacocalyceae
Chiococceae
Cinchoneae
Coffeeae
Coptosapelteae
Gardenieae
Subtribe Gardeniinae
Diplosporiinae
Hypobathrieae
Naucleeae
Pavetteae
Vanguerieae

inferior and usually bicarpellary and bilocular. Corolla sympetalous, the stamens adnate to it, usually attached at or near the throat. Inflorescence of solitary flowers or quite dense; dioecy and heterostyly widespread. The pollination syndrome of lepidopterophily is notably prevalent (ROBBRECHT & PUFF 1986, for *Gardenieae* and related tribes), the corolla-tube then being narrow and the limb either flat and spreading or reflexed.

Type of pollen presentation. Secondary pollen presentation takes place on the style ('Pseudo-Staubblatt'), the anthers being introrse. BREMEKAMP (1966) used the expression 'Ixoroid pollination' to denote it; in this he was followed by VERDCOURT (1976) and others, despite the fact that the same system is found in other families. BREMEKAMP stated that the part of the style that serves as a receptaculum pollinis (the 'pollen presenter' in our terms) is always covered with short hairs, but this is not often the case. However, the pollen presenter is usually ridged; in the examples of *Cinchonoideae* that I have examined the style is glabrous but has ten ridges, creating grooves corresponding to the ten thecae of the anthers. In addition, pollen presenters that are neither pilose nor ridged occur in the family.

BURCK (1884) made the interesting observation that secondary pollen can occur in dioecious species, the male plants of these having a well-developed style that functions only as a pollen presenter. Secondary pollen presentation and heterostyly are combined in one instance (species of *Tricalysia*, Tribe *Gardenieae*, subtribe *Diplosporinae* – ROBBRECHT 1988).

Examples of Secondary Pollen Presentation in Rubiaceae

Subfamily *Rubioideae*

Tribe *Rubieae*

Phuopsis (GRISEB.) J. D. HOOK.

P. stylosa (TRIN.) B. D. JACKSON. (YEO, unpublished 1989). **STR s. ISS exposed. SEQU pln. APPL a. RWD n. SYN** psycho. **VIS** –.

This herb from temperate Iran has ascending stems to about 30 cm and most of the flowers borne in dense hemispherical terminal clusters. The bright pink corolla has a very slender tube 6–9 mm long and a 5-lobed star-shaped limb 5 mm across. The style-head may still carry pollen when the style has reached its final length, being exserted about 5 mm from the throat, but one can also find pollen-carrying styles that are only about half extended or less. The style-head is 2.5–3 times the diameter of the style and the outer surface which carries the pollen is papillate at all stages dealt with here. The papillae appear smooth, highly polished and facetted; angles between the facets may catch the light in such a way as to give a false impression that small hairs are present. The external papillosity stops just short of the small cleft at the apex of the style-head which represents the stigmatic lobes. Within the lip-like edges of the lobes there is a further pair of lip-like cushions. In old flowers the lips diverge somewhat and pollen can usually be found on them, whereas by this time all of the flower's own pollen has gone from the outer papillate surface. Conversely, one can find young style-heads heavily loaded with the flower's own pollen, while no pollen can be found in the stigmatic cleft. Just before the flower opens it is possible by gentle squeezing to make the corolla pop open. The pollen-loaded style-head is then usually found to be above the throat, having been enclosed in the cavity formed by the valvate corolla-lobes. However, in the youngest buds that can be popped one may find the style in the throat; the pale green anthers may then be either undehisced and not yet equalled by the style, or just dehisced and leaning inwards over the style-head, the extreme tip of which now overtops them. When flowers are popped some pollen usually gets on to the corolla; in naturally opened flowers pollen is also sometimes found on the corolla. A considerable amount of pollen remains in the anthers after the flowers have opened. On the occasions when I inspected this plant I found a great preponderance of pollen-free style-heads (female stage) over pollen-carrying (male stage).

A quite different report of pollen presentation in *Phuopsis* is given by KERNER (1902: 265, 267) and repeated by MCCLEAN & IVIMEY-COOK (1956). KERNER says that the pollen is explosively released in a cloud, and there is a dramatic illustration of this in progress. This is incompatible with my observations because if it were true one would not find style-heads of naturally opened flowers covered with a heavy load of pollen. There seem to be three possible explanations, (1) error, (2) different plants of *Phuopsis stylosa* behave in different ways, (3) the corolla-lobes sometimes cohere longer than they should so that pressure from the growing style is what forces popping, and the sudden release of the style then causes some pollen to be thrown into the air. A photograph sent me by Dr CHR. WESTERKAMP shows naked and pollen-loaded styles, as found by me. Dr WESTERKAMP and Prof. ST. VOGEL (pers. comm., November 1990) think KERNER's account is an error.

Subfamily *Cinchonoideae* (*Coffeoideae*) (Tribes Arranged Alphabetically)

Tribe *Acranthereae*

Acranthera ARN. ex MEISSN.

The species in general. (BREMEKAMP 1947, 1966). **STR s. ISS exposed. SEQU pln. APPL** proboscis? **RWD –. SYN?. VIS** –.

The monograph of the genus (BREMEKAMP 1947) is unillustrated apart from some histological figures; it deals with 34 species. The flowers are homostylous and hermaphrodite. The corolla is glabrous within and the stamens are inserted at the base of the tube. The pollen is deposited on the upper part of the style, where there are rows of papillae. The anthers, which are basifixed, form a sheath round the style and have apical appendages which are connected with the top of the style; the pollen on the style can only be reached through windows between these connectives. The stigmas are subulate or semiconical and do not diverge; often they do not even separate from one another. Structurally, this sounds like a remarkable parallel with the flower of *Apocynaceae*, but it is not known whether it is functionally similar. Illustrations in KERR (1937, *A. siamense* (KERR) BREMEK.) and WONG (1982, *A. pulchella* (RIDL.) K. M. WONG) show a strictly funnel-shaped corolla (slightly tapered lower tube and conic-campanulate upper tube); the tips of the style and adherent anthers reach to about the middle of the expanded part. In *A. siamense* ridges on the style project through the slits between the anthers. It is not clear what is the distribution of pollen on the presenter, nor how the stigmatic surface becomes exposed.

Tribe *Aulacocalyceae*

Secondary pollen presentation is the rule in this tribe of 4–6 genera, as in all five tribes of the *Gardenieae* alliance (of which this is one) treated by ROBBRECHT & PUFF (1986).

Tribe *Chiococceae*

This is one of the tribes in subfamily *Ixoroideae* as delimited by BREMEKAMP (1966), which is defined by possession of secondary pollen presentation. I have no further information.

Tribe *Cinchoneae*

Some genera have secondary pollen presentation (VERDCOURT 1976). See also *Coptosapelteae*, below.

Tribe *Coffeeae*

Both genera, *Coffea* and *Psilanthus*, show secondary pollen presentation (ROBBRECHT & PUFF 1986). *Psilanthus* is peculiar in that the style is very short and the pollen presenter is always well below the anthers (D. BRIDSON, pers. comm.).

Tribe *Coptosapelteae*

The same applies as for *Chiococceae*, except that BREMEKAMP named the genera composing the tribe (they were *Coptosapelta* and *Crossopteryx*, but some authors retain one or both of these in *Cinchoneae*).

Tribe *Gardenieae*

Gardenieae, subtribe *Gardeniinae*

ROBBRECHT & PUFF (1986) assign 53 genera here, most of which definitely have secondary pollen presentation.

Burchellia R. BR.

B. bubalina (L. f.) SIMS. (YEO, unpublished 1985). **STR s. ISS exposed. SEQU pln? APPL a? RWD n. SYN** ornitho. VIS sunbirds (*Aves-Nectariniidae*).

The corolla is about 3cm long, tubular, with a reduced limb, orange-red in colour and thick and firm in texture. The distal part of the style, including the stigmatic region, is clavately thickened but somewhat bilaterally compressed in the plane of the stigmatic cleavage; below the thickening is a zone of loose hairs but the part where pollen is deposited is glabrous and bears 10 ridges. The stigmatic cleavage extends about half way down the thickened part and has a ridge close on either side of it. Flowers examined in an advanced bud stage had the furrows between the stylar ribs well filled with pollen, but a considerable amount of pollen remained in the anthers. The plant that provided the flowers (at the Cambridge Botanic Garden) was suffering from some disorder, and the flowers did not open in the 1985 season. The zone of secondary pollen presentation extends approximately to the tip of the style and therefore overlaps the stigmatic cleavage but probably in normally developed flowers the anthers are so placed that their pollen is not deposited in it. The anther connective extends over the backs of the thecae and beyond them distally as a small appendage. In essentials the arrangements in this African shrub are very like those of *Mitriostigma*, on which fuller observations were made. Bird visits were seen by VOGEL (1954) and by PUFF (ROBBRECHT & PUFF 1986).

Catunaregam WOLF (*Randia* sensu auctt.: p. p.)

C. spinosa (THUNB.) TIRVENGADUM (*Randia dumetorum* (RETZ.) POIR.) (BURCK 1884, illus.). [BURCK used the names *R. dumetorum* LAM., *R. longispina* DC., and *R. spinosa* (L. f.) POIR.?, which may all represent the same species (D. BRIDSON, pers. comm.).]. **STR s. ISS exposed. SEQU pln. APPL a. RWD n. SYN** psycho?, melitto? **VIS** –. (Fig. 46A).

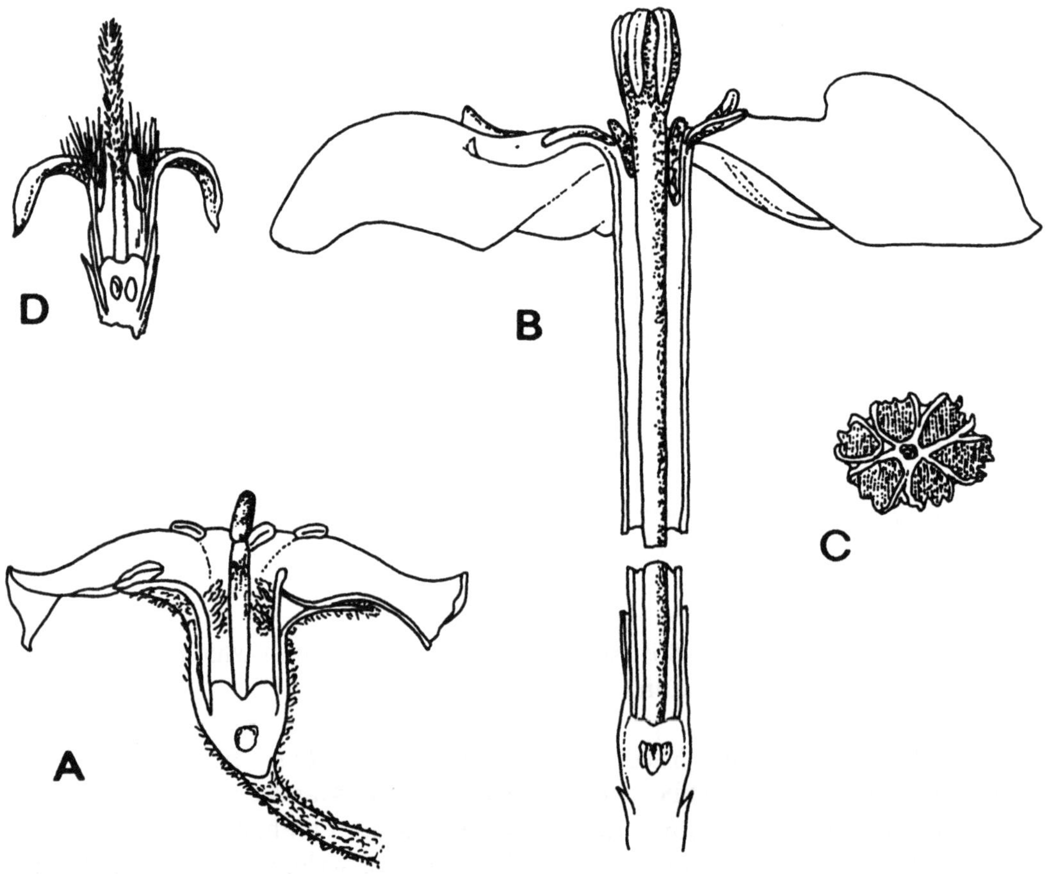

Fig. 46. *Rubiaceae*, tribes *Gardenieae* (*A–C*) and *Hypobathrieae* (*D*). *Catunaregam spinosa*: *A* longisection of male flower showing style head which functions only as a pollen presenter. *Gardenia resinifera*: *B* longisection of flower, showing anthers in the throat and ribbed style-head; *C* transection of style-head; pollen is presented on the raised tracts, while the grooves between these are apparently stigmatic (meniscus of stigmatic fluid visible). *Scyphostachys coffeoides*: *D* longisection of flower; the stylar brush sweeps pollen from the anthers situated in the corolla-throat. (P. F. Yeo, after Burck 1884.)

The plants are dioecious (perhaps also hermaphrodite – D. Bridson, pers. comm.) and the flowers are salverform with a rather short and wide tube. Colour is not mentioned. The style-head of the male appears approximately cylindrical. After the pollen has been deposited on the style-head of the male flower the head becomes sticky 'at the periphery'; Burck considers that this can only be an adaptation for gumming the pollen to the body of the visitor.

Ceriscoides (W. J. Hook.) Tirvengadum

C. campanulata (Roxb.) Tirvengadum. D. Bridson (pers. comm.) suggests that this may be the plant described by Burck as *Gardenia blumeana* (see below).

Gardenia Ellis

In this genus the pollen forms tetrads.

G. stanleyana Hook. (Burck 1884, illus.). **STR s. ISS exposed. SEQU pln. APPL a. RWD n. SYN** sphingo. **VIS** hawkmoths (*Lepidoptera-Sphingidae*).

This is a typical hawkmoth flower, with a tube not less than 15 cm long, abundant nectar and a very agreeable odour. The style has a slightly thickened and fusiform apex with a gently twisted furrow on either side throughout its length. The anthers are 4 cm long. When the flower opens the pollen is adhering round the style-head in a compact mass, except for a zone at the apex; here the furrow is slightly more open and its tissue secretes a very sticky liquid in which pollen grains can germinate. BURCK had four plants under observation in the botanic garden at Bogor and they never fruited despite insect visits and artificial pollination, which suggests that they represented a single self-incompatible clone.

G. resinifera KORTH. (BURCK 1884, illus.) (Fig. 46B, C) also has the sphingophil form but the style-head is obovate-truncate, with 6 longitudinal ridges on which the pollen is deposited. Between the ribs are six furrows with a papillate surface covered with a sticky secretion; these furrows are almost closed at first but widen gradually, making them more accessible to foreign pollen.

G. blumeana DC. (possibly = *Ceriscoides campanulata* – see above in this tribe). (BURCK 1884, illus.). A single cultivated male plant of this apparently dioecious species was available to BURCK. He illustrates a small, approximately funnel-shaped flower not conforming to any lepidopterophil syndrome. The ellipsoidal style-head has no furrow but 10 raised folds on which pollen is placed.

Griffithia WIGHT & ARN.

G. acuminata KORTH., *G. eucantha* KORTH., *G. fragrans* WIGHT & ARN. and *G. latifolia* TEIJSM. & BINN. (BURCK 1884, illus.). **STR s. ISS exposed. SEQU sim. APPL a. RWD n. SYN** psycho?. **VIS** –.

Pollen is presented on the style-head in the usual way. Some pollen falls into the throat and is caught by hairs. The furrow between the stigmatic lobes is wide, even in the bud stage, and some of the deposited self-pollen inevitably touches the papillae in the furrow. BURCK therefore considers the flower not to be dichogamous, and to allow the possibility of selfing.

Mitriostigma HOCHST.

M. axillare HOCHST. (YEO, unpublished 1985). **STR s. ISS exposed. SEQU sim. APPL a? RWD n? SYN** lepidoptero?, melitto? **VIS** –.

The flowers of this South African compact shrub or small tree are funnel-shaped and white but scentless. The calyx is insignificant, whereas the corolla-tube is 14 mm long and the limb about the same length (measured along the outside). The five rounded corolla lobes are free in the distal two-thirds of the expanded limb. The style is 23 mm long (from the base of the calyx); it has a fusiform distal enlargement, and the distal two-thirds of this bears 10 furrows. It is very shortly bifid at the apex, the cleft having swollen lips with a papillate surface which is presumably stigmatic. The style has been examined in bud just before dehiscence of the anthers and in anthesis, and proved to be glabrous throughout. Its furrowed part seems to be the last part to elongate, for in a bud 20 mm long it was only 2 mm long, whereas at anthesis it slightly exceeds the length of the anthers, which are about 5 mm long and virtually sessile. They dehisce in the very late bud stage and deposit their pollen on the grooved part of the style. When the corolla opens they diverge and lie between its lobes. Although the grooves on the style extend right up to the stigmatic lips, the deposit of pollen ends a short distance below this, so that spontaneous self-pollination seems to be ruled out. This is a difference from *Burchellia* but in that species it is suspected that pollen is presented before the stigmatic surface is receptive. As in *Burchellia* a considerable part of the pollen produced remains in the anthers. These observations were made at Cambridge Botanic Garden on a stock received from the Chelsea Physic Garden in 1968. Another stock, of unknown origin, is also grown at Cambridge, and this has a narrower throat and shorter style, making the arrangements for secondary pollen presentation rather less obvious.

Oxyanthus DC.

Secondary pollen presentation was recorded in 1884 by BURCK in a species of this African genus, *O. racemosus* (SCHUM. & THONN.) KEAY (as *O. hirsutus*), cultivated at Bogor, Indonesia. The corolla-tube was 20 cm long.

O. pyriformis (HOCHST.) SKEELS (*O. natalensis* SONDER). (YEO, unpublished 1985). **STR s. ISS exposed. SEQU pln. APPL a. RWD n. SYN** sphingo. **VIS** –.

A specimen of this South African shrub or small tree grows in the tropical section of the Cambridge Botanic Garden. The corolla-tubes are greenish and have a length of 8.5 cm and a width of 2 mm (external diameter 3/4 of their length from the base); the limb is white and reflexed, consisting of five narrow acute lobes with recurved margins. At the mouth of the tube the basal part of the limb forms a small disc (c. 6 mm in diameter). The flowers contain nectar but I have not detected any scent in them. The dorsifixed anthers are sessile at the mouth of the tube. The slender style projects 1.8 cm beyond this and ends in a fusiform thickening 2.5 mm long. The tip of this is acuminate and somewhat compressed; however, it does not appear acuminate because the excavation that makes it so, c. 0.5 mm long, is filled up with the flower's own pollen. This is deposited more heavily on the flattened sides than on the edges, which bear the narrow stigmatic cleavage and may remain free of self-pollen. The small size of the effective pollen presenter means that not very much pollen is offered; it forms a dense packet and is rather well stuck together. A large proportion of the pollen remains in the anthers, which are in this species much longer than the pollen presenter. In the advanced bud stage the inner epidermis of the corolla throat is deeply transversely wrinkled, probably to allow for subsequent eversion; below the wrinkled zone the tube is lined with soft irregular hairs. I have not seen the flowers with any separation of the stigmatic lobes. They are produced in cymes and stand erect and parallel, several together, and appear over a period of a few weeks in May or thereabouts.

Tocoyena AUBLET

T. brasiliensis MART. and *T. formosa* SCHUM. (SILBERBAUER-GOTTSBERGER 1972, illus.). **STR s. ISS exposed. SEQU pln. APPL a. RWD n. SYN** sphingo. **VIS** hawkmoths, including *Agrius cingulatus* (*Lepidoptera-Sphingidae*).

These two species are sympatric shrubs of the cerrado of Brazil, flowering at different seasons. The corolla-tube is 10–14 cm long and (*T. brasiliensis*) only 3 mm wide. The lobes are 15 mm long and white or yellow. The anthers are sessile in pits at the throat of the corolla. When the flower opens the considerably swollen and massive style-head is situated in the corolla-throat and is loaded with pollen; later it is carried a short distance out of the throat. Each half of the style-head bears three ribs. The anthers spread away from the style-head. The male phase of the flower lasts 2 or 3 days; the female stage is marked by the wide divergence of the broad stigmatic lobes and lasts a further 5 or 6 days. The colour of the corolla-limb changes to yellow with the onset of the female phase and the pollen shrinks and turns brown. The flowers are in groups of 10–15, with usually only 1 or 2 open at a time.

Experiments carried out on *T. formosa* showed that bagging the flowers prevented fruit-initiation, whereas all control (open-pollinated) flowers fruited. Fresh and shrunken pollen mixed with crushed stigmas from another plant germinated quickly and abundantly, whereas if it was mixed with crushed stigmas from the same plant germination was delayed until the second day and was less abundant, while pollen-tube growth was severely restricted. It appears that plants are self-incompatible. Self-pollen placed on the outsides of the tips of the stigmatic lobes in the male phase germinated, and did so more strongly as the flower aged. Foreign pollen could presumably do the same and might cause fertilization if received in this way.

Features of the syndrome of hawkmoth pollination in these species are the form and colour of the flower, the opening of the flower and the commencement of the female stage in the afternoon or evening and the emission of a strong scent, which is weaker in the daytime. Unlike some flowers pollinated by nocturnal *Lepidoptera*, these species are not less visible by day than at night. Unidentified hawkmoths were seen only at *T. formosa* by SILBERBAUER-GOTTSBERGER but the moth *Agrius cingulatus*, which has a proboscis-length of 16 cm, was later seen and photographed visiting *T. brasiliensis* (SILBERBAUER-GOTTSBERGER & GOTTSBERGER 1975).

Gardenieae subtribe *Diplosporiinae*.

A subtribe of seven genera according to ROBBRECHT & PUFF (1986), who illustrate styles of *Sericanthe* and *Cremaspora*, both of which are pilose throughout. The subtribe is shown as having secondary pollen presentation by ROBBRECHT (1988, p. 127).

Tricalysia DC.

Tricalysia species. (ROBBRECHT 1982, 1988).

The members of the small African section *Ephedranthera* combine secondary pollen presentation with heterostyly. Only the style is dimorphic; in long-styled flowers the stigmatic lobes are in the throat, at the same level as the anthers, and in the short-styled deep within the tube. ROBBRECHT does not describe pollen presentation.

Tribe *Hypobathrieae*

Secondary pollen presentation is the rule in this tribe of 23 genera according to ROBBRECHT & PUFF (1986); three genera in which the pollen-presenter is hairy are illustrated by these authors (*l. c.*: 86).

Kraussia HARV.

K. floribunda HARV. (SCOTT ELLIOT 1891). **STR s. ISS exposed. SEQU pln. APPL? RWD n? SYN** psycho. **VIS** mainly butterflies (*Lepidoptera*).

The corolla-tube is shortly and rather broadly tapered and the lobes patent or reflexed. The style-apex is clavate with longitudinal furrows. The anthers shed their pollen into the grooves while the flower is still in bud and when it is open the style is thickly covered with pollen. Later the stigmatic lobes separate. Visits of *Lepidoptera* (4 species) were abundant, though honeybees and small beetles were also visitors. Dr D. BRIDSON tells me that *K. speciosa* is similar; Fig. 121 in BRIDSON & VERDCOURT (1988) shows that this has a very short style, a long pollen presenter which is both pilose and furrowed and long stamens on very short filaments.

Hypobathrum BLUME

H. albicaule BAILL. (*Eriostoma albicaulis* BOIV. ex BAILL.). (BURCK 1884, illus.). **STR s. ISS exposed. SEQU sim. APPL a. RWD n. SYN?. VIS –.**

The flower is stellate-salverform with a very hairy corolla-limb. The anthers have quite large connective-appendages that lean over the apex of the style, which has parted lobes from the beginning. The pollen presenter is hairy. As the apices of the anther loculi just about reach to the stigmatic furrow there is a slight possibility of self-pollination.

Scyphostachys THWAITES

S. coffeoides THWAITES. (BURCK 1884, illus.). **STR s. ISS exposed. SEQU pln. APPL av? RWD n. SYN?. VIS** insects (Fig. 46D).

The flowers are very small and have masses of erect silky hairs emerging from the throat, forming tufts between the anthers. As the flower opens the style is almost completely sunken below the anthers. It soon grows to become exserted, brushing pollen from the anthers with its rough hairs and carrying it upwards; any falling grains are held by the hairs of the corolla.

Tribe *Naucleeae*

Secondary pollination is said by VERDCOURT (1976) to characterize this tribe. BREMEKAMP (1966) excluded the tribe from his *Ixoroideae*, stating that its members were previously but erroneously thought to have the 'Ixoroid pollination mechanism', but he is himself in error on this point (D. BRIDSON, pers. comm.).

Tribe *Pavetteae*

VERDCOURT (1976) and BRIDSON & ROBBRECHT (1985) state that the Ixoroid pollination mechanism is usually present in the tribe *Pavetteae*, in which 13 genera are recognized by ROBBRECHT & PUFF (1986). The latter confirm the occurrence of secondary pollen presentation in the two genera treated here and in the bird-pollinated *Captaincookia* HALLÉ, endemic to New Caledonia. In one genus of this tribe (*Cladoceros*) the style is distinctly below the anthers (cf. tribe *Coffeeae*) (D. BRIDSON, pers. comm.).

Ixora L.

BREMEKAMP (1966) proposed to recognize a subfamily *Ixoroideae* defined by possession of the 'Ixoroid pollination mechanism' and comprising all genera which display it. This is a firm, if implicit, assertion that *Ixora* (by definition included in the subfamily) has secondary pollen presentation, in agreement with the statement in the preceding paragraph. As shown in 'Floral Features' (above) it is also implied that the pollen presenting part of the style is covered with small hairs. Only one *Ixora*, the hybrid *I.* × *williamsii* SANDWITH, has been available for examination at Cambridge, and that clearly does not have secondary pollen presentation (YEO, unpublished observations made in May 1985). I have also looked at all the plates illustrating this genus in Curtis's Botanical Magazine ('Bot. Mag.') and none of them gives any hint that there might be arrangements for secondary pollen presentation. However, in 1990 the report below was published and in addition I saw stylar pollen presentation on plants labelled *I. coccinea* and *I. rosea* in the botanic garden of the Johannes Gutenberg University in Mainz. The stigmatic lobes clearly diverge in most species of *Ixora* (D. BRIDSON, pers. comm.) and the statement by KNUTH (1905) that in some species of *Pavetta* the stigmatic lobes can curl right back to touch the pollen-presenting zone of the style, so that autogamy can occur in the absence of good pollinator service, probably applies to *Ixora*, which was confused with *Pavetta* at that time (D. BRIDSON, pers. comm.).

I. platythyrsa BAKER (NILSSON & al. 1990, illus.). **STR s. ISS exposed. SEQU pln. APPL a. RWD n. SYN** phalaeno. **VIS** moths of families *Noctuidae* (species of tribe *Sarrothripinae*) and *Geometridae* (*Melimoessa catenata*) (*Lepidoptera*).

This plant grows 2–5 m high in dry semi-deciduous forest in Madagascar. The flowers are cream-yellow and open in the evening. The corolla tubes were 17 mm long and the rather narrow spreading lobes nearly 9 mm long. Anther-dehiscence and the loading of the pollen presenter take place on the day of opening. When the flowers opened the anthers were pressed back between the corolla-lobes and soon they were shed. An illustration shows neither hairs nor ridges on the male-stage style-head. At anthesis the style-head is just exserted from the throat. The stigmas diverged gradually and became receptive throughout the second day. Flowers lasted some days further, their duration depending apparently on water-availability. A light perfume was produced at night. The proboscis-length of the Noctuid moths was 20–23 mm and these could drain the flowers. It was therefore thought that these were the moths to which the flower was adapted. Nectar ascended some way up the corolla-tube which made some of it accessible to *Melimoessa* which has a proboscis length of 11.5 mm. This moth foraged earlier in the evening than the Noctuids, before the nectar level had been reduced by them. Pollen appears to be deposited in the facial cavity of the moths. The authors note that phalaenophily is infrequently reported from the tropics.

Pavetta L.

BRIDSON & VERDCOURT (1988) illustrate several species in which the pollen presenter is unridged and it alone, or the entire style, is finely pilose. SCOTT ELLIOT (1891) states that pollination arrangements are practically identical to those of *Kraussia* (see tribe *Hypobathrieae*, above), though that has a ridged and pilose pollen presenter.

P. elliottii K. SCHUM. & KRAUSE. (VERDCOURT 1958). **STR s. ISS exposed. SEQU pln. APPL a? RWD n. SYN? VIS** silver-striped hawkmoth *Celerio lineata* (*Lepidoptera-Sphingidae*).

I have seen illustrations only of other species of *Pavetta* and they show a syndrome of lepidopterophil characters; they are white-flowered but not as long-tubed as most sphingophil plants, so that in ignorance of the pollinator I would call them phalaenophil. Such flowers would, however, be accessible to hawkmoths. The flowers of *Pavetta* are tetramerous (BRIDSON & ROBBRECHT 1985). In the unopened bud of *P. elliottii* the clavate style is touching the anthers; it bears 7 lines of short hairs and these fit into the grooves of the anthers (between thecae?). When the bud opens the pollen is carried by the style-head and the anthers are empty. The style grows rapidly and becomes well exserted. The apex of the style is slightly cleft and the lobes open out to expose the stigmatic surface. There is abundant nectar at the base of the corolla-tube. It seems likely that it is the head of the insect that makes contact with the style-head.

Tarenna GAERTN.

BURCK (1884) reported on this genus under the name *Stylocoryne webera* (*Webera* is in fact yet another generic synonym which BURCK must have mistaken for a specific epithet). Here the style but not the style-head has fine hairs, whereas the pollen-presenting part of the style-head has 10 folds into which the (thecae of the) anthers fit exactly in the bud.

Tribe *Vanguerieae*

Secondary pollen presentation is the rule in this tribe (BREMEKAMP 1966; VERDCOURT 1958, 1976). The style-head is distinctive; the stigmatic branches are wider than usual and it appears that they are fused along their edges and the resulting tube is then turned inside out. Depending on their length and breadth the branches form either a short tube ('skirt') or a mushroom-like cap, in each case descending outside the tip of the style proper (Fig. 47E). The stigmatic region is lobed in agreement with the number of locules in the ovary (D. BRIDSON, pers. comm.).

Canthium LAM.

The construction of the style-head is as described for the tribe.

C. laeve TEIJSM. & BINN. (BURCK 1884, illus.). **STR s. ISS exposed. SEQU pln. APPL a. RWD n. SYN** melitto? **VIS** –. (Fig. 47A–D).

The flowers are similar in form to those of *Psydrax odorata* (see below) but the style-head is mushroom-shaped and the species is dioecious. The deposited pollen

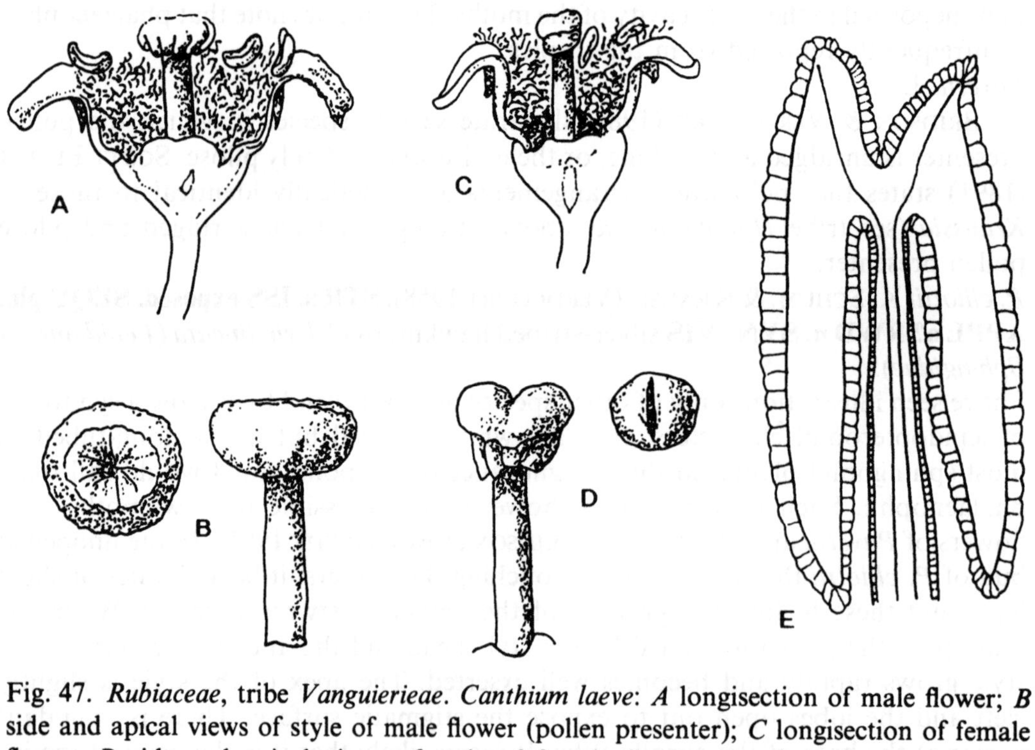

Fig. 47. *Rubiaceae*, tribe *Vanguierieae*. *Canthium laeve*: *A* longisection of male flower; *B* side and apical views of style of male flower (pollen presenter); *C* longisection of female flower; *D* side and apical views of style of female flower, showing pleated skirt and stigmatic furrow. *Keetia gueinzii*: *E* longisection of style-head, showing vascular strands and epidermis. (P. F. YEO, A–D after BURCK 1884, E after SKOTTSBERG 1945.)

forms a shining orange rim on the style-head and is removed little by little by numerous visitors (unspecified). The female flower, 6 mm in diameter, is smaller than the male (8 mm). Its style-head has a cleft across the upper surface and a pleated skirt descending beneath it; the male style-head has a central depression on the upper surface and no apparent skirt. The style is not exserted from the corolla.

C. inerme (L. f.) KUNTZE (SKOTTSBERG 1945, illus.).

This is a South African species similar to *Keetia gueinzii* (see below), but the unfurrowed style-head has a very deep stigmatic cleft and a very short skirt, and again the style is not or scarcely exserted.

Keetia PHILLIPS

The construction of the style-head is as described for the tribe.

K. gueinzii (SOND.) BRIDSON (*Canthium gueinzii* SOND.). (SKOTTSBERG 1945, illus.). **STR s. ISS exposed. SEQU pln. APPL a? RWD n. SYN** melitto or micromelitto. **VIS** –. (Fig. 47E).

This South African species is similar to *Psydrax odorata* (below) except that the style-head is relatively longer, with a much longer skirt, and its pollen-presenting surface is furrowed. The style in *Keetia* is exserted as in *Psydrax* (D. BRIDSON, pers. comm.).

Psydrax GAERTN.

The construction of the style-head is as described for the tribe (SKOTTSBERG 1945).

P. odorata (G. FORST.) DARWIN & A. C. SMITH (*Canthium odoratum* (G. FORST.) SEEM.). (SKOTTSBERG 1945, illus.). **STR s. ISS exposed. SEQU pln. APPL a? RWD n. SYN** melitto or micromelitto. **VIS** –.

This tree is widespread in the West Pacific region. The flowers are numerous in the inflorescence but small, being about 7 mm long, and they have a relatively short and broad corolla-tube and recurved corolla-lobes slightly longer than the tube. The flowers are mostly 5-merous but 4-merous and rarely 6-merous flowers occur additionally in the same plant in some populations. Pollen is deposited on the style-head in bud and after anthesis the style extends to at least twice the length of the corolla-tube. The style-head is nearly cylindrical and has a rather deep apical cleft occupied by the stigmatic surface. The pollen-presenting surface has neither hairs nor furrows. Flowers with barren anthers were sometimes found.

P. obovata (KLOTZSCH) BRIDSON (*Canthium obovatum* KLOTZSCH). (SCOTT ELLIOT 1891). **STR s. ISS exposed. SEQU pln. APPL a? RWD n. SYN** melitto or micromelitto. **VIS** bees (*Hymenoptera-Apoidea*).

This species apparently has a furrowed style head. It is the only member of the tribe for which I have found any visitor-records.

Discussion of Rubiaceae

The flowers of this family are simple in construction and show a prevalence of narrow-tubed corollas with a rotate limb, most of which must be adapted to pollination by *Lepidoptera*. The melittophil syndrome, with a funnel-shaped corolla and styles included in it, appears to be not uncommon, the ornithophil rather rare. As already mentioned, dioecy and heterostyly are frequent and dioecy may coexist with secondary pollen presentation (*Canthium*, *Gardenia*, *Catunaregam*). Secondary pollen presentation takes only one form in the family, but variation is found in the method of holding the pollen: ridges, furrows or hair-tracts are commonest, shallow cavities (*Oxyanthus*) and, possibly, adhesives (*Oxyanthus* and *Catunaregam*) occasional. Secondary pollen presentation is widespread in only one subfamily, though it is not entirely constant within that and apparently not constant even within tribes (VERDCOURT 1976, ROBBRECHT 1988, who records this inconstancy only in tribes *Pavetteae* and *Cinchoneae*). VERDCOURT (pers. comm.) suggests that the system could easily be lost. Thus, it is not yet possible to say whether secondary pollen presentation arose once or repeatedly in the *Rubiaceae*. The single occurrence in subfamily *Rubioideae* and the different ways of supporting the pollen on the presenter in the *Cinchonoideae* suggest that it might have originated repeatedly. [It is very probable that heterostyly arose many times (ANDERSON 1973)].

When the corolla-tube is narrow the position of a visiting insect's head when probing the flower is precisely determined in relation to the floral axis. How high the head is placed above the throat depends on the length of the legs if these are used to grasp the flower (and if the proboscis has the required minimum length) and on the length of the proboscis alone if the insect hovers. Effective pollinators will be restricted to those whose heads are brought to the same level as the style-head. Secondary pollen presentation could thus be involved in placing the pollen accurately on the appropriate part of the pollinator. The same effect could be achieved by having long stamens, exserted to the same distance as the style. However, these might provide a better alighting place for unwanted pollen-eating insects than the slender style. (In bee-pollinated *Proteaceae* where the reward is pollen presented on the style the style has a thickening, the alighting knob, a fact which suggests that at least the larger bees might have difficulty in clinging to a very slender style.) The role of secondary pollen presentation here may thus include pollen protection. Where the style is strongly exserted, the adoption of secondary pollen presentation might have been promoted also by some physical difficulty in accommodating long stamen-filaments (that would otherwise be required) in the corolla, with its narrow tube and frequently small limb. The large residue of pollen sometimes found in the anthers might even serve to deflect pollen-gatherers from the more important packet of pollen that is destined to be transported by the legitimate pollinators. However, little seems to be known in general about the proportion of pollen left in flowers at the end of the part of anthesis during which pollen-donation is possible. Substantial residues have been reported in the heterostylous *Primula vulgaris* (PIPER & CHARLESWORTH 1986). Alternatively, pollen may have a chance of being picked up from the anthers and of causing pollination, the plant thereby combining primary with secondary pollen presentation. In wider, funnel-shaped, flowers only precise pollen placement seems to be served by secondary pollen presentation.

Apart from a possible difficulty in accommodating long filaments, the basic design of the Rubiaceous flower seems to provide a more positive type of predisposition to develop secondary pollen presentation. Thus, according to ANDERSON (1973), protandry is characteristic of the family, and it is usually associated with an arrest in the elongation of the style during the male phase of the flower. Pollen is shed at or soon after [the beginning of] anthesis. If anther dehiscence became a little early, so as to take place while the stamens and style-head were still packed together in the corolla, secondary pollen presentation would result. The system would then be improved by further adaptation in stylar form and by the evolution of effective dissociation of the anthers from the pollen they had deposited. The extent of the later elongation of the style would come under selective pressure so as to bring the pollen to its adaptively most suitable position.

The situation in the apparently ornithophilous *Burchellia bubalina* is peculiar in that the style is more or less included in the corolla. Here secondary pollen presentation is difficult to account for in any way except as an inheritance from an ancestor with a different pollination syndrome [it belongs to subtribe *Gardeniinae*, in which secondary pollen presentation is universal (D. BRIDSON, pers. comm.)]. The tropical American Rubiaceous shrub *Hamelia patens* JACQ. has a similar flower though with a narrower and more cylindric tube. It is visited by hummingbirds and butterflies (THOMAS & al. 1986). The stylar head is furrowed as in *B. bubalina* and also papillate-hairy on the ridges but the stamens stand just away from it and do not spontaneously deposit pollen on it. If, however, the flower is handled, large masses of pollen adhere to the style-head (YEO, unpublished, 1985). The insertion of a hummingbird's bill would certainly cause such a transfer on the opposite side of the flower. If this has any significance at all it would only be in relation to a subsequent probing on the other side of the flower.

The type of secondary pollen presentation in this family is like that in some *Proteaceae*, but as far as is known the styles are not particularly stiff, and the *Rubiaceae* do not share the preponderance of ornithophily that is found in the *Proteaceae*. Resemblance to *Campanulaceae* is also apparent, but in that family pollen presentation is more likely to be within the corolla. Parallels between these two families are evident in the use of hairs or ridges to hold pollen on the presenter. One case where strongly projecting bristles sweep the pollen from the anthers is recorded in *Rubiaceae* (*Scyphostachys*); in other cases where pollen is placed in hair-tracts it seems likely that transfer takes place during a phase of apposition, the anthers then moving to the periphery of the flower (as repeatedly described by BURCK 1884) and the styles resuming growth afterwards. This is the method that seems to be adopted also by most *Campanulaceae*. *Rubiaceae* seem to be sufficiently isolated taxonomically from other families with secondary pollen presentation to rule out the possibility that this was either inherited from or passed on to any other family. Isolated resemblances to *Apocynaceae* occur in *Catunaregam* which may use adhesive to attach pollen to the visitor, and *Acranthera* in which the anthers form a cage round the pollen presenter.

Homomorphic self-incompatibility was thought by ANDERSON (1973) to be rare in *Rubiaceae*, though the evidence was meagre. However, indications of self-incompatibility are noted above and BAWA & BEACH (1983), in a study of 14 tropical *Rubiaceae*, found self-incompatibility in all four of the homostylous species in their sample. Inhibition of incompatible pollen took place in the style or even in the ovary, so the determination of incompatibility is probably gametophytic. Two features of the flower could, in different cases, prevent autogamy in self-compatible species with secondary pollen presentation: delayed receptivity of the stigma and restriction of the area of deposition of pollen. The pollen grains are usually binucleate but not infrequently trinucleate (BREWBAKER 1967) and the stigmas usually dry (HESLOP-HARRISON & SHIVANNA 1977).

The conflict between the numeric condition of the gynaecium and that of the androecium that occurs in *Apocynaceae* must occur also in the *Rubiaceae*, more especially where the androecium is pentamerous. How it is resolved is not always clear. Thus in *Burchellia bubalina* there are five ridges on either side of the decurrent stigmatic cleavage; if each furrow received the contents of one theca each end of the stigmatic cleavage would receive pollen, but if this were avoided 5 thecae would be required to place their pollen in 4 furrows. In *Oxyanthus* the stigmatic cleavage is left free of pollen, at least on one side, according to my drawing. With a tetramerous androecium, each side of a slightly flattened stylar head could receive pollen from two stamens; it is curious that in *Pavetta* VERDCOURT counted 7 rows of hairs. The problem is resolved in *Mitriostigma*, as the stigmatic cleavage is not decurrent and the pollen deposit does not even reach the distal extremities of the 10 ridges on the stylar head. The state of affairs in *Oxyanthus* suggests the possibility that the pollen depends on an adhesive for its attachment to the style, and that absence of adhesive from the neighbourhood of the cleavage could keep the latter free of pollen. However, as this is one of the genera with pollen in tetrads, Dr D. BRIDSON (pers. comm.) suggests that there could be some correlation of stickiness of the pollen with retention of tetrads. On the other hand, in *Hamelia* (mentioned earlier in this discussion) it was found that pollen would adhere to the style only in a region opposite the distal half of the notably long anthers (YEO, unpublished 1985).

It is interesting that explosive flowers exhibiting the sphingophilous syndrome occur in the tropical American Rubiaceous genus *Posoqueria* AUBLET (F. MÜLLER 1866). The flower-limb is zygomorphic and the five corolla-lobes are imbricated in the same way as the petals in *Leguminosae-Papilionoideae*. Tripping is caused by contact with a limited area of two lateral, outwardly bowed stamen-filaments after the flower has opened; the resulting pollen-cloud has a width of many centimetres. Although the anthers dehisce before the corolla opens, there seems to be no element of secondary pollen presentation and there seems to be no link with this method of pollen-release and the form of secondary presentation in the family.

References for Rubiaceae

ANDERSON, W. R., 1973: A morphological hypothesis for the origin of heterostyly in the *Rubiaceae*. – Taxon **22**: 537–542.

BAWA, K. S., BEACH, J. H., 1983: Self-incompatibility systems in the *Rubiaceae* of a tropical lowland wet forest. – Amer. J. Bot. **70**: 1281–1288.

BREMEKAMP, C. E. B., 1934: A monograph of the genus *Pavetta* L. Feddes Repert. – Spec. Nov. Regni Veg. **37**: 1–13.

BREMEKAMP, C. E. B., 1947: A monograph of the genus *Acranthera* ARN. ex MEISN. (*Rubiaceae*). – J. Arnold Arbor. **28**: 261–308.

BREMEKAMP, C. E. B., 1966: Remarks on the position, the delimitation and the subdivision of the *Rubiaceae*. – Acta Bot. Neerl. **15**: 1–33.

BREWBAKER, J. L., 1967.

BRIDSON, D. M., ROBBRECHT, E., 1985: Further notes on the tribe *Pavetteae* (*Rubiaceae*). – Bull. Jard. Bot. Belg. **55**: 83–115.

BRIDSON, D., VERDCOURT, B., 1988: *Rubiaceae* (Part 2). – In POLHILL, R. M., (ed.): Flora of Tropical East Africa 415–747. – Rotterdam: Balkema.

BURCK, M. W., 1884: Sur l'organisation florale chez quelques Rubiacées (Suite). – Ann. Jard. Bot. Buitenzorg **4**: 12–87.

HESLOP-HARRISON, Y., SHIVANNA, K. R., 1977.
KERNER, A., 1902: The Natural History of Plants, transl. F. W. OLIVER, **2**. – London: Blackie.
KERR, A. F. G., 1937: *Psilobium siamense*. – Hooker's Icon. Pl. **25**: t. 3332.
KNUTH, P., 1905.
MCCLEAN, R. C., IVIMEY-COOK, W. R., 1956: Textbook of Theoretical Botany **2**. – London: Longmans, Green & Co.
MÜLLER, F., 1866: Über die Befruchtung der *Martha* (*Posoqueria*?) *fragrans*. – Bot. Zeitung (Berlin) **24**: 129–133.
NILSSON, L. A., RABAKONANDRIANINA, E., PETTERSSON, B., RANAIVO, J., 1990: 'Ixoroid' secondary pollen presentation and pollination by small moths in the Malagasy treelet *Ixora platythyrsa* (*Rubiaceae*). – Pl. Syst. Evol. **170**: 161–175.
PIPER, J., CHARLESWORTH, B., 1986.
ROBBRECHT, E., 1980: The *Hypobathreae* (*Rubiaceae-Ixoroideae*). 1. Delimitation and division of a new tribe. – Bull. Jard. Bot. Belg. **50**: 69–77.
ROBBRECHT, E., 1982: The African genus *Tricalysia* A. RICH. (*Rubiaceae-Coffeeae*) 2. *Ephedranthera*, a new section of subgenus *Tricalysia*. – Bull. Jard. Bot. Belg. **52**: 311–339.
ROBBRECHT, E., 1988: Tropical woody *Rubiaceae*. – Opera Bot. Belg. **1**: 1–271.
ROBBRECHT, E., PUFF, C., 1986: A survey of the *Gardenieae* and related tribes (*Rubiaceae*). – Bot. Jahrb. **108**: 63–137.
SCOTT ELLIOT, G. F., 1891.
SILBERBAUER-GOTTSBERGER, I., 1972: Anthese und Bestäubung der Rubiaceen *Tocoyena brasiliensis* und *T. formosa* aus dem Cerrado Brasiliens. – Oesterr. Bot. Z. **120**: 1–13.
SILBERBAUER-GOTTSBERGER, I., GOTTSBERGER, G., 1975: Über Sphingophilie Angiospermen Brasiliens. – Pl. Syst. Evol. **123**: 157–184.
SKOTTSBERG, C., 1945: The flower of *Canthium*. – Ark. Bot. **32A(5)**: 1–12.
THOMAS, C. D., LACKIE, P. M., BRISCO, M. J., HEPPER, D. N., 1986: Interactions between hummingbirds and butterflies at a *Hamelia patens* bush. – Biotropica **18**: 161–165.
VERDCOURT, B., 1958: Remarks on the classification of the *Rubiaceae*. – Bull. Jard. Bot. État **28**: 209–291.
VERDCOURT, B., 1976: *Rubiaceae* (Part 1). – In POLHILL, R. M., (ed.): Flora of Tropical East Africa 1–414. – Nairobi: Crown Agents of Oversea Governments and Administrations.
VOGEL, S., 1954.
WONG, K. M., 1982: Notes on *Gardenia* and *Acranthera* (*Rubiaceae*) from Peninsular Malaya. – Gard. Bull. Singapore **35**: 21–32.

Asteraceae

Systematic position: *Asteridae, Asterales*
Restriction of occurrence within the family: None

Distribution within the family. Secondary pollen presentation is general in the family and I know of know definite exceptions to this; however, in the present account the relatively few wind-pollinated groups are omitted.

Floral features. Flowers protandrous. Calyx of scales or numerous fine hairs (collectively the pappus, which often allows for wind-dispersal). Corolla actinomorphic or zygomorphic, tubular at base; limb three-lobed to (typically) five-lobed. Zygomorphic corollas are two-lipped or, more usually, one-lipped, with a slit on one side and a flat strap-shaped limb (ligule) with the lobes in the form of teeth at the apex. Nectary within the base of the corolla tube. Stamens attached to the corolla-tube, which is usually dilated at and above the point of their attachment. Filaments free; anthers introrse, united into a cylinder; connective usually with an apical appendage; bases of thecae also frequently appendiculate. Ovary inferior, single-seeded. Style two-branched; branches linear, initially appressed, so that they form a continuation of the style, later more or less divergent to expose the stigmatic surface on the adaxial side, extending from the base part-way or all the way to the apex. Most frequently the stigmatic surfaces are found in a strip on each margin of each branch. They are often at least partly visible, though unreceptive, even while the stigmatic branches are still appressed.

Throughout the family the floral unit is the capitulum or, occasionally, an aggregation of capitula. The capitulum is composed of many small flowers (florets) packed together on a flat or convex receptacle and surrounded by an involucre of bracts. The sequence of flower-opening in the capitulum is centripetal. The capitulum may contain only tubular (actinomorphic) corollas, only ligulate (zygomorphic) corollas, or two types, tubular and usually actinomorphic in the centre (constituting the disc) and enlarged and zygomorphic round the edge (collectively the 'ray') and then either ligulate (one-lipped) or more or less two-lipped.

Type of pollen presentation. There are three common types of pollen presentation: (1) on a stylar brush (pollen presenter) projecting beyond the anther-cylinder ('Pseudo-Staubblatt'), (2) by a stylar piston mechanism causing pollen to appear on the apex of the anther-cylinder ('Nudelspritze'), (3) a combination of these two arrangements, extrusion first, followed by emergence of the style-branches bearing the remainder of the pollen (Fig. 48). Rarely presentation is on the corolla lobes.

Essential to the functioning of the system is the correct positioning of the distal part of the style immediately before dehiscence of the anthers (Fig. 48, stage 3). This depends partly on the type of pollen presentation and partly on the structure of the parts involved. It is mentioned at appropriate places in the following account. Owing to the flimsy state of the thecal walls it is difficult to tell by dissection when the anthers dehisce; according to THIELE (1988: 46) they do so at the beginning of anthesis.

In many members of the family the stamen filaments are irritable; on being touched they arch outwards at the middle. In some cases only the filaments touched respond, causing the anther-cylinder to lean towards the side touched; in others all filaments respond, drawing the anther-cylinder downwards, so that some or all of the pollen is presented. A third type of response is leaning to one side but in a random direction. In reviewing this subject SMALL (1917b) stated that hairs on the filaments of irritable stamens had previously been regarded as sensory in function ('trigger-hairs') but that some species without these hairs show irritability (see also introduction to tribe *Cardueae*).

Fig. 48. Diagrams of the three main methods of pollen presentation in *Asteraceae*. Hairs shown are all concerned with pollen presentation; numbering of columns indicates successive developmental stages. Upper row: piston mechanism. Middle row: stylar brush mechanism. Lower row: piston mechanism initially, followed by emergence of pollen-laden stylar brush. *ac* = connective appendage of anther, *at* = theca of anther, *cl* = corolla-limb, *ct* = corolla-tube, *f* = filament, *s* = style, *sb* = style-branch. (THIELE 1988.)

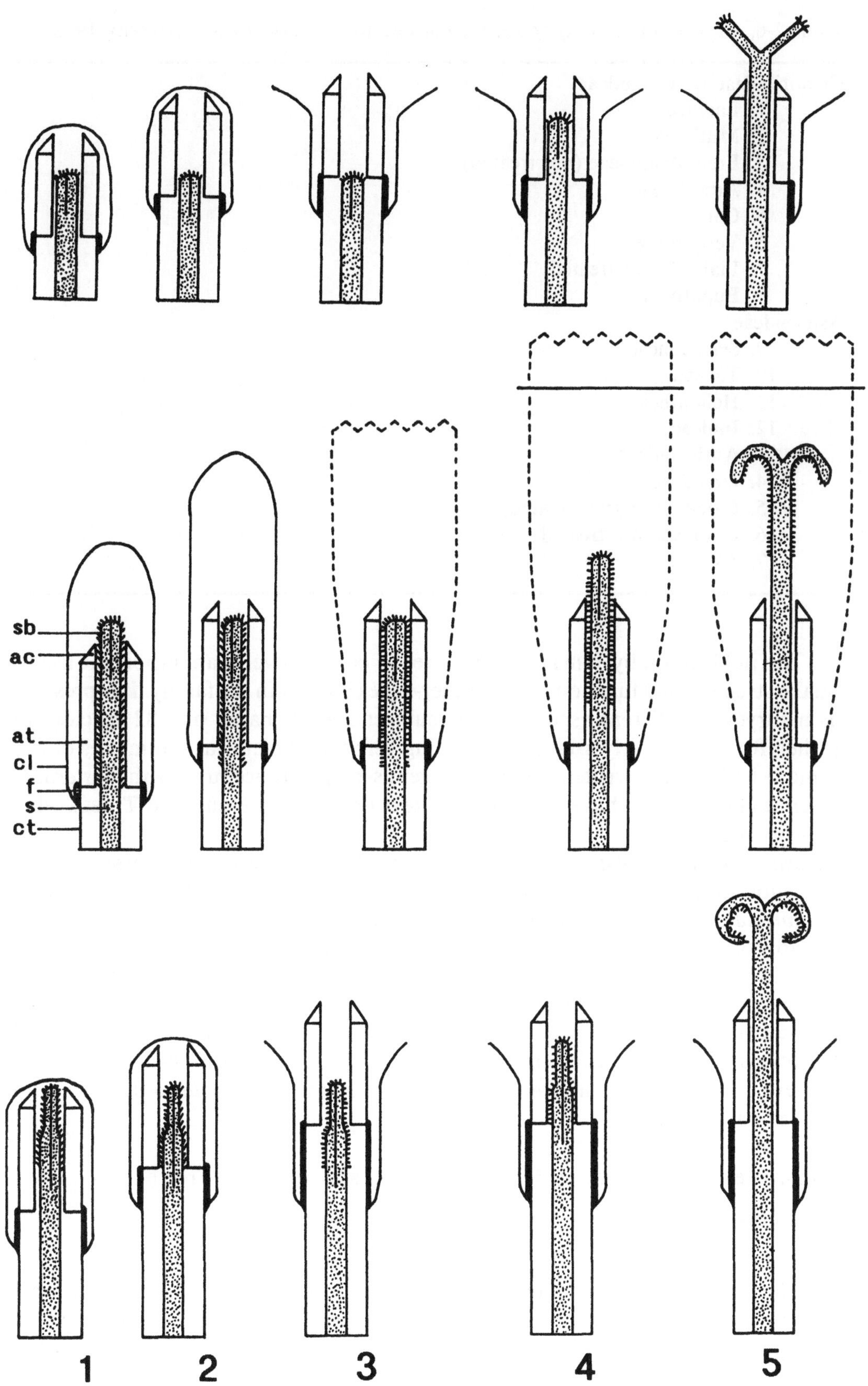
sb
ac
at
cl
f
s
ct
1
2
3
4
5

Table 2-6. Classification of Asteraceae according to C. JEFFREY (in HEYWOOD, 1978)

Cichorioideae (Lactucoideae)	
	1. Lactuceae
	2. Mutisieae
	3. Eremothamneae (not treated)
	4. Arctotideae
	5. Cardueae
	6. Vernonieae
	7. Liabeae (not treated)
	8. Eupatorieae
Asteroideae	
	9. Senecioneae
	10. Tageteae
	11. Heliantheae
	12. Inuleae
	13. Anthemideae
	14. Ursinieae
	15. Calenduleae (not treated)
	16. Cotuleae (not treated)
	17. Astereae

The publication by THIELE (1988), already cited above, is an important treatise on the structure and function of the anther-style complex in the family. It is presented as a survey of different aspects of the subject matter, based on 182 species, and is therefore particularly useful in providing verbal and pictorial summaries of each topic. I had already written the present genus-by-genus account of the *Asteraceae* when it appeared but have amplified my text or added references to THIELE where appropriate.

Classification of the Asteraceae. The system adopted here for the classification of this enormous family is shown in Table 2–6. Of the two subfamilies recognized, *Cichorioideae* has dorsifixed, and *Asteroideae* basifixed, anthers. (Changes in classification of the latter, compared with older systems, are that tribe *Helenieae* has been disbanded as unnatural, and some genera have been moved to new positions, for example *Arnica* from *Senecioneae* to *Heliantheae*.) Other differential characters of the two subfamilies are inconstant but *Cichorioideae* are the most variable in the forms of floret included in the capitulum. The system of pollen presentation, however, is more often uniform within tribes of the *Cichorioideae* than in those of the *Asteroideae*.

Examples of Secondary Pollen Presentation in Asteraceae

Subfamily *Cichorioideae*

Tribe 1, *Lactuceae*

The tribe is very uniform in its pollen presentation (KNUTH; JUEL 1908), which is shown diagrammatically in Fig. 48, middle row. All the florets in the capitulum are ligulate. The upper part of the style and the whole of the outer surface of the style-branches bear short hairs. This tract emerges from the anther-cylinder bringing with it the whole of the pollen produced by the floret. Immediately before the flower opens its position coincides with that of the thecae of the anthers so that it can be loaded with pollen. Later the style-branches diverge and can receive pollen. The stigmatic surface of the branch occupies the whole of its inner surface. The same system, with considerable differences of detail, is found also in tribes *Arctotideae*, *Vernoniae* and *Eupatorieae* of this subfamily.

Some members of the *Lactuceae* have irritable stamens and in these the anther-cylinder bends towards the side touched. Some genera are noted for the fact that the capitula have only a short daily period of opening (*Cichorium*, *Tragopogon*) and, indeed, may open on one day only (*Cichorium*, THIELE 1988: 36, 37). However, in several genera I have noted that the male stage of each floret is passed through rapidly in the first morning that the floret opens (*Cicerbita*, *Cichorium*, *Crepis*, *Tolpis*), whereas the female stage apparently lasts for the remainder of the period of anthesis of the capitulum; this is a few hours in *Cichorium* but apparently two days in *Cicerbita macrophylla* (WILLD.) WALLR. and several days in *Tolpis barbata* (L.) GAERTN. (YEO, unpublished 1985; plants in cultivation at Cambridge).

The great majority of species have erect yellow heads and are visited by a wide variety of insects, but especially by bees (*Hymenoptera-Apoidea*) for both pollen and nectar, hoverflies (*Diptera-Syrphidae*) mainly for pollen, and butterflies (*Lepidoptera*) for nectar. However, some have blue or purple heads and in some cases the capitula are few-flowered and nodding, suggesting specialization to long-tongued bees. The two cases formally cited below serve as examples of the prevalent form and colour. The structure is illustrated by drawings of *Tolpis barbata* (Fig. 49).

Hieracium L.

H. umbellatum L. (MÜLLER, illus.). **STR s. ISS exposed. SEQU pln. APPL vl. RWD n, p. SYN** entomo. **VIS** social and solitary bees (*Hymenoptera-Apoidea*), hoverflies and conopids (*Diptera-Syrphidae* and *-Conopidae*), butterflies (*Lepidoptera*).

The pollen presenter is 6 mm long and is composed of the style-branches (2.5 mm) and the upper part of the style. For mechanism see description of tribe.

Leontodon L.

L. autumnalis L. (MÜLLER, illus.). **STR s. ISS exposed. SEQU pln. APPL vl. RWD n, p. SYN** entomo. **VIS** social and solitary bees (*Hymenoptera-Apoidea*), various flies (*Diptera-Syrphidae*, *-Bombyliidae* and *-Muscidae*).

The pollen presenter is constituted similarly to that of *Hieracium*. For mechanism see description of tribe.

Tribe 2, *Mutisieae*.

Mutisia L. f.

The pollen presentation system here resembles that of tribe 5, *Cardueae*, rather than that of the preceding and following tribes.

M. coccinea ST. HIL. (SAZIMA & MACHADO 1983, illus.). **STR s. ISS rel(dose)-trigg. SEQU pln. APPL a. RWD n. SYN** ornitho. **VIS** four spp. of hummingbirds (*Aves-Trochilidae*), one sp. of bee *Exaerete smaragdina* (*Hymenoptera-Apoidea*).

The capitula of this plant are 6.5 cm long and 2.5 cm wide. The marginal florets are ligulate and female, and they secrete no nectar, whereas the central ones are tubular, bilabiate, hermaphrodite and nectariferous. In the male stage stimulation

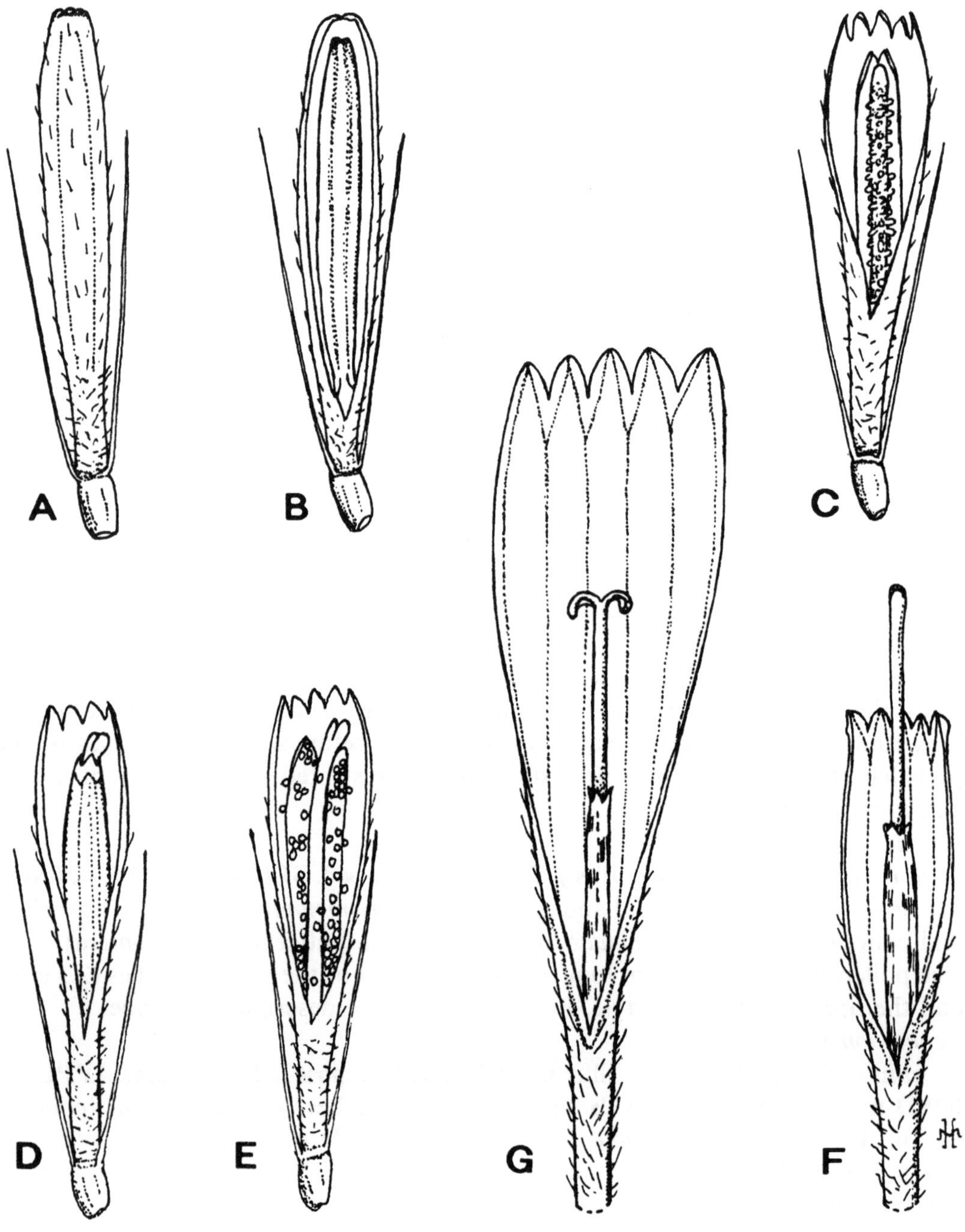

Fig. 49. *Tolpis barbata* (*Asteraceae-Cichorioideae*, tribe *Lactuceae*). *A* closed flower-bud, showing ovary, pappus (calyx) and corolla. *B* the same, cut open to expose anther-cylinder. *C* corolla and anther-cylinder opened to expose style; size of stylar papillae exaggerated. *D* floret beginning to open, with tip of style emerging from anther-cylinder. *E* the same, with anther-cylinder opened to show style (stylar papillae omitted) and pollen. *F* male stage of flower which has apparently misfunctioned, as anther-cylinder reported full of pollen. *G* female stage. (M. HICKEY.)

of the filaments causes them all to contract rapidly, so that pollen is forced out of the anther-cylinder. The cylinder quickly returns to its original position and is able to respond to further stimuli. If the stimulus is applied artificially the pollen accumulates on the tip of the anther-cylinder. In the female stage the filaments contract permanently, so that the style is exposed without further growth and remains at the level at which pollen was previously presented. The illustration suggests that the connective-appendages are nearly half as long as the thecae and fused for most of their length. The floral parts are partly red and partly orange or yellow. The colour and nodding posture of the capitulum, the length of the florets, the firmness of the anthers, and the large quantity of dilute nectar that is produced are all characteristics which suit the plant to pollination by hummingbirds, which are the principal visitors, although one long-tongued bee succeeds in causing some pollination. Plants are self-incompatible.

Gerbera L.

G. jamesonii J. D. HOOK. (SMALL 1917). **STR a, s. ISS rel(dose)-irrit. SEQU pln. APPL av. RWD n, p. SYN** entomo. **VIS** –.

The stamens are irritable and bend towards the side of the flower on which they are touched, but the outer florets are much more active than the inner. Half the pollen is expelled at the first touch and the rest at the second (compare *Perezia*, immediately following).

Perezia LAGASCA.

P. multiflora (HBK.) LESS. (SMALL 1917). **STR a, s. ISS expl depos. SEQU pln. APPL av. RWD n, p. SYN** entomo. **VIS** –.

This is similar to *Gerbera jamesonii* (immediately preceding) but all the pollen is expelled at the first touch in an explosive movement that is sometimes so energetic that the pollen is thrown over the edge of the capitulum. I have guessed that the method of issue is 'explosive deposit' rather than 'explosive cloud'.

Tribe 4, *Arctotideae*

This is one of the three tribes in which the whole of the pollen production is presented exposed on the style (see introduction to tribe 1, *Lactuceae*).

Arctotheca WENDL.

A. calendula (L.) LEVYNS (*Arctotis calendulacea* L.). (JUEL 1908, illus.). **STR s. ISS exposed. SEQU pln. APPL vl. RWD n, p. SYN** entomo. **VIS** –.

The capitulum has both disc and ray, only the disc florets being hermaphrodite. The upper part of the style is abruptly and cylindrically thickened for a distance equal to the length of the anthers; it emerges from the anther-cylinder with all the pollen on it. The pollen-carrying surface is hairy. The style-branches are very short and are receptive on their inner faces (there is no decurrent stigmatic margin). The style is irritable, not the stamens, and it bends towards the side touched.

Arctotis L.

A. venusta T. NORL. (*A. stoechadifolia* auct. non BERGIUS; *A. grandis* auct. non THUNB.). (JUEL 1908).

This is the same as *Arctotheca calendula*. (JUEL referred to *A. stoechadifolia* but it is presumed that he was referring to the plant widely cultivated under that name, which is *A. venusta*.)

Gazania GAERTN.

G. rigens (L.) GAERTN. (JUEL 1908).

This is like *Arctotheca* and *Arctotis* but the pollen presenting part of the style is less clearly marked, the style-branches are not so short, and the style is not irritable.

Tribe 5, *Cardueae*

In this group the pollen is usually released by the style's acting at first as a piston and afterwards as a direct pollen presenter; JUEL (1908) states that pollen-presentation takes place on the tip of the anther-cylinder but it is certainly not limited in this way. Compared with other tribes with this system in either subfamily of the *Asteraceae* the form of the style apex (including stigmatic branches) is different. In the *Cardueae* a considerable length of the distal part is always covered with papillae or very short hairs, and the proximal end of the papillate zone is delimited by a ring of slightly longer hairs, forming a brush greater in diameter than the rest of the style and as great as the internal diameter of the anther-cylinder (Fig. 50E, F). The florets are usually very long with narrow corolla-lobes, the sinuses between which are slightly unequal. In three genera that I have examined (*Carduus*, *Cirsium* and *Centaurea*), the final phase of elongation of the corolla takes place at the lower end of the tube just before the floret opens (at night in *Cirsium*: PIGOTT 1968). At an earlier stage the tip of the style can be found within the tip of the anther-cylinder, too high up for its sweeping hairs to be able to push pollen up from the bases of the anthers (Fig. 48, stage 1; KNUTH 2: Fig. 206C). It must therefore be inferred that in the last phase of corolla-elongation before anthesis (Vorbereitungsphase of THIELE 1988) the style does not grow, so that the anthers are carried forward to a point where their bases are ahead of the tip of the style, and that the style makes further growth only as the flower opens and enters on its male stage. [THIELE (1988: 17, 18) reported on this point for 12 tribes; always there was either a cessation or, as in *Cardueae*, retardation of growth]. Since according to THIELE anther dehiscence does not take place until the beginning of anthesis (see 'Introduction' to *Asteraceae*) the thecal walls can isolate the pollen from the pollen-presenting zone of the style during the phase of advancement of the anthers (relative retraction of the style).

In each of the three above-mentioned genera I found instances in which, by time the female stage is reached, the pollen had been cleared completely or almost so from the anther-cylinder but also instances in which large quantities, perhaps nearly all, remained behind; there are therefore circumstances, perhaps unfavourable weather conditions, in which the pollen presentation system fails to function properly; these occurrences will not be mentioned further. (THIELE 1988: 35, noted, for the family as a whole, an influence of the weather on failure of all pollen to be presented; she also found that this occurred especially in non-native ornamental species).

The position of the pollen presenting part of the style at the beginning of anthesis is different in different groups, as explained below. Irritability of the stamens in this tribe usually involves retraction of the whole anther-cylinder but bending towards the side touched also occurs (SMALL 1917b). The filaments usually bear short thick hairs. Contact with either the hairs or the intervening epidermis stimulates contraction. A loss of turgor in the cortical cells, combined with elasticity of the epidermal cuticle, is what causes contraction. After such an event turgor is restored over a period of from 10 to 25 minutes and the cuticle again becomes stretched. Contraction can be induced at least ten times in succession (PESACRETA & al. 1991).

The *Cardueae* are particularly suited to and visited by long-tongued insects such as bumble-bees and *Lepidoptera*. The flower colour is frequently purple but also sometimes blue or yellow.

The majority of the genera of *Cardueae* can be assigned to two groups that were recognized by JUEL (1908). (1), the thistle and saw-wort genera: these have a relatively weakly constructed anther-cylinder (JUEL 1908), with the connective-appendages short and free from one another (YEO, report below), and a male stage of the floret lasting from morning to midday (JUEL 1908), at the beginning of which the lower extremity of the pollen presenter (with its ring of hairs) is within the thecal part of the anther-cylinder (YEO, report below) and at the end of which the female stage begins, so that anthesis is completed in one day (PIGOTT 1968); movements of the anther-cylinder resulting from irritability of the filaments tend to be small in amplitude (YEO, report below). Group (2) is composed of the knapweeds: these have a firmer anther-cylinder (JUEL 1908) with connective-appendages more than half as long as the thecae and very firm (JUEL 1908; YEO, report below) and fused into a tube for more than half their length (Fig. 51; MÜLLER: 346–347; YEO, report below) so as to form a pollen magazine; the male stage typically lasts all day (JUEL 1908) and begins with the lower extremity of the pollen presenter (and consequently the pollen) in a position already above the thecae and within the pollen magazine (MÜLLER: 346, Fig. 115; YEO, report below); the female stage occurs on the second day of anthesis (e.g. LACK 1982); stimulation of the filaments often causes a large movement of the anthers (MÜLLER 1883; YEO, report below).

Apart from these two groups there are several genera that have their own peculiarities. The genera of the *Cardueae* are therefore dealt with here in the following order: group (1), thistles and saw-worts – *Cirsium*, *Carduus*, *Onopordum*, *Notobasis*, *Tyrimnus* and *Serratula*; group (2), *Centaurea* (knapweeds) and *Cnicus*; group (3), the other genera, namely *Xeranthemum*, *Arctium*, *Carthamus*, *Echinops* and *Carlina*. When more is known members of group (3) may be found to have further groups centred on them.

[The base of the pollen presenter, which is marked by the slightly longer hairs, is sometimes said to be thickened. Thus, according to DITTRICH (1977) it is thickened in *Carduus*, *Cynara* and *Rhaponticum* – corresponding to my group (1) – but not in *Carlina* and *Echinops* – my miscellaneous group, (3). I did not find such thickening in group (1) and its presence is denied by E.-M. THIELE (pers. comm. and 1988), yet it is shown in our Fig. 50F. In any case, the ring of hairs at the base of the pollen presenter creates a zone effectively thicker than the distal part of the pollen presenter.]

Cardueae, Group 1: connective-appendages short

Cirsium MILL.

C. arvense (L.) SCOP. (MÜLLER, illus.). **STR a, s. ISS rel(dose)-irrit?** and **growth?. SEQU pln. APPL vl. RWD n, p. SYN** entomo. **VIS** many short- and long-tongued bees and sphecid wasps (*Hymenoptera-Apoidea* and *-Sphecoidea*), some *Hymenoptera-Chrysidoidea*, *-Ichneumonoidea* and *-Tenthredinoidea*; many *Diptera*, especially *Syrphidae* and *Muscidae*; some butterflies (*Lepidoptera*) and several *Coleoptera*. (Fig. 50).

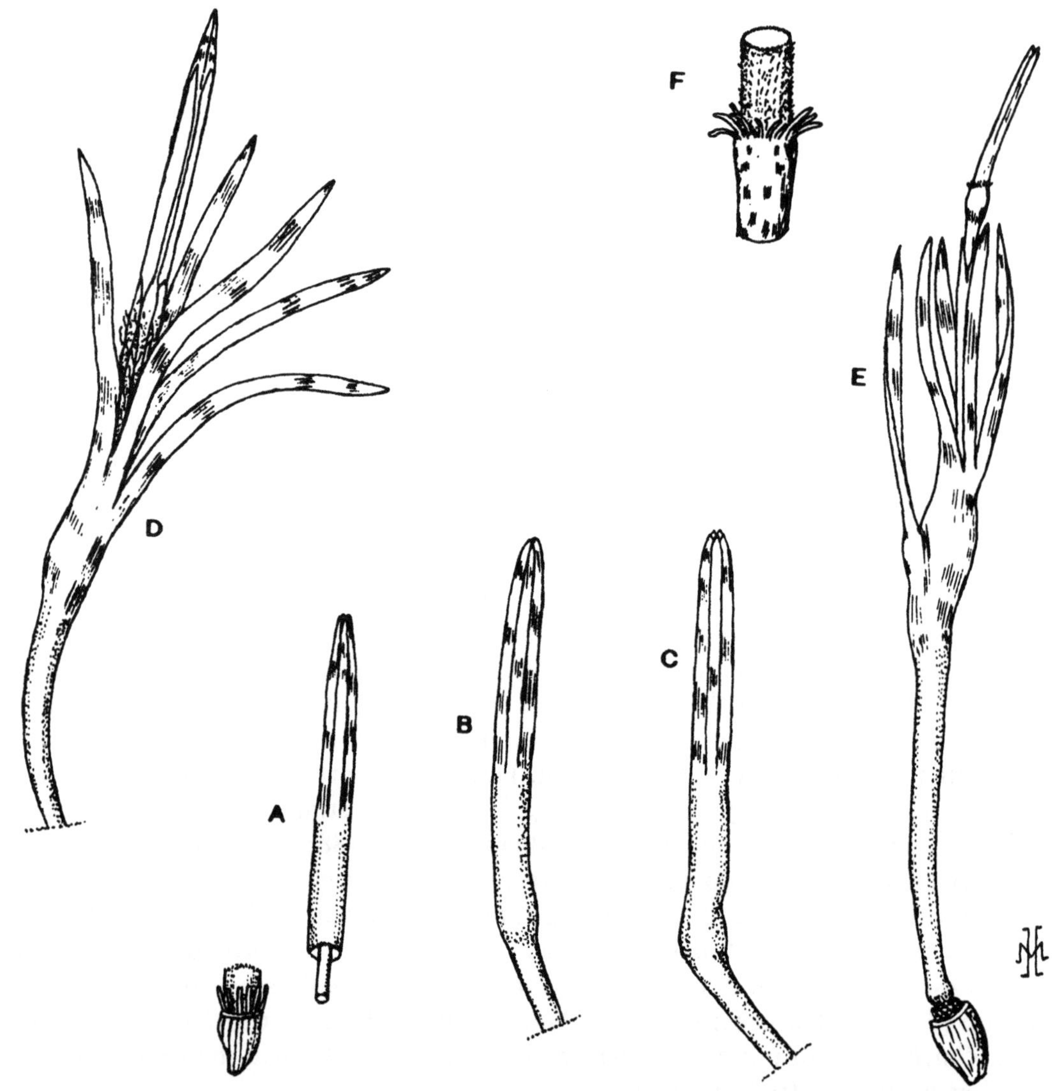

Fig. 50. *Cirsium arvense* (*Asteraceae-Cichioroideae*, tribe *Cardueae*). *A* young flower bud, with base broken off and displaced and pappus cut short; main portion of corolla 10 mm long. *B*, *C* corollas at successively later stages, up to 14 mm long; basal part of tube leans increasingly away from centre of capitulum. *D* corolla and stamens at male stage when corolla is 16.5 mm long; note inequality of incisions between corolla-lobes and hairiness of filaments and that connective appendages are about 1/4 length of thecae. *E* flower at female stage with style protruding from anther-cylinder (pappus cut off; whole floret 18 mm long). *F* enlargement of part of style bearing the hair-ring that sweeps pollen from anther-cylinder. Parts shaded with stippling unpigmented, those shaded with striations purple. (M. HICKEY.)

The florets spread outwards above the involucre and form a capitulum 2 cm wide. The sympetalous part of the corolla is 9–13.5 mm long (including the dilated throat 1–1.5 mm long) but the nectar rises up to the throat and can be reached by a wide range of visitors including those with short tongues. The flowers are very fragrant (YEO, pers. obs.). In the first stage of anthesis the style pushes pollen out of the anther-cylinder and then emerges itself, thickly covered with pollen. This stage lasts from morning to midday (JUEL 1908). In the second stage the stigmatic branches do not separate but their edges become lightly revolute for the full 2 mm of their length. Plants are self-incompatible (MICHAUX 1989).

C. vulgare (SAVI) TEN. (*Cnicus lanceolatus* (L.) WILLD.). (MÜLLER; YEO, unpublished 1987). **STR a, s. ISS rel(dose)-irrit** and **growth. SEQU pln. APPL avl. RWD n, p. SYN** melitto. **VIS** honeybees *Apis mellifera*, bumblebees *Bombus* and *Psithyrus*, leaf-cutter bees *Megachile* (*Hymenoptera-Apoidea*), long-tongued hoverflies *Eristalis* spp. (*Diptera-Syrphidae*), and some butterflies (*Lepidoptera*).

The capitula are rather larger than those of *C. arvense* and the corollas measure about 22 mm on first opening, and 25 mm in the female stage. The narrow tube is 12–16 mm and the enlarged sympetalous part 4–6 mm long. Nectar, even if it reaches the top of the narrow tube, is thus much less accessible to insects, and only the longer-tongued types seem to visit the flowers. I found the anther-cylinder to be about 5.5 mm long and the connective-appendages to be 1.3 mm long and free for most of their length. The filaments are irritable in the male stage of the flower and pollen exudes when they are touched; in the absence of insect visits it accumulates on the outside of the connective-appendages. In the later part of the male stage the style emerges with a considerable quantity of pollen on it and it thus appears to play a significant role in directly presenting the pollen. The ring of larger hairs at the base of the pollen presenter is not annular but in the form of two deep arcs meeting at the bases of the stigmatic bands. I have noted that early in the male stage the anthers become slit, presumably in the process of pollen-robbery by bumblebees.

C. acaule SCOP. (KNUTH; PIGOTT 1968).

The corolla is up to 37 mm long and the limb as much as 15 mm; its incisions are very unequal, three short ones being 5–6 mm long and two long ones about 10 mm. The hairs of the pollen presenter, except those of the basal ring, are extremely small. In the female stage of anthesis the anther-cylinder is somewhat retracted relative to the corolla. PIGOTT records that some plants are female only. He found that florets in each successive concentric ring elongate overnight and open between 10.00 and 12.00 hours (as also in *C. arvense* and *C. canum* (L.) ALL. – JUEL 1908); after the expulsion of the pollen the female phase lasted several hours before the stigmas became discoloured (thus anthesis of each floret seems to be completed within the day).

Carduus L.

C. nutans L. (KNUTH; YEO, unpublished 1987). **STR a, s. ISS rel(dose)-irrit** and **growth. SEQU pln. APPL av. RWD n, p. SYN** melitto. **VIS** long-tongued bees, mainly *Bombus* and *Psithyrus* but also the larger species of *Lasioglossum* (*Hymenoptera-Apoidea*), butterflies (*Lepidoptera*).

The nodding heads when in flower present a surface up to 4 cm in diameter. I have indicated antero-ventral pollen application as that seems the most likely in view of the nodding posture of the capitulum. The florets are 20 mm long. In the male stage of anthesis the anther-cylinder projects about 4 mm beyond the corolla but in the female stage is the same length as the corolla. The main part of the pollen presenter is merely papillate (the length of the papillae being about twice the diameter of a pollen grain), but still quite distinct from the smooth proximal part of the style. The basal hairs of the presenter are also relatively small. Stimulation of the filaments at the male stage results in a slight movement and a modest extrusion of pollen. In my own observations, pollen was seen on the exserted pollen presenter only in a capitulum that had been left in water indoors. The male stage was observed in *C. crispus* L. to begin in the morning and end at midday (JUEL 1908).

Onopordum L.

O. acanthium L. (MÜLLER). **STR a, s. ISS rel(dose)-irrit? and growth? SEQU pln. APPL vl. RWD n, p. SYN** melitto. **VIS** mainly long-tongued bees *Bombus* spp., *Megachilidae* and *Anthophoridae* (*Hymenoptera-Apoidea*), hawkmoths (*Lepidoptera-Sphingidae*) and butterflies (*Lepidoptera*).

MÜLLER gives a fairly detailed account. The nectar rises up the corolla tube to the throat; the tubular part of the limb is 3–4 mm long and the lobes 6–8 mm. The anther-cylinder is 8–10 mm long and it projects 5–7 mm above the corolla; later the stigmatic branches, which remain fully appressed, occupy the same position. MÜLLER describes the energetic gathering of pollen by leaf-cutter bees (*Megachile*).

Notobasis CASS.

N. syriaca (L.) CASS. (JUEL 1908).

This is said to be similar to *Carduus* and *Cirsium* species in the type of presentation, duration of male stage (morning) and the weakness of the anther-cylinder relative to that of *Centaurea*.

Tyrimnus (CASS.) CASS.

T. leucographus (L.) CASS. (JUEL 1908).

The same applies as to *Notobasis*.

Serratula L.

S. coronata L. (?*S. wolffii* ANDRAE). (JUEL 1908).

The same applies as to *Notobasis* and *Tyrimnus*.

Cardueae, Group 2: connective-appendages long, united

Centaurea L.

C. jacea L. (MÜLLER, illus.). **STR a, s. ISS rel(dose)-irrit** and **growth?. SEQU pln. APPL vl. RWD n, p. SYN** melitto. **VIS** many bees (*Hymenoptera-Apoidea*), *Diptera* and *Lepidoptera*. (Fig. 51).

The connective-appendages are much longer in relation to the thecae than in the genera described so far (see introduction to *Cardueae*, and further under *C. collina*). When the floret opens the hair-ring at the base of the style-branches has reached the bases of the connective-appendages and pushed all the pollen up into the pollen magazine formed by them. The anther-cylinder exceeds the corolla by about the length of the connective-appendages. If the heads are not visited by insects the application of an artificial stimulus to the filaments causes a great quantity of pollen to be expelled, while the whole of the pollen presenter, previously concealed, emerges loaded with pollen. It seems likely that if insects were to visit the floret soon enough, less pollen would come out, as indeed shown in MÜLLER's figure; see Fig. 51A), and it might then be issued in repeated doses (see *C. cyanus* and *C. montana*). In the female stage the style and its branches far exceed the corolla. MÜLLER found that in addition to homogeneous capitula there were capitula with enlarged sterile marginal florets and central fertile florets which might be all male or all female. Between the homogeneous capitula and the two extreme heterogeneous

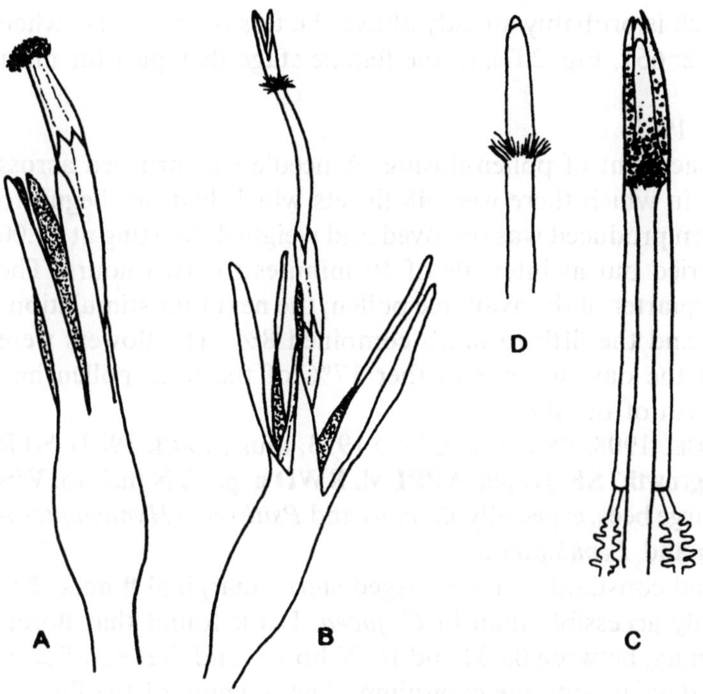

Fig. 51. *Centaurea jacea* (*Asteraceae-Cichorioideae*, tribe *Cardueae*). *A* flower in male stage; the upper part of the anther-cylinder is formed by large connective-appendages at least half as long as the thecae; some pollen has exuded from the anther-cylinder but the tip of the style has not yet emerged. *B* female stage, with style far exserted from the anther-cylinder. *C* anther-cylinder opened to show pollen and style-head; the stylar brush is level with the bases of the appendages. *D* style-head with its brush of sweeping hairs. (P. F. YEO, after MÜLLER 1883.)

types a series of intermediates was found. In describing this MÜLLER does not indicate how such capitula are distributed on the plant. The corolla of this species is purple.

C. cyanus L. (MÜLLER, illus. of style; KNUTH, illus. from MACLEOD). **STR a, s. ISS rel(dose)-irrit** and **growth? SEQU pln. APPL avl. RWD n, p. SYN** melitto. **VIS** large and small bees (*Hymenoptera-Apoidea*), various flies (*Diptera*), some noctuid moths (*Lepidoptera-Noctuidae*).

The corollas are deep blue, relatively few in number, and the capitula have a ring of very conspicuous outer sterile florets. On capitula kept indoors the response of the filaments to touch is particularly dramatic; MÜLLER says that the anther-cylinder is quickly drawn back 2–3 mm and then more slowly to an extent of 5–6 mm. When the flowers are exposed to insect visits such a strong response is not obtained, presumably because potential for such a large movement never builds up (see *C. montana*, below). The parts of the corolla in this species are relatively short; this makes it likely that short-tongued insects will visit it (as appears to be the case from the list of visitors) and that there would be a strong anterior component in the pollen loads of these visitors. As in other species of *Centaurea* the connective-appendages are more than half as long as the thecae of the anthers and are joined for most of their length. The style-branches are very short, with a pronounced hair-ring which is probably already above the tips of the thecae when the flower opens (KNUTH 2: 661, Fig. 207); in the female stage they part for most of their length.

C. montana L. (PERCIVAL 1965).

PERCIVAL gives a specific account of pollen-dosing. A needle was brushed across the florets of a capitulum in which there were 48 florets which had not begun to exude pollen, and the pollen produced was removed and weighed. Starting at 08.00 hours this action was carried out at intervals of 10 minutes for two hours. The first stimulation yielded a quarter of the available pollen; the next four stimulations yielded about 10% each, and the fifth to ninth combined 8%. The flowers were untouched for the rest of the day, when a further 17% of the total pollen had exuded as a result of growth of the style.

C. scabiosa L. (KNUTH; JUEL 1908; PROCTOR & YEO 1973, illus.; LACK 1982). **STR a, s. ISS rel(dose)-irrit** and **growth? SEQU pln. APPL vl. RWD n, p. SYN** melitto. **VIS** many medium-sized and large bees, especially *Bombus* and *Psithyrus* (*Hymenoptera-Apoidea*), various *Diptera* and *Lepidoptera.*

The capitula are purple and constantly have enlarged sterile marginal florets. The nectar is rather more easily accessible than in *C. jacea.* LACK found that florets began anthesis in the morning, between 08.00 and 10.00 hours, and that new florets opened each day for 3–5 days in any one capitulum. The opening of the flowers is quickly followed by the commencement of nectar-secretion and the availability of pollen in response to stimulation of the stamen filaments. The stigmatic branches are at most partially visible on the first day and apparently not receptive. They emerged fully, and were presumed to be receptive, on the second day. Nectar-secretion on the second day ceased any time up to 14.00 hours.

C. nigra L. (LACK 1982).

The capitula of this species are similar to those of *C. scabiosa* but over much of the British range of the species enlarged marginal florets are absent. LACK found that the stigmatic branches were fully exposed and apparently receptive on the afternoon of the first day of anthesis, and that nectar-secretion regularly ceased on the first afternoon.

C. collina L. (YEO, unpublished 1987). **STR a, s. ISS rel(dose)-irrit** and **growth. SEQU pln. APPL vl. RWD n, p. SYN** melitto. **VIS** –.

This species is yellow-flowered. In the late bud stage of the floret (when it is 16 or 17 mm long and probably due to open next day) the style with its branches reaches all the way to the tip of the anther-cylinder. Immediately before it opens, however, the corolla has lengthened to 26 or 27 mm, with the androecium protruding from its apex by 3 mm (showing that the elongation took place at the base of the corolla). Inspection of the whole capitulum in the morning suggests that this elongation takes place overnight, since intermediate lengths are not found. The thecal region of the anther-cylinder (6 mm long) is now full of pollen and the style-apex reaches only just to the bases of the thecae. The style apex is of the usual form in the tribe but its pollen presenting zone is only 2 mm long. In the next stage observed (which probably follows the preceding one the same morning) the corolla was 28 mm long and its lobes had parted. The pollen presenter was unchanged in length but proximal elongation of the style had brought it beyond the thecae to a point half way up the tube formed by the connective-appendages (the pollen magazine). There was no pollen proximal to the basal hair-ring of the pollen presenter but the region from here upwards was packed with it. The connective appendages were 4 mm long, horny, and firmly fused in the basal 2.5 mm, the free tips thus being comparable in length to the entire appendage in the thistles. Further developments are an increase in corolla-length to 30 mm and a lengthening of the style so that the pollen presenter begins to be permanently exserted; finally (perhaps on the second day of anthesis) the stigmas are receptive and the style exserted by 6 mm, a result associated with a withdrawal of the anther-cylinder by 3 mm relative to the corolla. In mature florets the enlarged part of the corolla tube (tubular part of the limb) is 6.5 mm long and the filaments are attached at its base.

The basal appendages of the thecae are bluntish, irregularly erose and made of almost crystalline cells. The filaments bear thick crystalline papilliform hairs, mostly antrorse, thinning out and becoming shorter distally. On some cut pieces of plant left in water indoors for 48 hours I probed the bases of flowers in the male stage and in one case got a prolonged and dramatic extrusion of pollen, borne along on the emerging style apex. Flowers that had already entered the female stage had a mass of pollen on the exserted pollen presenter and some more pollen resting at the tips of the connective-appendages.

In this plant I found some female flowers in which the style was of normal length, and the stigmatic branches parted for the whole of their length instead of the usual portion of less than 0.5 mm. These flowers had short corollas, 14 mm long, that lacked the swollen base of the limb, and between the corolla lobes fully developed polliniferous anther-cylinders, the pollen of which had no possibility of being presented. There were also some sterile flowers in the capitula, with 3–4-lobed corollas and an ovary, but no style and apparently no androecium.

Cnicus L.

C. benedictus L. (JUEL 1908; KNUTH, citing HILDEBRAND.)

JUEL lists this plant along with *Centaurea* and other genera that are stated to have a firmly constructed anther-cylinder and a male stage lasting all day. KNUTH states that neuter marginal florets are present but reduced in size, while the outer functional florets produce no pollen.

Cardueae, Group 3: deviant genera

Xeranthemum L.

X. annuum L. (JUEL 1908; KNUTH, citing SPRENGEL and HILDEBRAND.)

JUEL refers to this plant in the same way as he does to *Cnicus benedictus* but KNUTH states that the style-branches separate throughout their (short) length and are receptive to pollen on their inner surfaces (also that the capitula contain sterile marginal florets). This genus is classified alongside *Carlina* by DITTRICH (1977). (Judging by the diverse accounts quoted for *Carlina* by KNUTH, it needs further study and is not separately treated here; it agrees with *Xeranthemum* in that its style-branches are short. The pollen-sweeping hairs are similar to those in the rest of the tribe but as in *Arctium* [this tribe and group] they extend below the origin of the style-branches).

Arctium L.

A. minus BERNH. (MÜLLER, KNUTH). **STR a, s. ISS exposed. SEQU pln. APPL v. RWD n, p. SYN** melitto. **VIS** various bees, records of which are rather sparse but *Bombus* is probably the most important (*Hymenoptera-Apoidea*).

The capitula and flowers are rather small. The style branches are 1 mm long, and bear a stigmatic surface over their whole inner face which becomes fully exposed in the female stage. The hairs of the pollen presenter extend below the origin of the style-branches, thereby compensating for the shortness of the branches. According to JUEL (1908) *A. tomentosum* MILL. (*Lappa tomentosa* (MILL.) LAM.) is comparable in its pollen presentation with tribe *Lactuceae*, that is, pollen is not pushed out in advance of the emergence of the style branches but is entirely presented on the style; assuming the same applies to *A. minus* I have indicated the method of issue as by total exposure.

Carthamus L.

C. tinctorius L. (JUEL 1908). **STR s. ISS exposed. SEQU pln. APPL av. RWD n, p. SYN** melitto. **VIS** –.

No further information is available on this plant.

Echinops L.

The globose inflorescence in this genus is made up of single-flowered capitula, each with an involucre of many finely pointed bracts. JUEL (1908) states that the male stage lasts a whole day. The corolla has a swollen portion between the narrow basal tubular part (here extremely short) and the diverging lobes; however, whereas in thistles this part is sympetalous, in *Echinops* it is formed by the divergence and then convergence of the bases of the lobes and, as these are approximately linear, there are gaps between them where they are furthest apart.

E. sphaerocephalus L. (MÜLLER, illus.). **STR a, s. ISS growth. SEQU pln. APPL v. RWD n, p. SYN** melitto. **VIS** mainly the bees *Bombus* and *Lasioglossum* (*Hymenoptera-Apoidea*).

The inflorescence in this species is whitish. The corolla is only 5 or 6 mm long and is cleft nearly to the base, so that relatively short-tongued insects can reach the nectar. The anthers appear to have rather long connective-appendages (see *E. ritro*, below); the filaments are attached at the mouth of the tubular part of the corolla. The length of the stigmatic branches is inferred from the illustration as 1.5–2 mm, which is much shorter than the anthers. The hair-ring at their base is rather prominent; shorter hairs cover the remainder of their outer surfaces. Growth of the style after the beginning of anthesis at first pushes pollen out of the anther-cylinder; later, the style-branches themselves emerge and present the remainder of the pollen. In the female stage the style-branches separate completely; each is receptive to pollen over its whole inner surface.

E. ritro L. (YEO, unpublished 1987). **STR a, s. ISS growth. SEQU pln. APPL av. RWD n, p. SYN** melitto. **VIS** bumblebees *Bombus* (*Hymenoptera-Apoidea*).

My observations, which were made on scanty material taken when flowering was ending, largely agree with those of MÜLLER for *E. sphaerocephalus* and have in fact been used to interpret MÜLLER's drawings so as to amplify the account of that species. In *E. ritro* the flower is blue and measures about 10 mm in length. The corolla tube is 2 mm long and the swollen part formed by the corolla-lobe bases is the same length; the lobes diverge widely above the constriction and are twisted. The inner surface of the corolla lobes in the region of the swelling has a cushion-like thickening distally which terminates in a small semi-circular coronal scale. The edges of the scales are slightly erose and their decurrent extremities, visible in the slits in the swollen part of the corolla, are seen as a papillate margin there. The stamen-filaments are attached in the throat of the corolla and converge to pass through the neck at the distal end of the swollen part of the corolla; they are exserted 1–1.5 mm beyond this neck. The anthers have a length of 4.75 mm, made up approximately as follows: thecal tails (bearing forwardly directed hairs) 0.5 mm, thecae 3.25 mm, connective-appendages 1.75 mm (overlapping the thecae by 0.25 mm). The connective-appendages are in two parts, a slightly callused tip and, proximal to this, a hollowed part with a median connectival ridge; the above-mentioned overlap is possible because the thecae taper to a median point. The thecal walls are a little firmer than in thistles and knapweeds, and I found that the line of dehiscence is not along the middle of the theca but at or near its outer edge, where its wall meets the lateral abaxial extension of the connective. The style-branches are about 2 mm long, with a sinuous hair-ring at the base and shorter hairs on the outer surfaces distally. When mature they are fully separated and recurved. A flower in the early male stage was found to have the bases of the style-branches 1 mm above the bases of the anther thecae and loaded with pollen, though it had not cleared all pollen from the portions of the thecae proximal to it. On shoots that were left two days in water pollen was found to have emerged from the tips of the anthers in 5-ridged columns, the ridges corresponding to the slits between the connective-appendages. Mostly these columns concealed the emergent tips of the styles but in one case the latter was still within the anther-cylinder. Where florets had previously been in the male stage they had now reached the female stage but, insects being absent, pollen was still adhering to the outer surfaces of the style-branches. One immature flower, 8 mm in length, was opened; the bases of the stigmatic arms were positioned 3/4 of the way up the thecal part of the anther-cylinder, while the thecal walls themselves were essentially intact. It is probable that in younger buds the style reaches to the tip of the anther-cylinder, as in other *Cardueae*, and that in this flower the corolla was in the process of elongation that brings the bases of the thecae into alignment with the bases of the style-branches at the beginning of anthesis. Stroking of the filaments of open florets caused the anther-cylinder to lean towards the side touched but, in the few florets observed, no pollen was extruded as a result.

In summary, *Echinops* has structural peculiarities in the corolla and connective-appendages, but is similar to other *Cardueae* in the pollen presenting part of the style. JUEL (1908) classed it with *Centaurea* because it has a robust anther-cylinder but functionally it agrees better with the thistle group (my Group 1). However, if it turns out that movements in response to stimulation of the filaments do not result in the delivery of pollen, it could be classified functionally with, for example, the tribes *Astereae* and *Senecioneae* (subfamily *Asteroideae*) to be described below.

Tribe 6, *Vernonieae*

This tribe and the next are similar to each other and quite different from the *Cardueae*. Functionally it resembles the *Lactuceae* (q. v.) and *Arctotideae*. I have no accounts of individual species in this tribe, but have extracted information from THIELE (1988). Two species each of *Vernonia* SCHREBER and *Elephantopus* L. were included in her study, together with *Stokesia laevis* (HILL) GREENE. One species of *Elephantopus* showed daily simultaneous anthesis of the florets within the head, whereas *Stokesia laevis* and one species of *Vernonia* showed succesive anthesis through the day. In the developing flower the style nearly fills the anther-cylinder and there is no stage in which its growth in length is greatly retarded compared with that of the stamens. As the flowers open the pollen is loaded on to the sweeping hairs of the style, which cover the outsides of the style-branches and the upper part of the style. The pollen brush emerges completely. The stigmatic surface covers the whole width of the adaxial side of the style-branch but in *Stokesia* not the apex of the style-branch, which has sweeping hairs all round it.

Tribe 8, *Eupatorieae*

As in the *Vernonieae*, the entire pollen production of the flower is loaded on to the style and carried out of the anther-cylinder as the flower opens. The structural arrangements for this are quite different from those that achieve the same end in tribes *Lactuceae* and *Arctotideae*.

Eupatorium L.

E. cannabinum L. (MÜLLER, illus.; YEO, unpublished 1984) **STR s. ISS exposed. SEQU pln. APPL av. RWD n, p. SYN** psycho. **VIS** some bees (*Hymenoptera-Apoidea*) and some *Diptera* (*Syrphidae* – hoverflies, and families of the section *Calypterae*), but mainly butterflies (day-*Lepidoptera*) (Fig. 52).

The capitula of this tall perennial herb contain only four or five florets, which have long-tubed corollas, thus making the capitulum narrowly cylindrical. The small number of florets is compensated for by the very large number of capitula which are arranged in large corymbs. The flower colour is pale pink. The corollas are actinomorphic, with a dilated limb slightly shorter than the slender part of the tube. The limb is itself mainly tubular, the lobes occupying about a quarter of its length. The stigmatic area of the style-branch is a finely papillate strip along each edge, restricted longitudinally to the basal quarter. The remainder of each branch is terete and totally covered with coarse papillae. The thecae of the anthers are slightly shorter than the coarsely papillate parts of the style-branches, which lie opposite them in the advanced bud and at that stage receive all their pollen. In the male stage of anthesis the pollen-carrying parts of the style-branches have emerged fully and are widely divergent, while the stigmatic part is still within the anther-cylinder. In the female stage the style-branches emerge completely and the basal stigmatic parts diverge throughout their length. The florets present a mesh of divaricating pollen presenters and, in the female stage, short lengths of stigma placed close to the mouth of the floret. These features would seem to make the flowers well suited to pollen transfer via the thread-like proboscides of butterflies.

E. purpureum L.

JUEL (1908) describes and illustrates the pollen presenters as coherent in the male stage.

Liatris GAERTN. ex SCHREB.

L. spicata (L.) WILLD.

JUEL (1908) states that this is the same as *Eupatorium purpureum*.

Ageratum L.

A. houstonianum MILL. (*E. mexicanum* SIMS).

JUEL (1908) states that this is the same as *Eupatorium purpureum*.

Subfamily *Asteroideae*

The two main methods of pollen delivery in this subfamily are by a piston mechanism exclusively (not seen in subfamily *Cichorioideae*) and a combination

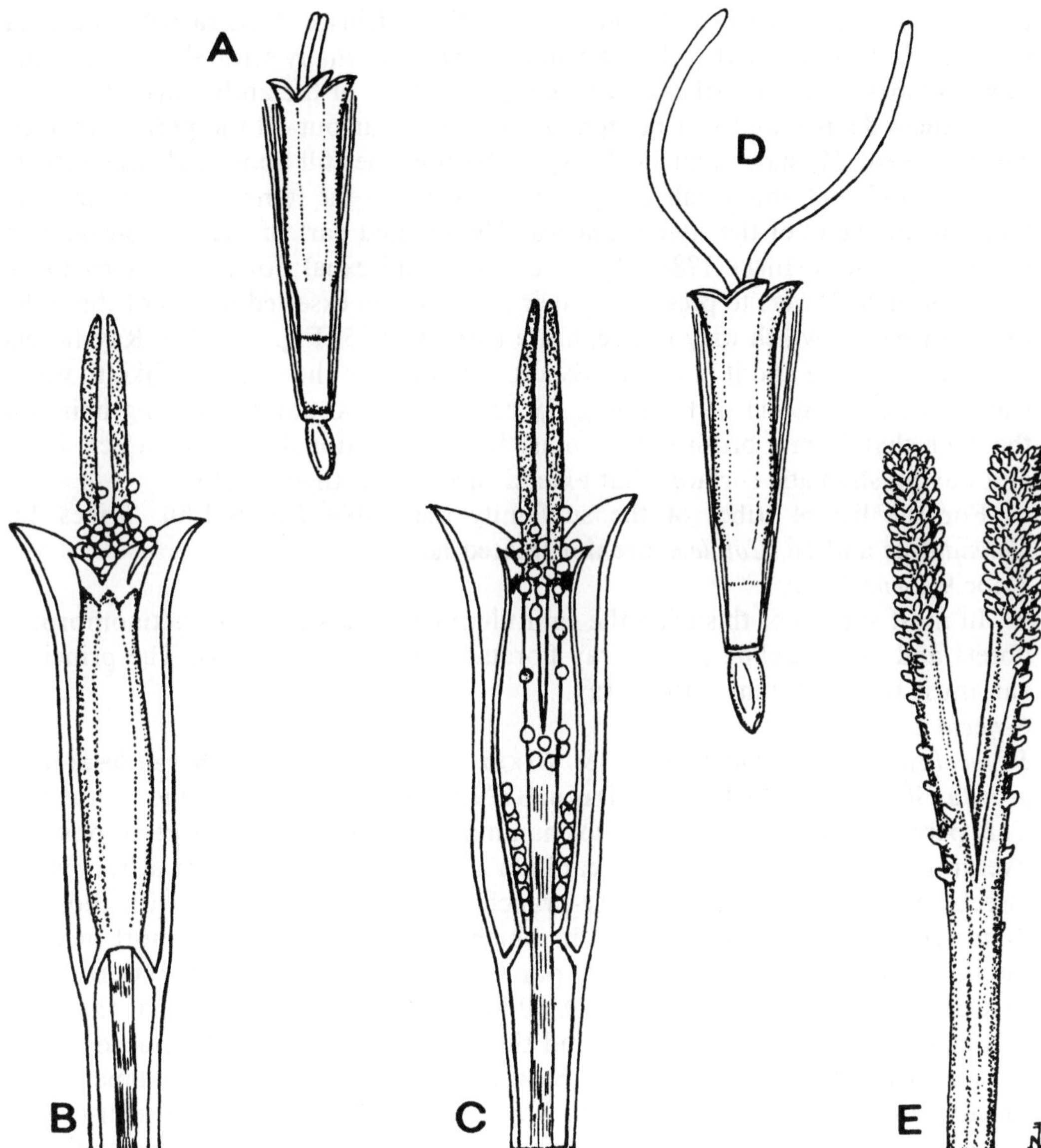

Fig. 52. *Eupatorium cannabinum* (*Asteraceae-Cichorioideae*, tribe *Eupatorieae*). *A* flower in early male stage, with corolla about 4 mm long. *B* similar to 'A' but corolla sectioned, showing anther-cylinder and presentation of some pollen. *C* the same with anther-cylinder sectioned; the presence of pollen in the anther-cylinder indicates malfunction. *D* flower in female stage with fully elongated style-branches. *E* the style in the region of the fork; pollen-presenting region densely papillate, stigmatic region below this sparsely papillate on the back and flat, epapillate, on the facing surfaces (these two regions are also differentiated in 'C'). (M. HICKEY.)

of the piston mechanism and presentation on a stylar brush (Fig. 48, upper and lower). Pollen extruded by piston activity is likely to be found clumped on the tip of the anther-cylinder until or unless it is removed by insects. Genera with delivery exclusively by piston activity have truncate style-branches with only a few short hairs at the apex; these truncate branch ends must be positioned no higher than the base of the anther-cylinder until the anthers dehisce. Genera with the dual mechanism have tapered style-branches clothed externally with short hairs; the tapered parts of the branches can be within the base of the anther-cylinder when the anthers dehisce and still be in a position to clear out all the pollen. In fact, THIELE (1988: 24) states that in this case the branches fill about half the anther-cylinder, and that about half the pollen is shed into the lumen of the latter and half into the hairs of the style-branches. The tapered part of such style-branches is not stigmatic (THIELE 1988). In genera with unisexual flowers it seems to be usual for male flowers to present all their pollen on an exserted region of the style, the branches of which may not separate (SMALL 1915: Fig. 2, XIII). Ray florets are usually female and it is said by SMALL that in these the styles are like those of tribe *Lactuceae* (SMALL 1915: Fig. 2, I), being long and narrow, having hairs on the back that form a brush (which must be non-functional), lacking apical hairs and having stigmatic surfaces that extend throughout their length.

For the list of tribes of the subfamily see Table 2-6 (p. 170). Tribes 15. *Calenduleae* and 16. *Cotuleae* are not treated here.

Tribe 9, *Senecioneae*

In most species of this tribe the capitulum comprises a disc of actinomorphic florets and an encircling ring of ray florets with ligulate corollas. The principal variant is the capitulum with no ray.

Senecio L.

S. jacobea L. (MÜLLER; HARPER & WOOD 1957). **STR a, s. ISS rel(dose)-irrit, growth. SEQU pln. APPL av. RWD n, p. SYN** entomo. **VIS** a wide range of insects.
This perennial herb is poisonous to mammals and often forms large colonies in overgrazed pastures. The yellow capitula are numerous and very attractive to insects. MÜLLER describes the style-branches by comparing them with those of *Tanacetum* (tribe 13, *Anthemideae*) in which they have only a few short hairs at the tip. Pollen presentation is of the purely piston type. HARPER & WOOD found that pollen presentation lasts from 08.00 hours to 17.00 hours and that there are two peaks, one about 10.00 hours and the other about 12.00, which account for 50% of a floret's pollen. However, SMALL (1917b) states that the filaments are irritable and contract simultaneously, which would be expected to cause discharge of pollen. To cover these two statements I have assumed that some pollen is extruded by growth and that some is released in doses in response to touch. From the account by HARPER & WOOD it seems likely that the floret becomes functionally female on the second day. (See further under *S. squalidus*.)

S. squalidus L. (PROCTOR & YEO 1973, illus.).
The illustration shows the disc floret at male and female stages and the ray floret. The truncate style-branches are clearly shown, with their few, short apical hairs. SMALL (1915) is quoted on the 'miserly' presentation of pollen following stimulation of the filaments.

Petasites MILL.

P. albus (L.) GAERTN. (MÜLLER; KNUTH: 579, 580, illus.). **STR s. ISS exposed. SEQU na. APPL av. RWD n, p? SYN** entomo. **VIS** *Diptera* and, especially, *Lepidoptera*.
The plants are dioecious. The style-branches in male florets are covered on their outer surfaces with papilliform, forwardly directed hairs and they emerge from

the anther-cylinder carrying a load of pollen. Pollen presentation is thus comparable with that in tribe *Lactuceae*. However, the lower part of the style is slenderer than the upper part which is of the same thickness as the combined style-branches.

P. hybridus (L.) GAERTN., MEYER & SCHERB. (KNUTH, quoting several authors). This species, which flowers in early spring and has dull purple capitula, visited mainly by social bees (*Apidae*), appears to be essentially the same as *P. albus*.

Tribe 10, *Tageteae*

Tagetes L.

JUEL (1908) reported only tentatively on this genus, saying that *T. signata* BARTL. may belong to the *Zinnia* type (next tribe) in which the pollen is presented on the corolla; he said that in *Tagetes* the pollen is brought out of the anther-cylinder at the beginning of anthesis by the emerging tips of the stigmas and that some of this is taken up by the hairy corolla-lobes.

Tribe 11, *Heliantheae*

Pollen presentation is commonly by a combination of extrusion and exposure on the exserted style. Exposure of all the pollen on the exserted style occurs in one species of *Bidens* and in *Gaillardia*. A more aberrant system is the use of the corolla for secondary pollen presentation. As there is evidence for this in the preceding tribe, the genera of *Heliantheae* in which it is reported are here dealt with first.

Arnica L.

A. montana L. (KNUTH: 631, partly quoting WARNSTORF). **STR c, s. ISS growth. SEQU pln. APPL av. RWD n, p. SYN** entomo. **VIS** *Diptera*, bees (*Hymenoptera-Apoidea*), *Lepidoptera*.

The orange-yellow capitula are 6–7 cm in diameter and solitary or few in number. Pollen is initially delivered on the tip of the anther-cylinder which is slightly exserted from the corolla, but according to the quotation from WARNSTORF, this falls on to the corolla lobes where it becomes lodged between marginal papillae (an occurrence not noted by KNUTH though he made his own observations on the plant). As, later, the style-branches become recurved and touch the corolla, this was seen as a means of allowing self-fertilization. JUEL (1908), who did not include *Arnica* in his study, claimed that pollen presentation on the corolla in *Asteraceae* occurred only in *Zinnia* (belonging to the present tribe) and perhaps *Tagetes* (previous tribe).

Zinnia L.

Z. haageana REGEL. (JUEL 1908; YEO, unpublished 1988). **STR c, a?, s. ISS exposed. SEQU po? APPL av. RWD n, p? SYN** entomo. **VIS** –.

Before the corolla opens the style pushes the pollen forward in the anther-cylinder, so that it emerges from five slits between the fleshy white connective-appendages in separate portions that are deposited on the grooved and strongly hairy inner surfaces of the corolla lobes, where they are exposed at the beginning of anthesis. The corolla lobes are scarcely divergent at this stage, and a proboscis thrust between them would inevitably pick up pollen from them. The hairs of the corolla lobes are coarse and tapered to a point. They are coloured yellow like the corolla and are directed slightly backwards. The anther-cylinder remains hidden within the corolla. The style-branches are truncate and their tips are densely covered by blunt spreading hairs; these decrease rapidly in length downwards and disappear one third of the way down the branch. I found buds in which all the pollen was free in the anther-cylinder and the tips of the style-branches were level with tip of the anther-cylinder. However, I also found the style-branches in this position in an open flower with pollen loaded onto the corolla lobes. It therefore seems as though there must be an intervening stage during which the anthers grow ahead of the style which then catches up, as in thistles. I am not entirely satisfied that this is the case, because the design of the stylar brush suggests that it is concerned with brushing rather than pushing pollen out. The timing of stigma receptivity was not reported by JUEL; I found that the corolla shrinks and goes brown early in the female stage of the floret. A peculiarity of the flower is that the stamens are attached at the base of the tube and the tube is not ventricose; this could be interpreted as indicating suppression of the part referred to as the tube and that the existing sympetalous region corresponds to the tubular part of the limb of other genera. I found some pollen in the lower part of the anther-cylinder after the distal part was cleared, and when the flower was in the female stage some on the style where it emerged from the anther-cylinder.

Z. elegans JACQ. (JUEL 1908).

JUEL found that some of the pollen is deposited on the corolla lobes but that some is found on the tip of the anther-cylinder.

Cosmos CAV.

C. atrosanguineus (HOOK.) ORT. (JUEL 1908, illus.). **STR a, s. ISS growth. SEQU pln. APPL av. RWD n, p? SYN? VIS –.**

The capitula are about 6–8 cm in diameter and the florets are blackish red. The anther-cylinder is rather well exserted from the corolla, and the pollen exudes from the slits between the connective-appendages so as to form a five-pointed star when viewed from above. In the female stage the strongly spreading style-branches are held at a lower level than that at which the pollen was previously presented. The male stage lasts all day.

Helianthus L.

H. annuus L. (JUEL 1908; SIMPSON & NEFF 1987, illus.). **STR a, s. ISS growth. SEQU pln. APPL av, pv. RWD n, p. SYN** entom. **VIS** bees (*Hymenoptera-Apoidea*), e.g. *Bombus* (*Apidae*), *Svastra* (*Anthophoridae*), *Megachiloides* (*Megachilidae*).

JUEL reported that initially part of the pollen is exposed on the tip of the anther-cylinder but the greater part is brought out afterwards on the emerging stigmatic branches. SIMPSON & NEFF made a detailed study of wild plants in Texas. The number of florets per capitulum was 200–250 and they opened over a period of 5 or 6 days; the number open in the capitulum was largest on the third day after flowering began. The corolla-limb is narrow, with very small lobes, and the anther-cylinder is strongly exserted. Newly open florets present about 60% of their pollen in the course of the morning as a result of stylar growth, culminating in the emergence of the tip of the style, with its branches still appressed. There is then a period of 5–6 hours during which little or no pollen is presented. In the evening there is a second period of pollen presentation caused by retraction of the anthers; at the end of this the style branches have started to separate and become receptive. The receptive (female) phase continues next day. A minority of plants have a different timetable, some proceeding rapidly to the female phase on the first afternoon, others delaying the second period of pollen presentation and the subsequent female phase until the second morning. Nectar is produced throughout the first day. The bee *Svastra* has a brush of specialized hairs beneath the tip of the abdomen with which it taps the capitulum to pick up pollen; *Megachiloides* also specializes in pollen of *Asteraceae*, whereas in this study the *Bombus* that visited the heads took nectar but rarely the pollen.

Rudbeckia L.

R. amplexicaulis VAHL. (JUEL 1908)

Pollen presentation is as in *Helianthus* (above).

Dahlia CAV.

D. variabilis DESF. (JUEL 1908)

Pollen presentation is as in *Helianthus* (above).

Bidens L.

B. tripartita L. (KNUTH, illus.). **STR a, s. ISS growth. SEQU pln. APPL av. RWD n, p? SYN** entomo. **VIS** hoverflies (*Diptera-Syrphidae*).

The tips of the style-branches are tapered to a point and are covered with hairs externally. Pollen presentation is therefore as in *Helianthus* (above, this tribe). The filaments are not irritable (SMALL 1917b) but the anther-cylinder becomes completely retracted into the corolla. *B. cernua* L. is similar (KNUTH).

B. cosmoides (A. GRAY) SHERFF. (GANDERS & NAGATA 1983, illus.). **STR s. ISS exposed. SEQU pln. APPL a. RWD n. SYN** ornitho. **VIS** –.

The capitulum of this Hawaiian species is nodding and borne on a stiff peduncle. The florets are reddish and the pollen, which is very sticky, is presented on the tips of the styles. These are exserted from the corolla and anthers by 20–25 mm, and give the head the character of a brush-blossom. Further features of the syndrome of ornithophily are the weakness of the scent and the large quantity of dilute nectar. The stigmatic branches diverge later. If the flowers are unvisited they curve sufficiently to come into contact with their own pollen, to which they are fertile. (This behaviour was commonly reported in European *Asteraceae* by observers in the nineteenth century, probably not always correctly; GANDERS & NAGATA state that some Hawaiian *Bidens* cannot do this). All Hawaiian *Bidens* are self-compatible.

Gaillardia FOUG.

G. aristata PURSH [possibly *G.* × *grandiflora* V. HOUTTE]. (JUEL 1908). **STR s. ISS exposed. SEQU pln. APPL av. RWD n, p?. SYN** entomo. **VIS** –.

The distal parts of the style-branches are roughly hairy and form a cone on which the entire pollen production is carried clear of the anther-cylinder. The stigmatic surface is confined to the lower parts of the branches. JUEL states that the filaments are not irritable but SMALL (1917b) found them irritable in *G. aristata* var. *grandiflora* [*?G.* × *grandiflora*].

Tribe 12, *Inuleae*

Pulicaria GAERTN.

P. dysenterica (L.) BERNH. (MÜLLER; YEO, unpublished 1984). **STR a, s. ISS growth. SEQU pln. APPL av. RWD n, p. SYN** entomo. **VIS** various *Diptera*, *Hymenoptera* (especially solitary bees, *Apoidea*), *Lepidoptera*.

According to MÜLLER the stigmatic branches are covered on their upper third with hairs directed obliquely upwards. The anther-cylinder has triangular connective-appendages bordered by thick unicellular hairs and these are said to hold the pollen that has been swept out of the anther-cylinder. Fully developed sweeping hairs are found on the styles of the female ray florets. The whole of the inner surface of the style-branches bears stigmatic papillae. My own observations do not agree with this. The inner surface of the style-branches has a central epapillose zone at the base, obliterated only distally. I could find no hairs at the tips of the anthers, but the tips of the corolla lobes have densely packed thick hairs on the adaxial surface. I found that pollen presentation appears to be purely by extrusion (piston method). This is in agreement with what JUEL (1908) says about the *Inuleae* in general.

Tribe 13, *Anthemideae*

It is in this tribe that a pure piston mechanism is best developed, the style-branches being truncate. Genera of this type are dealt with first; certain species of *Colletes*, a short-tongued bee (family *Colletidae*), seem to forage exclusively on them. Exposed presentation on the style occurs in another group of genera, dealt with second.

Achillea L.

A. millefolium L. (MÜLLER, illus.; KNUTH, illus.). **STR a, s. ISS growth. SEQU pln. APPL av. RWD n, p. SYN** entomo. **VIS** very varied, including *Coleoptera*, *Lepidoptera*, *Diptera* and *Hymenoptera*, the latter including not only short-tongued bees but some solitary wasps. (Fig. 53).

Achillea is characterized by having numerous few-flowered capitula in a corymbose arrangement. In this species they are white. The truncate style-branches have papillae at their tips and they push pollen out of the anther-cylinder, on the tip

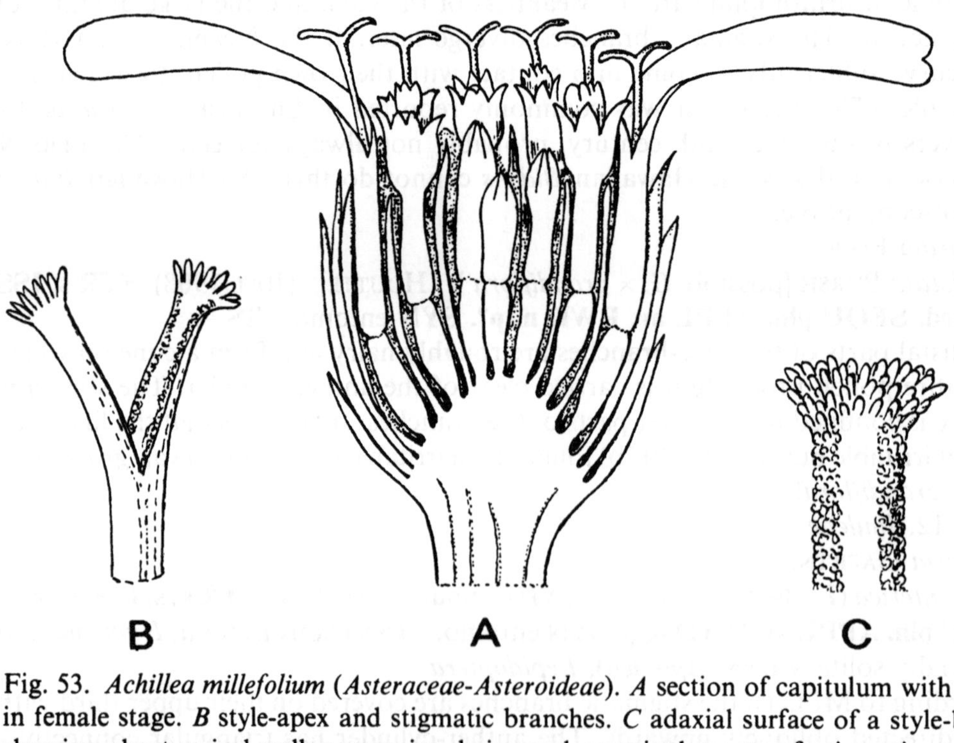

Fig. 53. *Achillea millefolium* (*Asteraceae-Asteroideae*). *A* section of capitulum with florets in female stage. *B* style-apex and stigmatic branches. *C* adaxial surface of a style-branch showing short apical pollen-sweeping hairs and marginal tracts of stigmatic papillae. (M. HICKEY, after WARMING 1895.)

of which it accumulates. At this stage the tips of the anthers project beyond the corolla. By the time the female stage of the floret is reached the anthers have been withdrawn and the divergent stigmatic branches are at the level at which previously the pollen was presented. SMALL (1917b) states that the filaments respond to touch by causing the anthers to lean towards the side touched. It is not clear whether this movement promotes the extrusion of pollen.

Leucanthemum MILL. (*Chrysanthemum* L., p. p.)

L. vulgare LAM. (*Chrysanthemum leucanthemum* L.). (MÜLLER, illus.; KNUTH). **STR a, s. ISS growth. SEQU pln. APPL av. RWD n, p.** SYN entomo. **VIS** very varied, much as in *Achillea millefolium* (above).

The capitula are large, with a densely flowered yellow disc and white rays. Pollen presentation proceeds similarly to that in *Achillea millefolium*; the irritability of the stamens is of the same type but weak (SMALL 1917b).

Tanacetum L.

T. vulgare L. (MÜLLER; KNUTH). **STR a, s. ISS rel(dose)-irrit. SEQU pln. APPL av. RWD n, p. SYN** entomo. **VIS** a wide range of *Hymenoptera*, chiefly bees (*Apoidea*) and wasps (*Sphecoidea* and *Vespoidea*), a smaller range of *Diptera* and *Lepidoptera* and a few *Coleoptera*.

The style-branches again appear to act as a piston and not as presenters. SMALL (1917b) found distinct irritability of the filaments causing a simultaneous response in all, so that the anther-cylinder was retracted. It is to be presumed that this will cause expulsion of pollen.

Antennaria GAERTN.

A. dioica (L.) GAERTN. (KNUTH). **STR s. ISS rel(dose)-irrit. SEQU pln. APPL av. RWD n, p. SYN** entomo. **VIS** a few *Lepidoptera*, a few *Diptera* (*Syrphidae* and *Calypterae*), one wasp (*Hymenoptera-Sphecoidea*).

This species is dioecious, with small rayless capitula. The style-branches of male flowers are densely covered with hairs on the outside so are probably not acting purely on the piston principle; when the filaments are touched the anther-cylinder is retracted.

Leontopodium (PERS.) R. BR.

L. alpinum CASS. (KNUTH, illus.). **STR s. ISS exposed. SEQU pln. APPL av. RWD n, p. SYN** entomo. **VIS** a few *Diptera*

The capitula are small and rayless, and a group of several is gathered with several stem-leaves to form a pseudanthium. The central florets are male and the marginal female. In the former the style-branches are strongly hairy and carry the pollen out of the anther-cylinder; having no stigmatic function they do not separate. In the female florets the style-branches are much slenderer.

Tribe 14, *Ursinieae*

Genera in this tribe were formerly placed in *Anthemideae*. The flower is distinguished by the fact that the style-branches are arranged on the radii of the capitulum, instead of tangentially, as in the rest of the family.

Ursinia GAERTN.

U. anthemoides (L.) POIR. subsp. *versicolor* (DC.) PRASSLER. (YEO, unpublished 1989). **STR a, s. ISS growth. SEQU pln. APPL av. RWD n, p. SYN** melitto. **VIS** –.

The capitulum has sterile ray florets and actinomorphic bisexual disc florets. The triangular connective-appendages of the anthers are longitudinally reduplicate and connivent, so that pairs of neighbouring edges form five ridges. At the beginning of anthesis the corolla lobes part just sufficiently to allow the tip of the anther-cylinder to emerge. Then the pollen starts to appear in the five slits, eventually forming a mass that conceals the appendages. Next the corolla-lobes spread and the style-branches emerge and separate until their tips are approximately horizontal, carrying great masses of pollen away, though a little gets onto the stigmas. The tips of the style-branches are truncate, as in *Anthemideae*, but the apex bears spherical epidermal cells in the centre and longer cells round the edge, forming hairs that spread away at right angles to the axis of the branch. Each style-branch bears two longitudinal stigmatic cushions, taking up all its width at the base but being somewhat separated distally. These developments occur in the morning and two or three rows of florets open each day. Most of the day's pollen production seems to be cleared away by the following morning. When I probed a number of young florets on heads that had been in water for four days, there was an abrupt expulsion of pollen through the cracks between the anther-appendages in one floret.

Tribe 17, *Astereae*

Aster L.

A. linosyris (L.) BERNH. (*Chrysocoma linosyris L.*). (MÜLLER, illus.; YEO, unpublished 1984). **STR a, s. ISS growth. SEQU pln. APPL av. RWD n, p. SYN** entomo. **VIS** bees (*Hymenoptera-Apoidea*), *Lepidoptera*, *Diptera*.

This species is unusual in lacking ligulate ray florets. The capitula are yellow. In the female stage the style-branches diverge from the base but higher up they are angled so that they converge again and touch at the tips. Pollen-sweeping hairs are borne on the outer surfaces above the point of inflexion; I found these hairs to be graded in size, decreasing from base to apex of the hairy zone, and overall smaller than is suggested by MÜLLER's figure. It seems probable that part of the pollen is presented on the apex of the anther-cylinder, though I could hardly find any here, and the remainder on the style-branches after these have emerged from the cylinder. The stigmatic zone is linear on either side of each style-branch and terminates where the pollen-sweeping hairs begin. This species is self-incompatible (C. D. PIGOTT, pers. comm.). (See also *A. simplex*).

A. simplex WILLD. (*A. lanceolatus auctt.*). (YEO, unpublished 1984). **STR a, s. ISS growth. SEQU pln. APPL av. RWD n, p. SYN** entomo. **VIS** probably a wide range of insects but especially *Diptera-Syrphidae*.

Here, as in most species of *Aster*, the capitula are radiate; the disc is yellow, turning to wine-coloured, and the rays usually white. Structurally the pollen presentation mechanism is the same as in *A. linosyris* but I found it easier to observe. At the beginning of the male stage of the floret the connective-appendages are connivent but soon they are forced into an erect position by pollen thrust up from below. At this stage the style is acting as a piston and its apex is well up inside the anther-cylinder and embedded in a mass of pollen. After emergence the style-branches had the configuration described for *A. linosyris* and, as in that species, I think they function partly as direct presenters of pollen (this is confirmed for the tribe by THIELE 1988: 24, 25; see also *Bellis*, this tribe, below). I found very little pollen on them (or on the anther-cylinder) but I think this was due to insect activity. The configuration of the style-branches after emergence from the cylinder suggests that if insects were scarce a floret might be presenting and receiving pollen simultaneously. It also suggests that ultimate self-pollination by recurvature of the branches is ruled out. (See also *Bellis*, this tribe, below.)

Bellis L.

B. perennis L. (MÜLLER). **STR a, s. ISS growth. SEQU pln. APPL av. RWD n, p. SYN** entomo. **VIS** many *Diptera*, some bees (*Hymenoptera-Apoidea*), some *Lepidoptera*, some *Coleoptera*.

Here there is a definite statement of what is suspected in the two species of *Aster* already described, thus the hairs of the stigmatic brush 'both sweep the pollen out of the anther-cylinder as the style elongates, and afterwards hold it until it is removed by insects'. In the purely female florets the hair-tufts of the style branches are absent and the stigmatic zones extend to the tip of the branch. SMALL (1917b) found that in the middle of the day there was strong irritability of the filaments, causing the anther-cylinder to lean towards the side touched.

Solidago L.

S. virgaurea L. This is similar to *Aster* and *Bellis*; MÜLLER states that in female flowers the hair-tufts of the style-branches are vestigial rather than absent but here too the stigmatic strips extend to the apex.

Discussion of Asteraceae

Possible functions of 2PP

The functions that might be performed by secondary pollen presentation in *Asteraceae* vary according to the details of the system. The types of pollen presentation are summarized in slightly different ways in Tables 2-7 and 2-8.

When pollen is presented on the style-branches its presentation is nearly always in a position corresponding to that of receptive stigmas. In the compositaceous capitulum, however, with its many closely packed flowers, only height above the surface of the disc could be of significance and usually it will be difficult to attribute importance to position even in this one dimension. In cases where some of the pollen is extruded from the anther-cylinder it is often presented at a lower level than that at which the stigmas are exposed, though sometimes this difference may be reduced by retraction of the anthers rather than growth of the style between the male and female stages.

Protection of pollen could be a benefit of the system if it is issued only during pollinator-visits. This requires (a) presentation on the tip of the anther-cylinder (at least in part) and (b) that the filaments have irritability of the type that retracts the anther-cylinder symmetrically. This is found particularly in *Cardueae* (subfamily *Cichorioideae*) but is also scattered in subfamily *Asteroideae*. This arrangement also has the effect of delivering the pollen in separate doses. Gradual expulsion of the pollen by growth of the style will also ensure that not all the pollen of a floret is carried away by one visitor, provided the interval between visits is not too long. The value of this is questionable (see section on 'Aspects of the Composite Capitulum and Floret'). The possible importance of protection of pollen from theft should be viewed in the light of the fact that, as observed in some other families, large amounts of pollen fail to get presented. Thus, HEYN & SNIR (1986) found in *Senecio vernalis* WALDST. & KIT. and *Calendula arvensis* L. that

Table 2-7. Occurrence of the three basic types of pollen presentation in all genera of Asteraceae here reported on

	Stylar Pollen Presenter	Piston	Piston first, then Stylar Presenter
Cichorioideae			
1. Lactuceae			
Tolpis	+		
Hieracium	+		
Leontodon	+		
2. Mutisieae			
Mutisia		+	
Gerbera		+	
Perezia		+	
4. Arctotideae			
Arctotheca	+		
Arctotis	+		
Gazania	+		
5. Cardueae			
Cirsium			+
Carduus			+
Onopordum			+
Notobasis			+
Tyrimnus			+
Serratula			+
Centaurea			+
Cnicus			+
Xeranthemum			+
Arctium	+		
Echinops			+
6. Vernonieae			
Vernonia	+		
Elephantopus	+		
Stokesia	+		
8. Eupatorieae			
Eupatorium	+		
Liatris	+		
Ageratum	+		
Asteroideae			
9. Senecioneae			
Senecio		+	
Petasites	+		
10. Tageteae			
Tagetes		+?	
11. Heliantheae			
Arnica			+
Zinnia			+
Cosmos		+?	
Helianthus			+
Rudbeckia			+

Table 2-7. (*Continued*)

	Stylar Pollen Presenter	Piston	Piston first, then Stylar Presenter
Dahlia			+
Bidens tripartita			+
Bidens cosmoides	+		
Gaillardia	+		
12. Inuleae			
Pulicaria		+	
13. Anthemideae			
Achillea		+	
Leucanthemum		+	
Tanacetum		+	
Antennaria			+?
Leontopodium	+		
14. Ursinieae			
Ursinia			+
17. Astereae			
Aster			+
Bellis			+
Solidago			+

18.5% of the capitulum's total pollen production fell into the corolla, without any possible use. This was attributed to adaptation of the architecture of the floret and, especially, of the structure of the style and anthers, to proterogyny [of the capitulum] and pollen presentation. Does this mean that selection for pollen success is not strong, or does it indicate that despite selection the plant has not found an efficient mechanism for avoiding this kind of wastage? In the former case protection against theft would not be critical; in the latter it might be all the more important. The fact that unsuitable weather conditions may sometimes be the cause of the loss, as suggested for some of my observations and for those of THIELE (1988), seems to show that the system may readily fail outside a narrow climatic range, but HEYN & SNIR's work was carried out in the plants' natural habitat.

In the one described case of explosive deposition (*Perezia*, tribe *Mutisieae*) the factors already discussed are likely to operate in the same way as in other groups; the forceful delivery of the pollen may provide better adhesion to the pollinator.

Plants that expose all their pollen at the beginning of anthesis, such as members of tribe *Lactuceae* and the genus *Eupatorium* (*Eupatorieae*), neither protect their pollen nor issue it in doses; in *Eupatorium* it is not even presented at the same level as that at which the stigmas are exposed, although this is a state of affairs which is obviously acceptable in flowers adapted to pollination by the thread-like proboscides of *Lepidoptera*.

Since identity of level in both presentation and reception looks rather unimportant in the family, it becomes difficult to see any function at all for secondary pollen presentation in the *Lactuceae*. This raises the possibility that the prime conditioner of secondary pollen presentation in the *Asteraceae* is that it makes for compactness (see also Chapter 3, 2.).

A further problem is offered by *Zinnia haageana*, since we do not know why it should transfer its pollen to the corolla lobes in the bud stage. This plant might be described as having tertiary pollen presentation!

Structure and function

The genera covered by my descriptive accounts and listed in Table 2-6 are those that I have examined myself and those that are reasonably adequately described in papers on their floral biology or in the reference works of MÜLLER and KNUTH. The styles and sometimes the stamens of a great many species are illustrated in the tribal surveys in HEYWOOD & al. (1977). These taxa have not been incorporated here, though the type of pollen presentation can in many cases be inferred from the form of the styles.

Structures involved in the different mechanisms have been summarized in the the descriptions of the family, subfamilies and (where necessary) tribes and in Table 2-8. The mechanism affects principally the shape and indumentum of the the stigmatic branches and (sometimes) upper part of style (THIELE 1988: 106–121), and the position of the style at the beginning of anthesis (THIELE 1988: 130, 131). The details of stylar structure are characteristic for most of the tribe *Cardueae*

Table 2-8. Occurrence of all pollen presentation types in Asteraceae

	All exposed on style or style-branches	Part on anther-cylinder (piston) part on style-branches	All on anther-cylinder (piston)	Deposited on corolla lobes	Part on anther-cylinder, part on corolla lobes
Cichorioideae					
1. Lactuceae	+				
2. Mutisieae	+		*Perezia*?		
4. Arctotideae	+				
5. Cardueae	*Arctium*	+			
8. Eupatorieae	+				
Asteroideae					
9. Senecioneae	♂ *Petasites*	+			
10. Tageteae				*Tagetes*?	
11. Heliantheae	*Bidens cosmoides*, *Gaillardia*	+	*Cosmos*	*Zinnia*	*Arnica*?
12. Inuleae			*Pulicaria*		
13. Anthemideae	♂ *Leontopodium*	♂ *Antennaria*?	+		
17. Astereae		+			

and yet the most usual method of presentation is the same as that in other tribes (e.g. *Heliantheae*) which do not have the same structure.

The proximal appendages of the thecae of the anthers are described and discussed by SMALL (1915). These are usually triangular laminae, sometimes tapering into an arista which is simple or plumose. Aristate appendages may have a reduced laminal component. Laminal appendages may be connate with those of neighbouring anthers. SMALL thought that their function was to extend the anther-cylinder downwards so as to embrace the style apex and so permit the anthers to dehisce shortly before the style apex surpassed the bases of the thecae proper. Without them SMALL thought that pollen would be wasted either by dropping out of the cylinder before the latter was plugged by the style or by being bypassed because the style would need to enter the anther-cylinder a short way in order to plug it. The shapes actually depicted for the thecal appendages leave me unconvinced and I think further study is needed. THIELE (1988: 72) suggests that they might be functionless.

SMALL (1915) considered that the entire organization of the capitulum of the *Asteraceae* was concerned with economy of materials. His view of the function of the thecal appendages was part of this thesis. SMALL's (1915) paper was entitled 'The pollen presentation mechanism in the Compositae' but it is disappointing in this respect because he catalogues (with illustrations) the different stamen types and style types separately and never considers how any particular androecium interacts with any particular style. He diverts his attention to evolutionary speculations, the basis of which is difficult to make out. Later, SMALL (1917a) provided better drawings of the styles, and indicated which style types were combined with which stamen types, but since the drawings are on different pages it is very laborious to collate them, and still not very instructive. He claimed to have shown (l.c.: 214) that complexity of the styles and that of the stamens was more or less complementary, and likewise (l.c.: 213) their efficiency. THIELE (1988: Chapter 5) did make the necessary comparisons, taking into account the details of the stigmatic surface and the way in which it was exposed, the details of the sweeping hairs and the method of pollen presentation ('pollen sweeping type'). This showed that many different combinations occur, but some were absent, and there were a few cases where one factor was always correlated with another. Thus with piston-type pollen presentation, sweeping hairs were always confined to the tips of the style-branches. With exposed pollen presentation the sweeping hairs usually cover the whole of the back of the style-branches and sometimes also part of the stylar shaft, and with the intermediate (combined) type the sweeping hairs usually cover only part of the backs of the style-branches. The causes of many other points of variation are still not understood.

In considering the functioning of the capitulum as a whole in the process of pollination the possibility that constraints are imposed by the requirements of seed-setting and seed-dispersal must be kept in view.

Aspects of the composite capitulum and floret

Pollen grains in *Asteraceae* are all, as far as is known (BREWBAKER 1967), trinucleate. The stigmas are dry and papillate (with unicellular papillae) or sometimes epapillate (HESLOP-HARRISON & SHIVANNA 1977; THIELE 1988). Self-incompatibility, when present, is of the sporophytic type (determined by the genotype of the pollen parent, not of the grain itself). The *Asteraceae* therefore conform to a general rule that sporophytic self-incompatibility is correlated with the dry papillate stigma type and the trinucleate condition of the pollen grain (HESLOP-HARRISON & SHIVANNA 1977).

It was suggested by BURTT (1961) that, as a consequence of the separation of ovules, each with its own style, the capitulum of the *Asteraceae* functions as a 'system for the exploration of genetic recombinations' [the antithesis of which is "sweepstakes pollination" where a large mass of pollen from a single source pollinates a large proportion of the ovules (CRUDEN 1977)]. The presence of many uniovulate gynaecial units will itself promote such exploration even in a capitulum that lasts only one day, as is apparently the case in *Cichorium*. The numerous androecial units in the capitulum presumably help to cause dispersal of pollen in a corresponding way. When the flowering period of a capitulum extends over more than one day, with florets opening in succession, these effects will be enhanced. The efficiency of the system in this respect would be impaired, on the other hand, by geitonogamy; factors governing this are the presence or absence of self-incompatibility systems and the synchronization within the population of the temporal pattern of pollen and stigma presentation in a single floret. In the present review we have noted two versions of the latter (see *Lactuceae* and *Cardueae*): in one the male stage lasts one day and is followed by a female stage lasting one or more days: in the other the male stage occupies the early part of the day and the female stage follows later the same day. The first pattern will impose no restriction on geitonogamy, so that outbreeding will depend heavily on the presence of an incompatibility system. [It seems probable that most tribes include both self-incompatible and self-compatible species, as in *Anthemideae* (HEYWOOD & HUMPHRIES 1977; THIELE 1988: 139) and the tribes for which reports are cited by BURTT (1961)]. The second pattern will severely restrict geitonogamy in populations of more than a few individuals. On the other hand, it requires that individual pollinators must remain active and retain viable pollen over a period that overlaps both phases of anthesis. [Trinucleate pollen has no storage viability (BREWBAKER 1967) so this period will inevitably be short.] Some interference with geitonogamy will also be offered if the peripheral florets of a capitulum (those that open first) are female only, the more so if the plant bears few capitula. Selfing without the intervention of insects must often be possible in self-compatible species, as THIELE (1988) found that compatible pollen can germinate from a moment shortly before anthesis.

In view of the way 'pollen-dosing' is built into the capitulum, it hardly seems necessary that it should also be built into the behaviour of the individual floret. Thus when pollen is issued in doses, it could be that this is more important for economy of pollen-provision than for any possible effect on the breeding system. The notion that the *Asteraceae* are economical with pollen is accepted by LLOYD (1979) and he suggests that this economy is a factor permitting a proportion of the florets to be female only.

Evolutionary and systematic considerations

It can hardly be supposed that we have yet much insight into the origin and subsequent evolution of the *Asteraceae*, but a defensible outline has been presented by JEFFREY (1977), based on the form of the corolla. The centripetal maturation sequence of the florets in the capitulum indicates an indeterminate type of inflorescence unit (spike or raceme). In such units, when not condensed into a capitulum, the horizontally disposed, bilabiate zygomorphic corolla is a biologically effective type and is found in other modern sympetalous families. Bilabiate corollas, usually disguised by the narrowness of their lobes, are in fact widespread in the *Asteraceae*. Since a realistically bilabiate form is unsuited to inclusion in a dense capitulum it is difficult to see why they should be there except as an inheritance from an ancestral group. They are commonest, most variable and most prominently developed in tribe *Mutisieae*. This is a tribe containing many small genera that show distributions suggesting a relictual status (CABRERA 1977 – who, it must be admitted, simultaneously proposed a recent origin for the tribe). Furthermore, as it contains a wide range of different corolla types it is easy to derive the corollas of neighbouring tribes in the classification from those of *Mutisieae*, for example those of the *Lactuceae* (uniform) and *Cardueae* (variable, though less so than in *Mutisieae*). The actinomorphic disc floret then has to be regarded as derived from the zygomorphic by neoteny. The extreme zygomorphy of the florets of *Lactuceae* can be seen as a possible alternative solution to the problem of collecting ancestrally zygomorphic flowers into a capitulum.

The pollen presentation mechanism in the *Asteraceae* is probably highly flexible in evolution, as indicated by the distinctive style types that occur in male flowers, which are often suggestive of those found in other tribes. It is thus difficult to suggest evolutionary trends. The *Mutisieae* again are here probably more diverse than most other tribes (see illustrations in CABRERA 1977); the fact that *Mutisia* has strong similarities with an important group of the *Cardueae* agrees with the neighbouring position of these two tribes in the classification. However, BREMER (1987), while confirming the diversity of the *Mutisieae*, found that the subtribe *Barnadesiinae* (not treated in this review) was apparently the most primitive group, coming out as the sister group to the rest of the family.

The tribes of the *Cichorioideae* in general tend to display distinct and uniform pollen presentation structures, whereas those of the *Asteroideae* show a smaller structural range overall but enough plasticity in detail to allow them readily to switch from one presentation class to another. The subfamily *Cichorioideae* may thus be regarded as representing an 'experimental' phase of evolution and thus perhaps a more ancient one. Since all *Asteraceae* have connate anthers and secondary pollen presentation, and since all the different versions are easily derivable from each other, there is no reason to suppose that these characters arose more than once in the family.

If we follow JEFFREY's (1977) thinking with regard to the ancestry of the *Asteraceae* we have to look for a group with bilabiate flowers in an indeterminate inflorescence as ancestral. The present-day group meeting this requirement, and having fewest objections to relationship when its characters are considered overall (DAHLGREN 1983, BREMER 1987), is *Lobeliaceae* (*Campanulaceae-Lobelioideae*). It also shares with *Asteraceae* the connate anthers and the use of the style to present the pollen. However, it approaches different groups of *Asteraceae* in different characters. Thus, the style is employed as a piston, as it is at least partly in *Cardueae* and also probably in some *Mutisieae* but not in *Lactuceae*. On the other hand, in their latex vessels the *Lobeliaceae* approach the *Lactuceae*, since in both groups these are ducts, whereas in *Cardueae* they are cells or cell-clusters (cf. WAGENITZ 1976). In any case the *Lobeliaceae* are separated from *Asteraceae* in having the odd petal on the side of the flower next to the parent axis instead of on the opposite side.

The *Campanulaceae* in the narrow sense are morphologically the second nearest family to the *Asteraceae*, and they show a tendency to develop capitula, which is not seen in the *Lobeliaceae*. They also provide an interesting parallel with the *Lactuceae*, since in both groups the pollen is consistently presented fully exposed on the style. This parallel, however, is combined with a contrast in the symmetry of the corolla, *Campanulaceae* being fully actinomorphic, whereas *Lactuceae* represent the extreme of zygomorphy in the *Asteraceae*. Since in addition *Campanulaceae* do not usually have connate anthers it would seem that the parallel is no more than a coincidence in two indirectly related groups which have similar patterns of evolution. Thus the resemblance between *Lactuceae* and *Campanulaceae* in pollen presentation does not provide an opening for support of suggestions that the *Lactuceae* had an origin separately from the rest of the *Asteraceae*. Detailed arguments against this are given by WAGENITZ (1976).

The small family *Calyceraceae*, which has flowers in capitula and is usually considered to be closely related to *Asteraceae*, does not have secondary pollen presentation (HANSEN 1992). SKVARLA & al. (1977) found resemblances to the pollen of *Asteraceae* in that of several other families that have been considered as possibly related to them, but no resemblances were found in the *Campanulaceae* (one genus) and *Lobeliaceae* (two genera), which are themselves dissimilar (DUNBAR 1975a, b). In pollen grain nuclei, *Lobeliaceae* resemble *Asteraceae* more than *Campanulaceae* do, since trinucleate pollen has been found in three genera, whereas in *Campanulaceae* binucleate pollen prevails. Both families have exceptions, however (BREWBAKER 1967). The family that has dry stigmas like *Asteraceae* is *Campanulaceae* (HESLOP-HARRISON & SHIVANNA 1977).

LEINS & ERBAR (1989, 1990) draw attention to the fact that asynchrony of growth of style and androecium is a common feature of *Asteraceae*, *Lobeliaceae* and *Goodeniaceae*. This is of functional importance in getting the stylar pollen presenter or piston into position for pollen-loading. As other details of the mechanisms differ so much among these families this character can hardly be seen as evidence for the phylogenetic proximity of these families.

Among many genera whose placement in the system is problematical is one, *Echinops*, that is described in this review. The characters that make it difficult to include this genus in the *Cardueae* are listed by WAGENITZ (1976) and BREMER (1987), while further morphological information about it is supplied by DITTRICH (1977). DITTRICH in fact treats *Cardueae* in our sense as a subfamily within which *Echinops* (together with *Acantholepis* LESS.) constitutes one of three tribes. He regards the style as importantly distinct, which I do not. Of the features seen by me, the distinctness is in the depth of incision in the corolla, the form of the lower part of the corolla-limb (not represented in DITTRICH's illustration) and the details of the connective-appendages (but I have not looked sufficiently closely at those of other genera, and comparison ought to be extended to cells and tissues).

References for Asteraceae

BREMER, K., 1987: Tribal interrelationships of the *Asteraceae*. – Cladistics **3**: 210–253.

BREWBAKER, J. L., 1967.

BURTT, B. L., 1961.

CABRERA, A. L., 1977: *Mutisieae* – systematic review. – In HEYWOOD & al. (1977) **2**: 1039–1066.

CRUDEN, R. W., 1977.

DAHLGREN, R., 1983: General aspects of angiosperm evolution and macrosystematics. – Nordic J. Bot. **3**: 119–149.

DITTRICH, M., 1977: *Cynareae* – systematic review. – In HEYWOOD & al. (1977) **2**: 1001–1015.

DUNBAR, A., 1975a: On pollen of *Campanulaceae* and related families with special reference to the surface ultrastructure. I. *Campanulaceae* Subfam. *Campanuloidae*. – Bot. Not. **128**: 73–101.

DUNBAR, A., 1975b: On pollen of *Campanulaceae* and related families with special reference to the surface ultrastructure. II. *Campanulaceae* Subfam. *Cyphioidae* and Subfam. *Lobelioidae*; *Goodeniaceae*; *Sphenocleaceae*. – Bot. Not. **128**: 102–118.

GANDERS, F. R., NAGATA, K. M., 1983: Relationships and floral biology of *Bidens cosmoides* (*Asteraceae*). – Lyonia **2**: 23–31.

HANSEN, H. V., 1992: Studies in the *Calyceraceae* with a discussion of its relationship to *Compositae*. – Nordic. J. Bot. **12**: 63–75.

HARPER, J. L., WOOD, W. A. , 1957: *Senecio jacobaea* L., in Biological Flora of the British Isles, no. [63]. – J. Ecol. **45**: 617–637.

HESLOP-HARRISON, Y., SHIVANNA, K. R., 1977.

HEYN, C. C., SNIR, S., 1986.

HEYWOOD, V. H. (ed.), 1978.

HEYWOOD, V. H., & al. 1977.

HEYWOOD, V. H., HUMPHRIES, C. J., 1977: *Anthemideae* – systematic review. – In HEYWOOD & al. (1977) **2**: 851–898.

JEFFREY, C., 1977: Corolla forms in *Compositae* – some evolutionary and taxonomic speculations. – In HEYWOOD & al. (1977) **1**: 111–118.

JUEL, O., 1908.

KNUTH, P., 1908.

LACK, A. J., 1982: Competition for pollinators in the ecology of *Centaurea scabiosa* L. and *Centaurea nigra* L. II. Observations on nectar production. – New Phytol. **91**: 309–320.

LEINS, P., ERBAR, C., 1989: Zur Blütenentwicklung und sekundären Pollenpräsentation bei *Selliera radicans* CAV. (*Goodeniaceae*). – Flora **182**: 43–56.

LEINS, P., ERBAR, C., 1990: On the mechanisms of secondary pollen presentation in the *Campanulales-Asterales*-complex. – Botanica Acta **103**: 87–92.

LLOYD, D. G., 1979.

MICHAUX, P., 1989: Reproductive and vegetative biology of *Cirsium vulgare* (SAVI) TEN. (*Compositae*: *Cynareae*). – New Zealand J. Bot. **27**: 401–414.

MÜLLER, H., 1883.

PERCIVAL, M., 1965: Floral Biology. – Oxford: Pergamon.

PESACRETA, T. C., SULLIVAN, V. I., HASENSTEIN, K. H., DURAND, J. M., 1991: Thigmonasticity of thistle staminal filaments. – Protoplasma **163**: 174–180.

PIGOTT, C. D., 1968: *Cirsium acaulon* (L.) SCOP., in Biological Flora of the British Isles, no. 111. – J. Ecol. **56**: 597–612.
PROCTOR, M., YEO, P., 1973.
SAZIMA, M., MACHADO, I. C. S., 1983: Biologia floral de *Mutisia coccinea* ST. HIL. (*Asteraceae*). – Revista Brasil. Bot. **6**: 103–108.
SIMPSON, B. B., NEFF, J. L., 1987: Pollination ecology in the arid southwest. – Aliso **11**: 417–440.
SKVARLA, J. J., TURNER, B. L., PATEL, V. C., TOMB, A. S., 1977: Pollen morphology in the *Compositae* and in morphologically related families. – In HEYWOOD & al. (1977) **1**: 141–248.
SMALL, J., 1915: The pollen presentation mechanism in the *Compositae*. – Ann. Bot. (Oxford) **29**: 458–470.
SMALL, J., 1917a: The origin and development of the *Compositae*. Chapter II. The pollen presentation mechanism. – New Phytol. **16**: 198–221.
SMALL, J., 1917b: The origin and development of the *Compositae*. Chapter III. Irritability of the pollen presentation mechanism. – New Phytol. **16**: 253–276.
THIELE, E.-M., 1988.
WAGENITZ, G., 1976: Systematics and phylogeny of the *Compositae* (*Asteraceae*). – Pl. Syst. Evol. **125**: 29–46.
ZAVADA, M. S., 1984: The relation between pollen exine sculpturing and self-incompatibility mechanisms. – Pl. Syst. Evol. **147**: 63–78.

Hydrocharitaceae

Systematic position: *Alismatidae, Hydrocharitales*
Restriction of occurrence within the family: *Hydrilloideae*

Distribution within the subfamily. Known only in the genus *Blyxa*.
Floral features (of *Blyxa*). Flowers unisexual, as is usual in the family; plants dioecious. Male flower with three sepals (free but connivent into a tube), three linear petals and nine stamens. Female flower with a long slender hypanthium above the inferior ovary, terminating in three connivent sepals; petals three, thread-like; style with three stigmatic branches.
Type of pollen presentation. Pollen presentation is on the petals ('Pseudo-Staubblatt').

Examples of Secondary Pollen Presentation in Hydrocharitaceae

Blyxa THOUARS
B. octandra (ROXB.) THWAITES. (COOK, LÜÖND & NAIR 1981, illus.). **STR c. ISS exposed. SEQU na. APPL** feet. **RWD** water (for *Diptera*), perch (for *Odonata*). **SYN** special. **VIS** blowflies and houseflies (*Diptera-Calliphoridae* and *-Muscidae*); damselflies and dragonflies (*Odonata-Zygoptera* and *-Anisoptera*).
This plant is a rooted submerged aquatic growing in India; plants detached from the substratum float and can continue to flower. In bud the anthers of the three whorls of stamens are superposed in three tiers. The linear petals of the male flowers are lightly held together by interlocking papillae along their edges. They are thickened along the middle and papillate over their inner surface; they also have three longitudinal furrows on this side. In the two hours before the flower opens they increase in length from about 5 to about 15 mm. The anther walls are

so thin that it is difficult to tell how the anthers dehisce, but during elongation of the petals the pollen is transferred to the latter. These then diverge and present the pollen. The female flower has a calyx about twice as long as that of the male and takes longer to open than the male. The most conspicuous part of the female flower is the group of three stout linear and divergent styles. The petals of the female flower are apparently vestigial, being limp and clinging loosely to the styles. After the flowers have opened water is secreted on the petals of the male flowers and on the styles; on the male flowers this wets the pollen. *Diptera* drink the water and regularly fly from flower to flower, carrying pollen on their feet. *Odonata* that rest on the flowers are also apparently capable of transferring pollen. The petals and stigmas are white and the petals of the male flowers appear to mimic the stigmas, so that male and female flowers look similar, apart from the yellowness of the pollen. The secreted liquid was not chemically analysed but appeared to be water.

Discussion of Hydrocharitaceae

Blyxa is remarkable for the way it combines different floral functions in one organ, and allocates functions common to flowers of both sexes differently in the male and the female flower. In the male flower the petals combine display, secretion and pollen presentation, while in the female the stigmas combine display, secretion and pollen reception. Its other curious attributes, namely that the petals of the male flower are mimetic on the stigmas of the female, and its habit of secreting water, and doing so in a watery environment, are incidental from our point of view. The need of the visiting *Diptera* for water is understandable, because these insects use saliva to liquefy dry foods. Secondary presentation of pollen exposed on the petals, and deposited there in the bud, is unusual. *Zinnia*, in *Asteraceae*, is comparable.

Cook, Lüönd & Nair (1981) do not comment on the furrows on the petals of the male flowers, but these might be the seat of the fluid secretion. These authors note that six of the probable total of eight species in the genus *Blyxa* appear (from herbarium studies) to have the same pollen presentation mechanism as *B. octandra*.

None of the usual functions that I ascribe to secondary pollen presentation operates in *Blyxa*. However, the placement of pollen and stigmas in comparable places in flowers of the respective sexes is achieved. Presumably also the three petals could more easily evolve into stigma-mimics than could the nine anthers. This might be the origin of secondary pollen presentation in this case.

Blyxa has trinucleate pollen (Brewbaker 1967).

References for Hydrocharitaceae

Brewbaker, J. L., 1967.

Cook, C. D. K., Lüönd, R., Nair, B., 1981: Floral biology of *Blyxa octandra* (Roxb.) Planchon ex Thwaites (*Hydrocharitaceae*). – Aquatic Bot. **10**: 61–68.

Xyridaceae

Systematic position: *Commelinidae, Commelinales*

Restriction of occurrence within the family: Reported in one of the four genera of the family

Distribution within the family. *Xyris*
Floral features. Flower actinomorphic, small, with three sepaloid outer perianth segments and three petaloid inner, in an arrangement recalling *Commelinaceae*. Androecium with outer whorl of three staminodes and inner whorl of three functional stamens.
Type of pollen presentation. Pollen is presented on the staminodes ('Pseudo-Staubblatt').

Examples of Secondary Pollen Presentation in Xyridaceae

Xyris L.
The staminodes are simple, bifid or quadrifid and tail-like or plumose. They are thought to collect pollen from the neighbouring anthers and present it to small visiting insects (mainly *Hymenoptera*), thereby 'facilitating pollination' (DAHLGREN, CLIFFORD & YEO 1985, illus.). I do not have the original source of this statement.

Discussion of Xyridaceae

In the illustrations referred to the anthers are well exposed in the flower, though in one case appressed to the perianth segments. It is not clear that the anthers could not themselves present pollen, but they are latrorse, which means they might load the staminodes at the bud stage. As the staminodes are highly developed it seems that they must be important in the functioning of the flower, so that the tentative indications of secondary pollen presentation may well receive confirmation when the flowers are studied further. On the other hand, they seem similar to staminodes found in *Commelinaceae*, which function as real or imitation food bodies for pollinators (VOGEL 1978).

The pollen is binucleate (BREWBAKER 1967).

References for Xyridaceae

BREWBAKER, J. L., 1967.
DAHLGREN, R. M. T., CLIFFORD, H. T., YEO, P. F., 1985.
VOGEL, S., 1978: Evolutionary shifts from reward to deception. – In RICHARDS, A. J., (ed.): The pollination of flowers by insects 89–96. – London: Academic Press.

Cannaceae

Systematic position: *Commelinidae, Zingiberales*
Restriction of occurrence within the family: This is a unigeneric family

Floral features. Perianth of two whorls, the outer actinomorphic, the inner slightly united at base; segments firm in texture, protective in the bud stage but contributing little to the attractive display. Androecium united at base with inner perianth whorl, of one stamen and up to 5 staminodes (number varying, even within one plant). Staminodes enlarged, petaloid, unequal, making the flower asymmetric; one staminode of the inner whorl recurved, forming a labellum. Stamen belonging to the inner whorl, petaloid, with one theca which is situated on the edge, well below the apex. Style flattened, petaloid, with stigmatic surfaces along its edges at the apex. Ovary inferior, trilocular, with many ovules. Observations on the

ontogeny, homologies and symmetry of the flower are reported by KIRCHOFF (1983) and KUNZE (1984).
Type of pollen presentation. Pollen is presented on the surface of the style ('Pseudo-Staubblatt').

Examples of Secondary Pollen Presentation in Cannaceae

Canna L.

Canna is found in the tropics and subtropics of both the Old World and the New. However, it is likely that it is native only in the latter. Very many of the described species have been reduced to synomymy, leaving a genus of 25, or perhaps only 7, species (ROGERS 1984). Some of the rejected taxa are of hybrid origin and some of these apparently originated in the Old World.
Canna species (BOUCHÉ 1833; others, named when quoted).
Presentation of the pollen on the style, as described below for *C. indica*, was reported by BOUCHÉ for the genus as a whole. BOUCHÉ refuted the idea, previously favoured, that this was an act of self-fertilization. He correctly identified as stigmas the cushions on the edge of the obliquely truncate style. He found that the species fell into two groups; in one the pollen placement was as shown below for *C. indica*, so that automatic self-pollination was impossible, whereas in the other, the pollen deposit overlapped the stigma and self-pollination occurred automatically. The two processes were named respectively 'foecundatio indirecta' and 'foecundatio directa'. BOUCHÉ listed 25 species (including *C. indica*) with 'foecundatio indirecta' and 10 with 'foecundatio directa'. Working in the Berlin Botanic Garden, he was able to say that taxa of the second group set seed in cultivation much more readily than those of the first, though these usually gave a good seed-set if artificially self-pollinated. He realised that in nature these needed insect visits or some other agency to enable them to set seed.

KRÄNZLIN (1912) accepted BOUCHÉ's conclusions in essentials and pointed out that other authors who wrote after BOUCHÉ refused to accept them owing to prejudice about the impossibility of automatic self-fertilization; he himself maintained that in species that BOUCHÉ credited with 'foecundatio indirecta' a slight overlap of the pollen deposit with the stigmas could occur, resulting in a degree of self-pollination. KRÄNZLIN noted that the honeybee (*Apis mellifera*) only touched the style of *C. indica* when the flowers were not fully open. He therefore thought that narrow-flowered species such as this were the most likely to be cross-pollinated. He concluded that in *C. iridiflora* RUIZ & PAVON, with pendent flowers, and *C. flaccida* SALISB. and *C. reevesii* LINDL., with flowers that are limp at anthesis, so that their members fall apart, the behaviour must be different, and that in the two latter fertilization except by autogamy in the bud is impossible (apparently because a honeybee would not touch the stigmas). This is despite the fact that he recognized that birds as well as insects might be pollinators, and that he cited observations by K. RECHINGER in Samoa of a Meliphagid bird visiting and pollinating *C. humilis* BOUCHÉ. I have examined published figures of *C. flaccida* and *C. reevesii*, showing that they have large trumpet-shaped flowers recalling those of *Costus*, and doubtless similarly flimsy, but for which it is difficult to accept a primary adaptation to autogamy (though according to BOUCHÉ *C. flaccida* has 'foecundatio directa'). ROGERS (1984, illus.) states that *C. flaccida* flowers are fragrant, nocturnal and ephemeral (which might explain why they are reported as flaccid, being seen by most botanists only in the morning!). (Dr W. J. KRESS, who has been studying pollination of *Zingiberales* in the field, tells me that he detects no scent in *C. flaccida*, nor in any other species of *Canna* – see, however, *C. brittonii*, below). *C. iridiflora* has very large pink, pendent, trumpet-shaped flowers, suggesting that it might be pollinated by a large hawkmoth. According to WINKLER (1930), there are many reports in the literature of birds visiting the flowers and, among insects, particularly bees and nocturnal moths.

C. indica L. (YEO, unpublished, 1981, and YEO in DAHLGREN, CLIFFORD & YEO 1985, illus.). **STR s. ISS exposed. SEQU sim. APPL ?. RWD n. SYN** ornitho. **VIS** –. (Fig. 54).
The long-lasting flowers of this species are about 6 cm long with more or less oblanceolate and porrect staminodes. The general colouring is flame-red but the rolled-up, tongue-like labellum, which has a channelled base, is yellow with red streaks. The labellum and stamen are about half the length of the other two or three staminodes. The stamen is conduplicate in bud, enfolding the style. Before

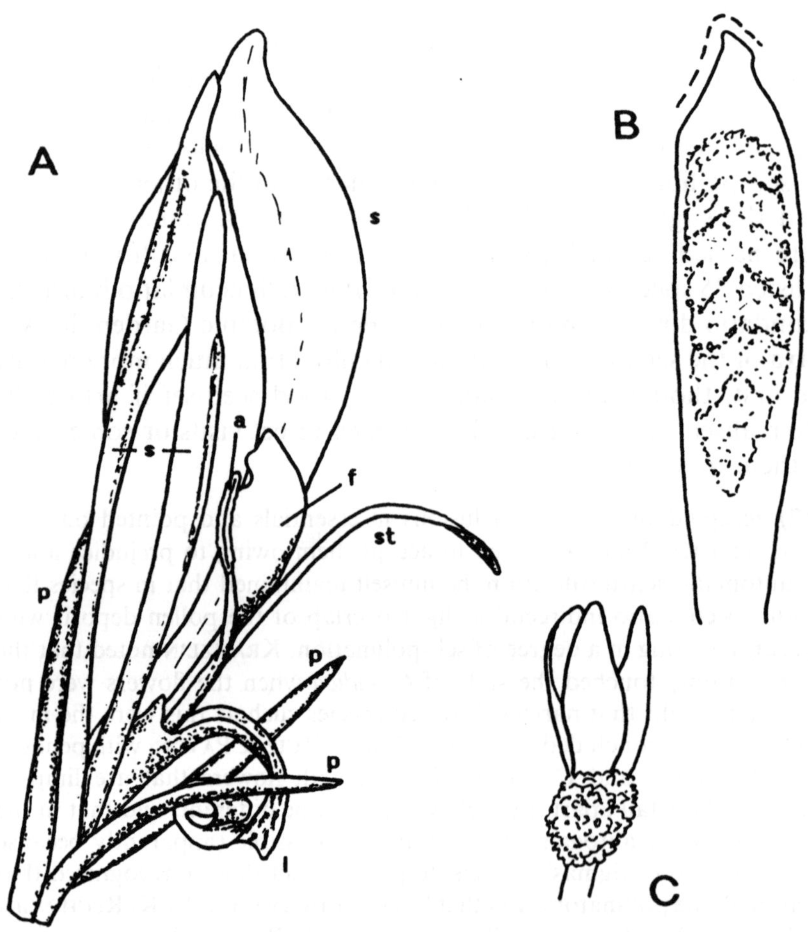

Fig. 54. *Canna indica* (*Cannaceae*). *A* flower in side view, ovary and sepals removed; overall length 63 mm. *B* ovary and sepals, from which remaining parts can be separated as a unit, as in 'A'. *C* distal portion (c. 12 mm long) of style with deposit of the flower's own pollen; edges next to dashed lines are stigmatic. *a* = anther, partly attached of 'f', *f* = expanded petaloid part of stamen-filament, *l* = the coiled and guide-marked staminode of the inner whorl known as the labellum, *p* = petal, *s* = staminodes, which are unequal in length but similar in shape, here seen at different angles, *st* = style. (P. F. YEO.)

the flower opens the anther dehisces and deposits its pollen on the style in a large patch. When the flower opens the anther has shrivelled somewhat and the distal part of the stamen has bent away from the style. The labellum probably acts as a nectar guide and the style stands over it in a position to be touched by a probing bird. The style is asymmetrically tapered at the apex and there are two unequal stigmatic areas on the edges of the tapered part; the pollen deposit stops short just before the longer stigmatic strip. The pollen is loosely deposited on the downward-facing side of the style, showing a herringbone pattern, and it is lightly coherent. If it is pushed with a needle to a different part of the style it does not adhere but drops off on being shaken. Therefore either the style has an area specially prepared to receive the pollen, or the anther provides an adhesive (?Pollenkitt) at the moment of dehiscence. If pollen is pushed up to the stigmatic strips it becomes infiltrated with stigmatic fluid. That the pollination syndrome is clearly ornithophilous nullifies the inferences of KRÄNZLIN cited under '*Canna* species' (above) based on observations of visits to the flowers by the honeybee.
C. brittonii RUSBY (VOGEL 1969, illus.). **STR s. ISS exposed. SEQU sim. APPL** al. **RWD n. SYN** chiroptero. **VIS** –.
This species grows to a height of 4–6 m and occurs at high altitude in cloud-forest in Bolivia. The inflorescence reaches well above the leaves and is spike-like, while the greenish white, firmly constructed, flowers are erect at the base but curved so that the entrance, which is funnel-shaped, is horizontal. The style measures 7 × 0.8 cm and is curved to the right so that it touches the left side of a visitor with its edge and inner surface. The surface bears a layer of pollen deposited in the bud from an anther that measures 30 × 4 mm. The nectaries are enormously enlarged. There is a herbaceous-soapy scent. (Observations are limited because the plant when found was at the end of its flowering period, but the chiropterophil syndrome is strongly developed.)

Discussion of Cannaceae

A single system of pollen presentation exists in species that appear to show a range of pollination syndromes, this range incorporating extremes of duration of individual flowers. The system does not give protection to the pollen, unless it makes it harder to find or harder to gather up; it does place pollen and stigma at approximately the same part of the flower and it seems to provide for the receipt and donation of pollen during the same pollinator visit. It is difficult to see what its function might be.

The placement of the pollen subapically on the style is a character in common with the related family *Marantaceae*. However, in *Marantaceae* the pollen is deposited on the opposite side of the style (KUNZE 1984). KUNZE suggests that a common tendency to neoteny of the stamen led to the independent origin of the respective pollination systems of the two families. However, this leaves out of account the question of possible adaptive significance of staminal neoteny.

The occurrence respectively of a totally passive system and a vigorously explosive system in the two related families is noteworthy. Further comparison with *Marantaceae* is found under the latter family and in Chapter 3, 11. (p. 230).

The pollen of *Canna* is binucleate (BREWBAKER 1967).

References for Cannaceae

BOUCHÉ, P. C., 1833: Mittheilung vieljähriger Beobachtungen über die Gattung *Canna*. – Linnaea **8**: 141–168.

BREWBAKER, J. L., 1967.

DAHLGREN, R. M. T., CLIFFORD, H. T., YEO, P. F., 1985.

KIRCHOFF, B. K., 1983: Floral organogenesis in five genera of the *Marantaceae* and in *Canna* (*Cannaceae*). – Amer. J. Bot. **70**: 508–523.

KRÄNZLIN, F., 1912: Cannaceae. – In ENGLER, A., (Hrsg.): Das Pflanzenreich **IV.47** (**56**. Heft): 13–15. – Leipzig: Engelmann.

KUNZE, H., 1984:

ROGERS, G. K., 1984: The *Zingiberales* (*Cannaceae*, *Marantaceae*, and *Zingiberaceae*) in the Southeastern United States. – J. Arnold Arbor. **65**: 5–55.

VOGEL, S., 1969: Chiropterophilie in der neotropischen Flora, neue Mitteilungen III. – Flora, Abt. B, **158**: 289–323.

WINKLER, H., 1930: *Cannaceae*. – In ENGLER, A., HARMS, H., (eds.): Die Natürlichen Pflanzenfamilien. 2nd edn. **15c**: 640–654. – Leipzig: Engelmann.

Marantacae

Systematic position: *Commelinidae, Zingiberales*
Restriction of occurrence within the family: None

Distribution within the family. In both thc tribcs that arc rccognizcd.
Floral features. Flowers usually borne in pairs, asymmetric, with the members of each pair being mirror-images of each other. Outer perianth members three, free, sepaloid; inner members three, usually fused, petaloid. Androecium connate at base with the inner perianth whorl when this is syntepalous; outer whorl rarely absent, otherwise of one or two petaloid staminodes that are usually broader than the lobes of the inner perianth, being thus the major contributors to the floral display; inner whorl comprising a monothecous stamen, usually with an enlarged petaloid filament, and two specialized staminodes, the cucullate staminode and the callose staminode, the latter having a petaloid lobe. Style thick and fleshy, fused at base with the perianthial/androecial tube. Ovary trilocular (tribe *Phrynieae*) or unilocular (tribe *Maranteae*) with not more than one seed per locule.

Plants are usually broad-leaved, caulescent and branched; they occur in the tropics of both hemispheres (for further details see DAHLGREN, CLIFFORD & YEO 1985, S. R. CHANT in HEYWOOD 1978, KIRCHOFF 1983, KUNZE 1984, and ROGERS 1984). The division of the family into the two tribes named above is artificial and is ignored hereafter.

Type of pollen presentation. Pollen presentation takes place on the style ('Pseudo-Staubblatt'). During an insect visit to the flower an explosive movement occurs, causing deposition of pollen on the visitor, fastened by an adhesive secretion. The same movement first allows the stigma to pick up pollen from the visitor. KUNZE (1984) reported observations on the ontogeny, homologies and symmetry of the flower. His experiments showed that the traditional view that the cucullate staminode holds the style under tension before triggering is erroneous. He found that there is no tension in the style before the flower is probed. The cucullate staminode bears a lobe which acts as a trigger and transmits pressure via a structure called the basal plate of the trigger to the style; the style then makes its abrupt movement, lasting 0.2 s. The inference is that when the pressure is received by the style an immediate change in turgor is induced and that this sets up the tension which is relieved by the movement of the style. Both of the staminodes of the inner whorl are involved in helping the transfer of pollen from the anther to the platform on the style that receives it. In the following account observations not in agreement with this account are ignored. KUNZE worked mainly with *Calathea*, *Maranta* and *Phrynium*.

Examples of Secondary Pollen Presentation in Marantaceae

Calathea G. F. W. MEYER.

Various species (KENNEDY 1978, illus.). **STR s. ISS expl depos. SEQU stg. APPL v. RWD n. SYN** melitto. **VIS** the large bee *Eulaema* (*Hymenoptera-Apoidea*: *Apidae*, subfam. *Euglossini*).

KENNEDY worked principally on the 'closed-flowered' species of *Calathea* which are peculiar in that the pollinator has to bite the flower to allow it to open; after this action the bee withdraws, extends its proboscis and probes the flower for nectar. The style is thick and distally curved upwards. During development of the flower the stamen deposits its pollen in a depression on the outside of the curved distal region of the style. The probing insect touches the lobe on the cucullate staminode which acts as a trigger. The style then coils more tightly in a forceful movement, during which any pollen from a previously visited *Calathea* flower is picked up and the flower's own pollen is deposited on the insect. The action is very precise: the apical stigma of the style scours the insect's proboscidial fossa (the cavity on the underside of the head into which the proboscis can be folded when out of use); then the increasingly sharply curved style-apex continues its movement within the fossa, so that its pollen-containing depression (see *C. bachemiana* below) rubs the fossa and leaves behind its pollen. The last refinement is that the leading border of the pollen-presenting depression is coated with an adhesive secretion; this is pushed into the proboscidial fossa after the stigmatic apex of the style has scoured it and before the flower's own pollen is offered up, so that the latter is secured to the bee's chitinous surface. KENNEDY reported that open-flowered species of *Calathea* functioned in the same way except that they do not need to be bitten open. The proboscidial fossa is used for pollen deposition by at least two other New World genera of *Marantaceae*, and the principal pollinators of the family in the New World are *Euglossini*.

C. bachemiana MORREN. (YEO, unpublished 1986). (Pollen presentation details are as shown above for the genus except for **VIS** which is not known.) Fig. 55.

This is an open-flowered species. The flower is white and the tube formed by the inner perianth segments and androecium is nearly 3 cm long. The petaloid part of the callose staminode is about as large as that of a staminode of the outer whorl. The lever-like trigger-arm on the cucullate staminode appears solid but is formed from a membranous lobe that is rolled up for thickness and rigidity. It was inferred that the style in the untripped flower is for the most part straight; however, the apical part, comprising slightly more than the length occupied by the pollen-presenting surface, is set at right-angles to the remainder. After tripping, the part just proximal to the angle is also strongly curved. The tip of the style is obliquely truncate and part of the apical surface is occupied by an asymmetric stigmatic

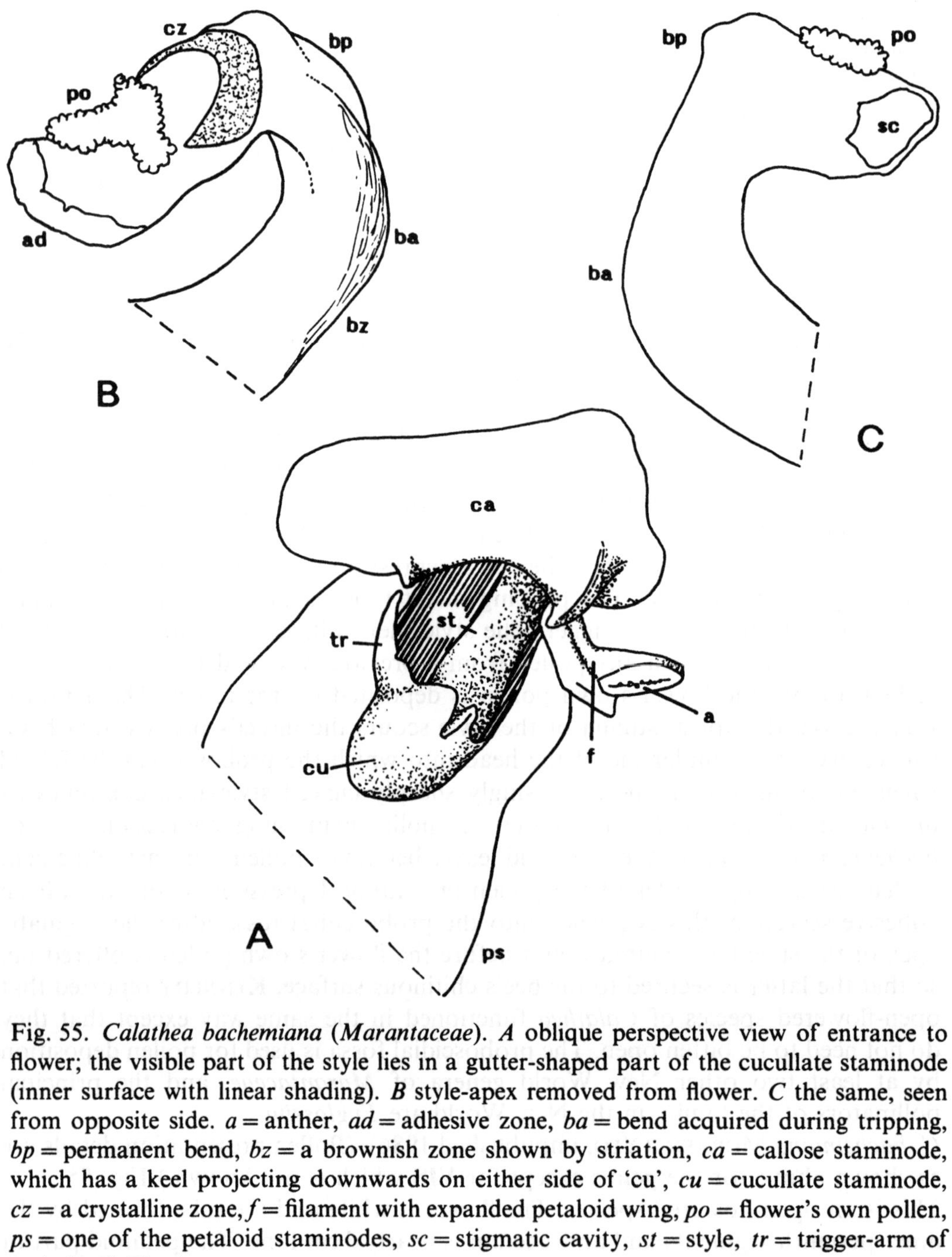

Fig. 55. *Calathea bachemiana* (*Marantaceae*). *A* oblique perspective view of entrance to flower; the visible part of the style lies in a gutter-shaped part of the cucullate staminode (inner surface with linear shading). *B* style-apex removed from flower. *C* the same, seen from opposite side. *a* = anther, *ad* = adhesive zone, *ba* = bend acquired during tripping, *bp* = permanent bend, *bz* = a brownish zone shown by striation, *ca* = callose staminode, which has a keel projecting downwards on either side of 'cu', *cu* = cucullate staminode, *cz* = a crystalline zone, *f* = filament with expanded petaloid wing, *po* = flower's own pollen, *ps* = one of the petaloid staminodes, *sc* = stigmatic cavity, *st* = style, *tr* = trigger-arm of 'cu'. (P. F. YEO.)

hollow; presumably the rim of the hollow acts as a scraper. The platform on which pollen is presented is a saddle-shaped, not lens-shaped depression, a fact which makes it easier to see how it can give up its pollen. The surface of the 'saddle' is smooth and its elevated distal edge bears the adhesive secretion. The artificial triggering of an intact flower caused the pollen to be deposited on the limb of the callose staminode, and the number of grains appeared to be well under 100; a very few may have been left within the cucullate staminode, and several were found along the closed-up dehsicence line of the anther. When a flower was tripped after the removal of the upper 'petal' and the limb of the callose staminode, the pollen remained on the pollen presenter in a neat dollop. The elaborate moulding of the style-apex and the cucullate staminode presumably ensures that the pollen-mass is not in contact with anything before tripping and is not scraped by any part of the flower during tripping.

Marantochloa Griseb.

M. purpurea (Ridley) Milne-Redhead. (Yeo, unpublished 1986). (Pollen presentation details are as shown above for *Calathea* except for **VIS** which is not known.)

The style was found to be very similar to that of *Calathea bachemiana*, whereas the cucullate staminode lacks a trigger-arm. The adhesive is in the same place; in bud it was found to be contained within a very thin membrane. The release of the pollen from the anther takes place at a very late stage in floral development.

Thalia L.

T. geniculata L. (Rogers 1964, illus.; Davis 1987; Classen-Bockhoff 1991, illus.). **STR s. ISS expl depos. SEQU stg. APPL v. RWD n. SYN** melitto. **VIS** the large bee *Xylocopa* (*Hymenoptera-Apoidea*: *Apidae*); the hummingbirds *Chlorostilbon canivetii* and *Amazilia tzacatl* (*Aves-Trochilidae*).

Classen-Bockhoff shows that the flowers of a pair, lacking pedicels, are held close together and that they open simultaneously; they thus form a single symmetrical blossom with two entries. Rogers illustrates the flower in detail; the inner perianth members are free from each other and more or less free from the androecial tube. In the USA the plant occurs outside the range of the Euglossine bees; various insects visit the flowers but only *Xylocopa* trips them. In Costa Rica the principal pollinator was *Xylocopa* (two species). In addition hummingbirds visited the flowers, two species accounting for 40% of all visits and causing some tripping; 5% of visits were by butterflies that did not trip the flowers. On plants in cultivation Rogers found that the flowers were very easily tripped. Davis found that the flowers lasted approximately through one daytime and became progressively more sensitive. It was later in the day that hummingbirds achieved a high tripping rate. Davis obtained evidence that self-tripping occurred after dark and caused pollination. Rogers tripped the flowers artificially and found self-pollen on stigmas afterwards; the column was allowed to strike the callose staminode, as in self-tripping, and it is not clear how the flower's own pollen gets on to the stigma. Classen-Bockhoff, working in Java and Germany, found that in Java *Xylocopa* species also visited the flowers and that the pollen is deposited asymmetrically on the mandibles and maxillary galeae. This is because tripping results in a lifting and twisting movement of the style (elucidated by high-speed cinematography) instead of the usual simple coiling movement; the tip of the style finally lies across the flower. Most of the movement is accomplished in the space of about 0.03 seconds (an order of magnitude faster than found in *Maranta* by Kunze). The sensitive area of the style is restricted to a patch adjacent to the basal plate of the trigger of the cucullate staminode. Before tripping the style is twisted towards the callose staminode; its apex bears a membranous flange, which functions in guiding the visitor's proboscis, and a large channelled appendage bent towards the base of the flower. At the end of the movement the channelled appendage reaches the nectar-store and seems to take up nectar by capillarity, thus bringing it nearer the mouth of the flower, presenting it secondarily to the insect (transfer of nectar to a secondary, more accessible, chamber, was found by KUNZE in *Ctenanthe* of this family).

Discussion of Marantaceae

This family seems to be primarily adapted to insect pollination (DAHLGREN, CLIFFORD & YEO 1985) and it seems likely that its pollination system arose in connection with bee-pollination and possible that the place of pollen transfer was primitively the proboscidial fossa. Secondary pollen presentation in *Marantaceae* achieves protection of pollen in the bud and exact positioning of the pollen on a particular part of the insect pollinator. When the place of deposition is the proboscidial fossa pollen protection after removal from the flower is also achieved since, as far as is known, this is totally inaccessible to the insect. In the case of *Calathea* the precision of pollen placement and its inaccessibility recall pollination in the orchids. The arrangements allow pollen production to be minimized. The small amount of pollen, however, should also be considered in the light of the presence of only one or three ovules in the ovary.

A possible penalty of such precision in pollen-transfer is that the system breaks down. This seems to be a factor in the breeding system of this family, for ANDERSSON (1986) states that fruit set in allogamous *Marantaceae* is generally low and that many of the sixteen species of *Maranta* subgenus *Maranta* are mainly autogamous.

The use of the style as pollen-presenter and of an adhesive presented on the style exactly parallels the situation in *Polygala* species described by BRANTJES (1982); *Marantaceae*, however, have 'one-visit' flowers. KENNEDY (1978) states that open-flowered *Calathea* species are subject to theft, but this, of course, is nectar-theft, since the thieves do not trip the flowers.

In the organization of the flower and the use of the style to present the pollen the *Marantaceae* are similar to *Cannaceae*, but the latter lack any explosive mechanism, and the value of their pollination arrangements is less obvious.

A third closely related family, the *Zingiberaceae*, lacks secondary pollen presentation but is also specialized in its pollen-transfer arrangements. The inner perianth segments form a floral tube which is often long and slender and is parted into three lobes at its mouth. The androecium is attached at this level and consists of an outer whorl of two petaloid staminodes and an inner whorl of two petaloid staminodes and one fertile stamen. The two sets of staminodes are variously developed and varyingly combined to produce flowers that are vexillate, labellate, bilabiate or funnel-shaped. The stamen has a dithecous introrse anther. The peculiarity of the family is that the stamen is adaxially channelled so as to embrace the distal part of the style, thus fixing the position of the anther and stigmas in relation to each other. The details vary considerably but in the labellate flowers of *Hedychium* and *Roscoea*, for example, the anther-connective is itself channelled and the stigma appears to terminate the anther, while in the vexillate *Brachychilum* the 3-lobed stigma emerges near the base of the anther, backed by a laterally expanded connective. In the latter case the anther is ascending so that, as in the other two genera, an approaching pollinator is likely to brush the stigma a moment before contacting the thecae. The flowers can thus probably receive and donate pollen during a single pollinator-visit, but can also receive more than one visit. In principle, the organization of these flowers could provide a stepping-off point for the development of the flowers of the *Cannaceae* and *Marantaceae*, though it must not be overlooked that the style in these families is robust, whereas in *Zingiberaceae* the part embraced by the stamen is extraordinarily thin and limp. *Zingiberaceae*, like *Cannaceae*, have numerous ovules.

The present account, together with scanty records cited by East (1940), indicates that self-compatibility is likely to be the rule in both *Marantaceae* and *Zingiberaceae*. (This is supported by unpublished work of Dr W. J. Kress – pers. comm.) The pollen is binucleate (two species of *Maranta*; Brewbaker 1967).

References for Marantaceae

Andersson, L., 1986: Revision of *Maranta* subgen. *Maranta* (*Marantaceae*). – Nordic J. Bot. **6**: 729–756.

Brantjes, N. B. M., 1982.

Brewbaker, J. L., 1967.

Classen-Bockhoff, R., 1991: Untersuchungen zur Konstruktion des Bestäubungsapparates von *Thalia geniculata* (*Marantaceen*). – Botanica Acta **104**: 169–256.

Dahlgren, R. M. T., Clifford, H. T., Yeo, P. F., 1985.

Davis, M. A., 1987: The role of flower visitors in the explosive pollination of *Thalia geniculata* (*Marantaceae*), a Costa Rican marsh plant. – Bull. Torrey Bot. Club **114**: 134–138.

Heywood, V. H. (ed.), 1978.

Kennedy, H., 1978.

Kirchoff, B. K., 1983: Floral organogenesis in five genera of the *Marantaceae* and in *Canna* (*Cannaceae*). – Amer. J. Bot. **70**: 508–523.

Kunze, H., 1984.

Rogers, G. K., 1984: The *Zingiberales* (*Cannaceae*, *Marantaceae*, and *Zingiberaceae*) in the Southeastern United States. – J. Arnold Arbor. **65**: 5–55.

Liliaceae

Systematic position: *Liliidae, Liliales*
Restriction of occurrence within the family: Reported in one genus

Distribution within the family. Albuca; in a more recent classification than the standard one used in this review *Albuca* is placed in super-order *Liliiflorae*, order *Asparagales* and family *Hyacinthaceae* (Dahlgren, Clifford & Yeo 1985).
Floral features. Perianth segments six, in two morphologically differentiated whorls; stamens six; ovary syncarpous, styles united.
Type of pollen presentation. Pollen is presented on stylar papillae ('Pseudo-Staubblatt').

Examples of Secondary Pollen Presentation in Liliaceae

Albuca L.
A. canadensis (L.) Leighton. (Westerkamp, pers. comm., April 1988, illus.). **STR s. ISS exposed. SEQU sim? APPL v, l? RWD n, p? SYN** melitto. **VIS** bees of family *Megachilidae*, and perhaps species of *Xylocopa* (*Hymenoptera-Apoidea*).
All species of this southern African genus have nodding, more or less bell-shaped flowers and inner perianth segments that are distinctly shorter and more truncate

than the outer. In the present species the flowers are yellow, the ovary and style are each columnar and the apex of each stamen is held in the hooded tip of the adjoining perianth segment. However, before the flower opens the pollen is deposited on a papillate zone of the style. Assuming that visiting bees cling to the gynaecium, they will touch the pollen deposit with their mid-ventral surfaces. If the pollen were not transferred to the style a bee would be unlikely to touch the anthers and stigmas with the same part of the body.

Discussion of Liliaceae

The arrangement described here is identical with that of *Campanula* (*Campanulaceae*). It would be interesting to know what the pollination mechanism is throughout the genus. *Albuca* is reported to have binucleate pollen (Brewbaker 1967).

References for Liliaceae

Brewbaker, J. L., 1967.
Dahlgren, R. M. T., Clifford, H. T., Yeo, P. F., 1985.

Addendum: Calyceraceae

This small family in *Asteridae-Dipsacales* also displays secondary pollen presentation (Cronquist, A., An Integrated System of Classification of Flowering Plants, New York, Columbia Univ. Press, 1981; Mabberley, D. J., The Plant-book, Cambridge, Cambridge Univ. Press, 1987) although this fact is not mentioned by Hansen (1992, see references for *Asteraceae*). The flowers are small, epigynous, long-tubed and borne in capitula. The pollen grains are binucleate. I have examined *Acicarpha tribuloides*, in which some flowers are functionally male. The anthers are united around the style at their bases and deposit pollen in the late bud stage on the thickened style-head. They then become splayed out. After the flower opens growth of the style carries the pollen upwards beyond the corolla. Later the style-head of functionally hermaphrodite flowers enlarges and becomes prominently papillate. Cronquist's description of pollen being pushed out does not seem to apply.

The function of secondary pollen presentation in this family can hardly be other than a space-saving one. The addition of just this one more small family brings out the prevalence of stylar pollen presentation in the families of the *Asteridae* that have inferior ovaries; I do not at present see any possible functional connection between the two characters.

Chapter 3

General discussion

1. Preamble

The study of secondary pollen presentation is the study of attributes involved in pollination success. These attributes could promote pollination success of two kinds. The first kind is achieved by leaving more progeny: genotypes giving more pollen should leave a greater number of progeny than those producing less pollen (i.e. the fitness of these genotypes is higher). Secondary pollen presentation is not concerned with increasing pollen supply but it would have the same effect if it ensured that a greater proportion of the pollen that is produced reaches receptive, compatible stigmas. The second kind is achieved by leaving progeny of better quality; this improves the fitness of the next generation. It depends on exercising a choice of mates, but in plants the scope for mate-choice is limited. However, the fitness of the next generation might be improved if, as a male, a plant can sample a genetically widened range of females, and this might be achieved by spreading pollen more widely in space and/or time. Some forms of secondary pollen presentation could have this effect. In either case, we have to look at the processes of pollen removal and pollen transfer, and these operate mainly through the individual flower, though not infrequently attributes promoting pollination success will be found in the inflorescence and sometimes in the form of the whole plant. Secondary pollen presentation takes effect through the individual flower.

In this 'General Discussion' I consider these and other possible functions of secondary pollen presentation; then I review the occurrence of some special measures that help to make it work; finally I consider its evolution. Readers of this chapter may find it helpful to refer to the discussions that conclude the family treatments in Chapter 2, where greater detail will sometimes be found.

2. Functions of secondary pollen presentation

There are a number of different forms of secondary pollen presentation. These are shown in the column headings in Table 3-4 (p. 227), which bring out the contrast between 'exposed' presentation and all the rest, showing that we can reduce secondary pollen presentation to two basic phenomena, the functions of which we can now consider separately.

A. Secondary pollen presentation involving protection and delivery systems

As a protection system evolves, so must also a delivery system. On the other hand, if the primary requirement had been for regulating the amount of pollen delivered on a single visit (whether for dividing it up or for ensuring that all of it is delivered during one visit) a protection for the pollen awaiting delivery would have had to evolve. Thus, either way, the one is the corollary of the other. Where all pollen is delivered during a single visit, 'explosive deposition' or 'explosive cloud' types of pollen issue (see Chapter 1) become possible. These explosive systems do not always involve secondary pollen presentation, but there is a high probability that they will, because of the need to have all the pollen in a position ready for immediate release on tripping. Thus secondary pollen presentation opens up one possible avenue for the evolution of explosive flowers.

Protection and delivery systems ('Nudelspritzen', see Chapter 1) may further increase pollination efficiency (and so minimize the required pollen production) by causing presentation of pollen in exactly the same place as it is received, which is 'the essence of biotic pollen transfer' (LLOYD & WEBB 1986), thereby assisting correct placement on the pollinator. Some systems go further, guiding or manipulating insect pollinators so as to place the pollen more precisely on a particular part of the body ('explosive deposition' types in several families and asymmetric 'release' types in *Leguminosae* and *Polygalaceae*).

B. Secondary presentation of pollen on exposed sites within the flower

Whereas in protection and delivery systems there is movement of floral parts in relation to one another during pollen delivery, and especially when pollinators are in contact with the flowers, 'exposed' presentation is characterized by its passivity. Here, the main feature is that pollen-presenting structures are physically different from anthers. (They are 'Pseudostaubblätter' or, occasionally, deposition sites for 'Pollenhaufen' – see Chapter 1.) The structures may have different mechanical properties from anthers, or their use may give rise to different spatial arrangements of the pollen. Thus the pollen from several anthers can be placed together on the style (*Proteaceae*, *Campanulaceae*, *Rubiaceae*), or the pollen from a single anther-cylinder can be divided up by presentation on five corolla lobes (*Zinnia*).

The most obvious function that could be served by 'exposed' types is one already noted for protection and delivery systems: approximation of the place of pollen presentation and pollen reception. The reason for the (rare) converse condition is not evident. To the extent that pollen-presenting styles of *Proteaceae* regulate the distance between the pollen and the nectar in bird-pollinated species, the 'exposed' system can play a part in exact placement of pollen on the body of the vector (as protective systems can). This is presumably also what is involved in the simple system of secondary pollen presentation in *Turraea* (*Meliaceae*), even though it is difficult to see why it is needed. Many *Proteaceae* have crowded flowers with far-exserted indurated styles. The stiffening of the styles is presumably an adaptation to bird-pollination. By transferring pollen to the styles the plant avoids the need to stiffen the stamens. This would save materials, though it is thought that photosynthate is not likely to be in short supply in these plants; it would also avoid undue congestion in the inflorescence. The possible deleterious effects of congestion within flowers have been emphasized by LLOYD & YATES (1982), who suggest that styles might obstruct pollen removal or accidentally remove newly donated pollen from an insect, and that anthers might similarly interfere with pollen reception. Alternatively, as will be discussed shortly, the system could have evolved for other reasons and merely been preadapted to bird-pollination (and small mammal-pollination) in a dense inflorescence. In *Campanulaceae*, as already mentioned, and as will be discussed further, secondary pollen presentation may have arisen to cope with the introrse anther dehiscence in a bell-like flower. The 'exposed' presentation system in *Rubiaceae* is often found in *Lepidoptera*-pollinated species with a long slender corolla-tube and far-exserted style; the corolla might be able to support only slender stamens which perhaps would not remain in the place corresponding to the stigmas. Letting the pollen be carried upwards on the style from anthers situated in the throat of the corolla would ensure its correct placement. There could also be difficulty at the bud stage in accommodating long filaments in very slender corollas. This is a problem of lack of space within the flower instead of in the inflorescence, as suggested for *Proteaceae*. However, not all members of the family with secondary pollen presentation are so built that they would need secondary pollen presentation for these reasons.

The case of *Dombeya* in *Sterculiaceae* is particularly puzzling. Why should not the staminodes have atrophied and the stamens (or some of them – there are 15) have become longer, so as to bring the pollen level with the stigmas? Even without this example, however, 'exposed' pollen presentation is much harder to understand than the concealed types. *Acrotriche* in the *Epacridaceae* presents a special case: pollen is presented on the corolla apparently to keep it clear of the copious nectar that fills the corolla-tube.

Although members of the family *Asteraceae* fall into both the above classes, the family needs separate consideration on account of the complications introduced by its having flowers in capitula. BURTT (1961) has pointed out that the organization of the capitulum ensures that it makes a series of separate pollen donations (and acceptances) over a relatively long period of time (usually several days) and also (BURTT 1977) that it is an evolving functional unit in its own right. As sequential pollen donation is built in to the capitulum pollen-dosing by the individual floret seems to be an irrelevance. However, among the *Cardueae* there is a clear protective role for secondary pollen presentation because it is linked with release of concealed pollen actuated by an irritability mechanism, so that pollen is delivered during pollinator visits. In many members of the family, however, the pollen is extruded by growth, independently of insect visits. LLOYD (1979) considers that both these arrangements promote economy in the use of pollen. However, the third method found in *Asteraceae*, 'exposed' presentation (not considered by LLOYD), is not protective; it is very difficult to see the function of secondary pollen presentation here. In many cases there could be an effect in controlling distance between the level of the nectar in the capitulum and the level of presentation and reception of pollen (cf. *Proteaceae*). However, this could easily be achieved without secondary pollen presentation. It is, therefore, tempting to explain it in the same way as for *Campanulaceae*. In this case, not only do we have introrse anthers, but we have miniaturized flowers. Perhaps secondary pollen presentation is here coping not so much with introrse anthers in an axial position in the flower but with an acute shortage of space, as suggested by VAN DER PIJL (1978: 82). The involvement of the crowding factor has already been suggested as relevant in *Proteaceae* and *Rubiaceae*. In *Asteraceae* there is lack of space both in the individual floret and in the inflorescence (capitulum). It has elsewhere been suggested that successive presentation (dichogamy) may be dictated by a confined space (LLOYD & YATES 1982: 91); successive presentation combined with secondary pollen presentation may further help floral functioning when space is limited.

I have alluded to pollen-dosing as a frequent feature of 'release' systems and as a property of the Asteracaeous capitulum. Although quantitative information is lacking, there are definite statements for some *Leguminosae* that during an insect-visit only a small proportion of the available pollen is released. Pollen is also released progressively in the *Campanulaceae*. Further, the occurrence of arrangements for reloading a pollen presenter (in *Pisum* and *Polygala vauthieri*) suggest that this is important and not merely incidental. As mentioned in Chapter 2 (p. 198) the antithesis of this is 'sweepstakes pollination' (exemplified by *Orchidaceae*): the first pollen package to arrive is derived from one flower and fertilizes all the ovules. The single-visit system, which we find in the explosive flower, is not the same as that, because the incoming pollinator may be carrying pollen from several previously visited flowers (or from none – a disadvantage to the plant, as a proportion of its flowers will be pollen-donors without being pollen-receivers).

The subject of pollen-dosing has been reviewed by HARDER & THOMSON (1989). Citing earlier authors on the importance of intrasexual selection to increase the proficiency of pollen donation, they point out that this could arise through changes in the removal, transport and deposition on stigmas of the pollen. Previous authors (e.g. LLOYD & YATES 1982) had suggested that increase in the number of visitors that can remove pollen from a flower will improve pollination success. HARDER & THOMSON explore how such regulation might promote maximum pollination in different circumstances and how it might reduce the uncertainty of successful pollen transport.

These authors studied pollen removal and dispersal in *Erythronium* and made mathematical models to show the optimal temporal deployment of pollen for various hypothetical conditions. They classify pollen allocation into two categories, packaging and dispensing. Packaging is the division of the plant's total pollen production into separate units that sequentially become available to pollinators. Examples are asynchrony of flower-opening (usually the case) and staggered anther-dehiscence within the flower. Dispensing is a floral behaviour that restricts the amount of the available pollen that a pollinator removes during one visit. Pollen release in plants such as *Lupinus*, *Lobelia* and *Campanula* is by dispensing. The models were based on two hypotheses of the pollen removal pattern, 'numerical', whereby the same amount of pollen is removed during each visit, and 'proportional', whereby the same proportion of the available pollen is removed at each visit. Which situation applies is not known for any plant, and it is thought that actual pollen removal can stand between the two. However, it is suggested that dispensing mechanisms would result in numerical removal. Numerical removal is presumed to be a feature of *Orchidaceae* and *Asclepiadaceae*. I think it should also characterize explosive flowers among which are a number with secondary pollen presentation.

The way the models might operate in natural situations was also considered. The experimental results with *Erythronium* showed that the proportion of removed pollen that reaches the stigmas of other flowers decreases as the amount of pollen removed from a donor-flower during an insect-visit increases. Pollination will therefore be more effective if the flower limits the amount of pollen removed at each visit (agreeing with the predictions of earlier authors). This tendency will be countered by any limitation on the number of insect visits. There might be an average number of visits received by flowers in a population which would set the limits. Given such a statistic the plant will do best if it allows all pollinators visiting it to participate in pollen dispersal, while ensuring the removal of most of the pollen by the time flowering finishes (i.e. to spread the pollen over the 'expected' number of visitors).

If the expected number of visits is variable (for example, according to position of the plant in the population) the optimal pollen packaging differs according to removal pattern ('numerical' versus 'proportional'), and with proportional removal it further differs according to the attenuation coefficient (which measures loss in transit, mostly by insect grooming). Many plants have inbuilt flexibility that copes with variability of pollinator visits. Thus, if the probability of visits is low the longer intervals between visits may result in larger doses of pollen when the pollinators arrive. HARDER & THOMSON think such flexibility is unlikely in dispensing mechanisms such as that of *Lupinus*, and on this view dispensing mechanisms should not be evolved unless a high frequency of visitation can be relied upon. However, my accounts of *Lobelia* and *Centaurea* suggest that flexibility is not ruled out in all cases of dispensing. Thus, subject to various constraints, pollen packaging and pollen dispensing are advantageous to the male function of the flower. However, in the majority of plants one would expect that the production of numerous flowers not opening simultaneously would provide sufficient subdivision of the pollen, in which case it is difficult to see why, on these grounds, there should be dispensing in addition. This difficulty does not apply to *Wahlenbergia albomarginata* (LLOYD & YATES 1982) which produces few flowers at any one time on each clonal clump, so some further subdivision of the available pollen by dispensing could be advantageous. In other cases, one should perhaps continue to view dispensing as related to protection. This does have a role in the HARDER & THOMSON models because larger doses suffer more loss in bee-grooming than smaller doses. The question is, why not produce less pollen in each flower and rely on packaging? Measurement of dispensing has yet to be attempted, and how far it is flexible remains to be investigated.

Explosive systems represent the case where only one package of pollen is issued by each flower and it would seem that it is immaterial whether removal is numerical or proportional. They probably give rather better pollen protection overall and are probably more specific in pollinator-selection than non-explosive release systems. Both the cloud type and the deposit type seem to be directed to putting pollen on to the vector in places from which it cannot be removed by grooming. Pollen pick-up and deposition during a single pollinator visit is efficient in that it maximizes the rate of pollen transfer (LLOYD 1979[1]) provided pollinators can recognize tripped flowers and avoid spending time visiting them.

The four possible functions of secondary pollen presentation listed in Chapter 1 still seem to be the only ones of widespread significance. They are harmonization of site of presentation and reception of pollen, protection of pollen, placement of pollen on the vector where it cannot misuse it, and pollen-dosing. Systems of type A (see earlier in this section) can fulfill all four functions; those of type B cannot provide protection and do not seem to put the pollen on inaccessible parts of the vector.

---

[1]He quotes BAKER & HURD (1968) in support of this point but with little justification.

However, other reasons for having secondary pollen presentation appear to apply in a few special cases. These are as follows:

1. to spread pollen more widely over the pollinator (as in *Myristica*, *Myristicaceae*, and perhaps *Cytinus*, *Rafflesiaceae*);
2. adaptation to special bees (in *Eucnide* and *Mentzelia*, *Loasaceae*);
3. use of hollow petals as pollination chambers which small flies are induced to enter to pick up pollen which is deposited there (in *Herrania*, *Sterculiaceae*);
4. a need to imitate the reward-providing stigmatic branches (in *Blyxa*, *Hydrocharitaceae*);
5. placement of pollen on the corolla lobes to remove it from flooding by the nectar which fills the corolla tube (in *Acrotriche*, *Epacridaceae*);
6. to enable mixing of pollen with an oily fluid from a gland on the anther which seems to be useful for sticking pollen to beetles and the beaks of birds (as in *Myrtaceae*).

The pollination-type in cases (1) and (2) is only slightly removed from the 'mess-and-soil' method.

Kerner (1873) reviewed the devices that protect pollen against premature removal by wind or premature moistening by rain and dew. All the types of secondary pollen presentation that involve enclosing the pollen afford such protection. Kerner recognized that floral arrangements could have more than one function, and I would expect that in most plants with concealed, secondarily presented pollen, protection against abiotic disturbance is merely a supplementary benefit.

3. Evidence for pollen-saving

If pollen-saving was an important reason for secondary pollen presentation one would expect that the transfer of pollen to the presentation system would be efficient.

There is evidence for both efficiency and inefficiency in this. I have noted only small amounts of pollen remaining in the anthers of *Petromarula* (*Campanulaceae*) and of *Marantaceae*. In some cases where apparently large amounts remain there is evidence for malfunction, in that the event is occasional and on other occasions very efficient transfer to the presenter takes place. This applies to some of my observations on *Asteraceae*, and to *Turraea* (*Meliaceae*), where I found transfer to be poor but F. White (pers. comm.) has found good transfer. Such effects are probably related to unfavourable environmental conditions. There is some evidence of this in the work of Stumpf (1988), where failures occurred with exotic plants in cultivation. The wastage that I observed in *Phyteuma* (*Campanulaceae*) might have been related to the age of the inflorescence or the lateness of the season.

Other instances of pollen being wasted are those of *Isotoma petraea* (*Lobeliaceae*), where some pollen is dropped as the style emerges from the pollen magazine, some *Rubiaceae* and *Scaevola plumieri*, where some pollen remains in the pollen cup, though this might be used in selfing if needed. Gross wastage was recorded for two *Asteraceae* by Heyn & Snir (1986) where 18.5% of the pollen fell, apparently spontaneously, into the corolla tubes.

Clearly this subject should be properly studied and comparisons made with wastage by retention in the anthers in plants with primary pollen presentation.

Table 3-1. Pollen-ovule ratios of primary and secondary pollen presenters, derived from Cruden, 1977

Class	p	c	s
3	171.25	168.5	152.70
4	830.08	796.6	616.38
5	6194.36	5859.2	3401.67

All members of group 's' are in family Asteraceae

'Class' is Cruden's outcrossing-index class (see text); '*c*' is Cruden's figure for the entire class, '*p*' is the mean of all results for primary pollen presenters, '*s*' is the same for secondary presenters

The only observation of the latter kind that I know of is in the study of heterostyly in *Primula vulgaris* by PIPER & CHARLESWORTH (1986). Enormous amounts of pollen remained in the anthers; over 2/3 in pin flowers, over half in thrums and about one-fifth in long homostyles. The 'explosive cloud' type of pollen release appears inherently wasteful but it may pay for itself if it places adequate quantities of pollen where it cannot be groomed off.

Another indication of pollination efficiency is a low pollen-ovule ratio (CRUDEN 1977). I believe that it will turn out to be quite low in *Marantaceae*. In *Apocynaceae* there are only 32 pollen grains per ovule in *Nerium oleander*, a value that is characteristic of obligately autogamous plants, though this species would have been classified as facultatively xenogamous by CRUDEN. The pollen/ovule ratio in *Vinca* is 2,500 to 3,500 which is outside the range that CRUDEN found in his facultatively xenogamous class but within that for the xenogamous class, though below the average of 5859 ± 937. All the plants with secondary pollen presentation used by CRUDEN belong to the *Asteraceae*; all have pollen/ovule ratios below the average for their outcrossing index (Table 3–1) and as the amount of xenogamy goes up from class 3 to class 5 so this difference increases. However, the difference is not statistically significant, on account of the large variance (C. A. GILLIGAN, pers. comm.) [a further test, taking account of the direction of the deviations, is possible but Dr GILLIGAN has not been able to carry it out].

4. Ways of preventing selfing

In the great majority of families with secondary pollen presentation the style is involved as part of the presentation mechanism (Table 3–3, p. 225). It might be that, since in these families the flower's own pollen is close to its stigmas, self-incompatibility would be advantageous. There are thirteen families with stylar presentation for which we have some information on the occurrence of self-incompatibility and self-compatibility. Table 3–2 clearly shows that self-incompatibility is not an essential part of the stylar presentation syndrome. The independence of the occurrence of self-incompatibility and stylar presentation is emphasized by the *Papaveraceae*, which show self-incompatibility in both the subfamilies that show secondary pollen presentation, although only one of them has stylar presentation. Further evidence in support of this view is to be found in the discussions of the families *Leguminosae*

Table 3-2. Presence of self-incompatibility and self-compatibility in families in which the style takes part in secondary pollen presentation

Family	Self-incompatibility	Self-compatibility
Papaveraceae	x	x
Leguminosae	X	x
Myrtaceae		x
Proteaceae	x	x
Polygalaceae		x
Apocynaceae	x	x
Campanulaceae	x	x
Lobeliaceae	x	x
Goodeniaceae		x
Rubiaceae	x	X
Asteraceae	X	x
Cannaceae		x
Marantaceae		x

A large symbol indicates the preponderant condition on present evidence (the indication for Rubiaceae refers to homostylous groups)

and *Proteaceae*. In *Rubiaceae* most occurrences of self-incompatibility are in the many heterostylous genera. There seem to be three other ways of preventing autogamy in stylar pollen-presenters: (1) sufficient separation between the stigmatic surfaces and pollen-presenting areas, (2) dichogamy and (3) ability to receive and donate pollen simultaneously on the occasion of the first pollinator-visit (the last will not stop all autogamy but will at least ensure that allogamy is not precluded by earlier autogamy). All these systems are widespread in plants with stylar presentation and it seems likely that they (rather than self-incompatibility) have the function of preventing interference with the breeding system that could result from stylar pollen presentation. Therefore, although all secondary pollen presenters are by nature at least partially allogamous their breeding system is otherwise apparently determined independently; here, as elsewhere, self-incompatibility is one of the regulators of the breeding system. It might seem that method (1) of preventing autogamy in secondarily pollen presenting plants conflicts with the function of harmonizing site of pollen presentation and reception within the flower. However, the pollen is always deposited close to the stigmatic region and dichogamy usually occurs as well. In many *Leguminosae* the spatial separation is supplemented by a physical barrier in the form of a peristigmatic hair-fringe.

5. Stigmatic membranes

Method (3) for preventing selfing is often the result of containment of the stigmatic fluid by a fragile membrane which is broken by the first visitor. This is known with certainty only in the *Papaveraceae* and *Leguminosae*. These are families in which the stigma itself is embedded in pollen at the beginning of anthesis. The membrane ensures that the flower's own pollen will not be wetted by the stigmatic secretion and that it will not germinate until a potential pollinator arrives. It allows the stigma to become functional from the moment foreign pollen is delivered to the flower. In some *Leguminosae* (*Lupinus*) there is no membrane; instead the flowers are protandrous, with the stigma not reaching a receptive condition until some time after the flower has opened.

6. Pollen longevity and secondary pollen presentation

Self-incompatibility is of two kinds, gametophytic and sporophytic, and among homomorphic (i.e. non-heterostylous) plants the second occurs in only two predominantly zoophilous families, *Brassicaceae* and *Asteraceae* (Gibbs 1986), and these have trinucleate pollen grains. Trinucleate pollen grains are much shorter-lived than binculeate. Pollen of *Chrysanthemum segetum* (*Asteraceae*) can last 5 h at 30 °C and about 18 h at 26 °C, whereas (binucleate) pollen of *Nicotiana alata* (*Solanaceae*) can last respectively 70 h and 400 h at these temperatures (Hoekstra & Bruinsma 1975). Even the short viability of *Chrysanthemum* pollen is not likely to be relevant to the transit time of vectors from blossom to blossom, since this will be much less. On the other hand, pollen viability is lost much more rapidly at very high humidity, as this allows respiration and the consumption of reserves. Hoekstra & Bruinsma (1975) found that at RH 97%, pollen of *Aster* lost germinability sharply after 4 h at 26 °C. Assuming that humidity is high within the anther-cylinder of *Asteraceae* and that oxygen is available, pollen would need to be dispersed within a few hours of anther dehiscence. This seems to be the case in *Asteraceae*, where florets change from the male to the female stage on either their first or second day. In *Lobeliaceae*, where pollen is also retained in the anther-cylinder, it has been found to be binucleate in one species of *Siphocampylus* and five of *Lobelia*, but trinucleate in one species each of *Isotoma* and *Lobelia* (in the binucleate *L. cardinalis* pollen may be presented over periods ranging from three to ten days). Close enclosure of pollen after release from the anthers also occurs in *Leguminosae*, *Papaveraceae*, *Goodeniaceae* and perhaps *Polygalaceae*; pollen is binucleate in the first four of these and binucleate or trinucleate in the last. It would be interesting to have information on RH and oxygen availability in the pollen magazines of plants in these families and on how excessive respiration is avoided within the anthers immediately before dehiscence. The conditions under which the pollen is held on a 'pseudostamen' (see Chapter 1) must be similar to those in the dehisced anthers of flowers that do not have secondary pollen presentation. Most of these have binucleate pollen but a few have both binucleate and trinucleate. It would perhaps be worth acquiring information on the duration of the phases of anthesis in the trinucleate representatives of these families.

7. Pollen flowers

Pollen flowers with secondary pollen presentation appear to be confined to *Leguminosae* and *Proteaceae*. The examples from the former are *Coronilla* excluding *C. emerus*, *Ononis spinosa*, *Lupinus luteus* and all members of tribe *Genisteae*; from the latter *Isopogon dubius*, *Petrophile biloba* and *Grevillea pilulifera*. Bees are the pollinators in the examples from *Leguminosae* and the first two of those from the *Proteaceae*, as is usually the case with pollen flowers (Vogel 1978). The rarity of pollen flowers with secondary presentation is probably no greater than in zoophilous flowers in general. Dr CHR. WESTERKAMP (pers. comm.) suggests that some nectarless flowers that are visited for pollen are really deceit flowers, and that insects that obtain their pollen may not be pollinators. This seems particularly likely to be the case with *Coronilla*.

8. Adhesives

Adhesives are used to fasten pollen to pollinators in *Myrtaceae*, *Polygalaceae*, *Apocynaceae*, possibly *Campanulaceae-Wahlenbergieae*, *Goodeniaceae*, *Marantaceae* and perhaps rarely in *Rubiaceae*. They might be needed either to defeat the effects of grooming, or to utilize inaccessible parts of the insects' bodies that are hairless. The first possibility seems the less likely, because a glue strong enough to resist grooming is not likely to relinquish the pollen grains when they touch the stigma. The question arises whether the stigma could be made tacky enough to cause this. Description of the physical and chemical properties of the adhesive is available only in the case of *Apocynaceae* (see Chapter 2).

9. Structures involved in secondary pollen presentation in relation to taxonomic position

Among the dicotyledons there appears to be a dichotomy between the subclass *Asteridae* on the one hand and the remaining subclasses on the other (the *Metachlamydeae* and *Archichlamydeae* respectively in the ENGLER system). One difference is in the frequency with which the style is involved. Table 3–3 lists different structural systems according to the family in which they occur. This shows that the style is used in all *Asteridae*, whereas among other subclasses there are five families in which the style is not used (*Myristicaceae*, *Sterculiaceae*, *Loasaceae*, *Epacridaceae* and *Rhizophoraceae*) and in addition there are genera that do not use the style in *Papaveraceae* and *Leguminosae*. A second difference lies in the organ that complements the style. In subclass *Asteridae* the perianth (corolla) does so only as a rare exception (in three families). In the Archichlamydeous subclasses the perianth complements the style in *Papaveraceae* – type A, *Rhizophoraceae*, *Polygalaceae* – type B and in all types of *Leguminosae*. (In *Papaveraceae* – type B, and *Polygalaceae* – type A, the perianth is important in pollen protection but does not come into contact with the pollen). A probable reason for this difference is that the separateness of the petals, allowing independence of movement, makes division of labour possible among them, so that the functions of display and pollen presentation are emphasized in different members. The relative unavailability of the perianth for the job of protection in the *Asteridae* may, indeed, be the reason why the style is more used in this group. A third difference is that the anthers of *Archichlamydeae* never form the cylinder of a piston mechanism (with one partial exception in *Leguminosae*); this arrangement is confined to the *Asteridae* (*Cyphiaceae*, *Lobeliaceae* and *Asteraceae* – type B).

The two groups are much alike in two other respects. Transfer of the whole of the pollen (or all the presented pollen) to an organ other than the anthers is implicit in cases where the 'Anthers' column in Table 3–3 is empty; it applies to 14 out of 18 units in the *Archichlamydeae* and to 8 out of 12 in the *Asteridae*.

Table 3-3. Structures used in secondary pollen presentation by taxonomic position (somewhat simplified)

Subclass, order, family	Perianth	Filaments	Anthers*	Style	Other
Magnoliidae					
Magnoliales					
Myristicaceae	+		+		
Papaverales					
Papaveraceae-A	+				
Papaveraceae-B				+	
Dilleniidae					
Malvales					
Sterculiaceae-A	+?	+	(+)		
Sterculiaceae-B	+				
Violales					
Loasaceae		+			top of ovary
Ericales					
Epacridaceae	+				
Rosidae					
Fabales					
Leguminosae-A	+			+	
Leguminosae-B	+	+			
Leguminosae-C	+	+	+		
Myrtales					
Rhizophoraceae	+		+		
Myrtaceae				+	
Proteales					
Proteaceae				+	
Sapindales					
Meliaceae				+	
Rafflesiales					
Rafflesiaceae	+				
Polygalales					
Vochysiaceae				+	
Polygalaceae-A				+	
Polygalaceae-B	+			+	
Asteridae					
Gentianales					
Apocynaceae				+	
Campanulales					
Campanulaceae-A				+	
Campanulaceae-B	+			+	
Cyphiaceae			+	+	
Lobeliaceae			+	+	
Goodeniaceae-A				+	
Goodeniaceae-B	+			+	
Rubiales					
Rubiaceae				+	

Table 3-3. (*Continued*)

Subclass, order, family	Perianth	Filaments	Anthers*	Style	Other
Asterales					
Asteraceae-A				+	
Asteraceae-B			+	+	
Asteraceae-C	+			+	
Alismatidae					
Hydrocharitales					
Hydrocharitaceae	+				
Commelinidae					
Commelinales					
Xyridaceae		+			
Zingiberales					
Cannaceae				+	
Marantaceae				+	
Liliidae					
Liliales					
Liliaceae				+	

*(+) = primary presentation
Subdivisions of families do not necessarily correspond with taxonomic subdivisions. Where more than one structure is indicated without qualification all are usually operating in a co-ordinated way in the same plant. For Leguminosae only the important types are given.

The style is used on its own in 5 cases out of 11 in *Asteridae* and 6 out of 18 in '*Archichlamydeae*'.

10. Type of pollen presentation in relation to taxonomic position

The simplest type of secondary pollen presentation is 'exposed'. This occurs in four isolated genera of monocotyledons and in a number of dicotyledonous families outside the *Asteridae* where secondary pollen presentation is rare (*Myristicaceae*, *Sterculiaceae*, *Loasaceae*, *Epacridaceae*, *Myrtaceae* and *Meliaceae*). In the *Proteaceae*, on the other hand, exposed presentation (on the style) is widespread. Among the *Asteridae* the *Apocynaceae* are a special case because secondary pollen presentation seems to be more or less incidental to a specialized pollination system and, although I have defined 'exposed' so that it describes their pollen presentation, the pollen is usually in fact well concealed. Among other *Asteridae* the *Campanulaceae* constitute a parallel to *Proteaceae*, having exposed stylar presentation as the norm. This is also the only system found in the *Rubiaceae*. In the *Asteraceae* 'exposed' pollen presentation (again using the style) characterizes the large tribe *Lactuceae* and has scattered occurrences elsewhere.

Systems for the non-explosive release of concealed pollen ('release'; first two columns of Table 3–4) are widespread and usually appear to involve the issue of the pollen in successive doses. Expulsion of such concealed pollen by growth is almost confined to the *Asteridae*. In the *Leguminosae* it is known only in *Clianthus*.

Table 3-4. Method of secondary pollen presentation by taxonomic position

Subclass, order, family	Release by pollinator	Release by growth	Explosive cloud	Explosive deposition	Exposed
Magnoliidae					
Magnoliales					
Myristicaceae					+
Papaverales					
Papaveraceae	+			+	
Dilleniidae					
Malvales					
Sterculiaceae					+
Violales					
Loasaceae					+
Ericales					
Epacridaceae					+
Rosidae					
Fabales					
Leguminosae	+	+[1]	+	+	
Myrtales					
Rhizophoraceae			+		
Myrtaceae					+
Proteales					
Proteaceae			+?	+	+
Sapindales					
Meliaceae					+
Rafflesiales					
Rafflesiaceae					+
Polygalales					
Vochysiaceae					+
Polygalaceae	+			+	
Asteridae					
Gentianales					
Apocynaceae					+
Campanulales					
Campanulaceae		+			+
Cyphiaceae		+			
Lobeliaceae	+	+			
Goodeniaceae	+	+			
Rubiales					
Rubiaceae					+
Asterales					
Asteraceae	+	+		+	+
Alismatidae					
Hydrocharitales					
Hydrocharitaceae					+
Commelinidae					
Commelinales					
Xyridaceae					+

Table 3-4. (*Continued*)

Subclass, order, family	Release by pollinator	Release by growth	Explosive cloud	Explosive deposition	Exposed
Zingiberales					
Cannaceae					+
Marantaceae				+	
Liliidae					
Liliales					
Liliaceae					+

Release by pollinator and release by growth are sometimes combined in the same species.
[1]Occurs in *Clianthus* (Chr. Westerkamp, pers. comm.; not covered in my description).

One might expect it to occur more often in this family, but the reason for its scarcity may be that the family has such a well-developed system of springy petals which can be actuated by the pollinator that there is no need for expulsion by growth which, after all, has a possible disadvantage in not being linked to pollinator visits. Among the *Asteridae*, expulsion by growth in *Cyphiaceae*, *Lobeliaceae* and many *Asteraceae* is caused by the style's moving through the anther-cylinder, whereas in *Goodeniaceae* it is caused by the growth of the style-tip within the stylar pollen-cup (indusium). In the *Lobeliaceae*, *Goodeniaceae* and *Asteraceae* there is frequently also an element of response to pollinator visits. In *Goodeniaceae* this takes place by pressure on the lips of the pollen cup, in *Lobeliaceae* through the non-living trigger-hairs of the anthers and in *Asteraceae* through irritability of the stamen-filaments. (It should be borne in mind that descriptions do not always say whether an element of dosing is involved or make clear the role of growth in expelling the pollen).

Of the explosive systems recorded, the scattering of a cloud of pollen occurs in *Rhizophoraceae* and explosive deposition in *Marantaceae*, neither of which families shows any other kind of secondary pollen presentation. There remain four families where explosive systems occur alongside others. In three of these, *Papaveraceae*, *Proteaceae* and *Polygalaceae*, it is reasonable to assume that they evolved out of passive or less vigorous systems widespread in their respective families. In the *Leguminosae* there is evidence of so much parallel evolution among the genera which release concealed pollen in response to pollinator visits that one should allow the possibility of the independent origin of the explosive systems (unless perhaps they depend physically on the previous establishment of a non-forceful mechanism).

Explosive systems that have to be tripped involve the development of tension between two organs (or sets of organs). Tension between the perianth and androecium occurs in *Pseudofumaria alba* (*Corydalis ochroleuca*, *Papaveraceae*), while in *Genista* and *Cytisus* (*Leguminosae*) it appears that tension exists in both the androecium and the style, in opposition to the keel petals. In *Polygala myrtifolia* (*Polygalaceae*) and the explosive *Proteaceae*, however, it is the style that is opposed to the perianth. In the peculiar *Rhizophoraceae* (*Bruguiera*) the tension is mostly within the petals, but there seems to be some between these and the stamens. There seem to be no explosive secondary pollen presentation systems in the *Asteridae* except for one marginal case in *Asteraceae*, although some *Lamiaceae* (see Chapter 1) and one Rubiaceous genus (*Posoqueria* – see discussion of this family in Chapter 2) have explosive systems of primary presentation. The monocotyledonous family *Marantaceae* provides a case where the perianth, and in this case also the petaloid staminodes, are united and yet the system is explosive, but the extent of the union (and hence the resemblance to the *Asteridae*) is slight; here the tension is between the style and the staminodes.

11. Evolution of secondary pollen presentation: constraint and canalization

On a family basis secondary pollen presentation is about equally abundant as heterostyly, occurring in 25 as against 24 families (Ganders 1979).

We have now established a number of possible functions for secondary pollen presentation, though there is doubt about its function in certain cases. Where we have found a function, we may at first think we have the answer to the question, why present pollen secondarily? This is not so, however, because all the things that secondary pollen presentation can do can also be done by plants with primary pollen presentation. Thus the 'release' systems of *Leguminosae* are similar in their effect to the 'valvular' system that is common in the family, in which the pollen-bearing anthers themselves emerge from the keel of the flower when a visit takes place. Similar protective systems are widespread. Likewise, explosive deposition takes place in, for example, some *Lamiaceae* and in *Stylidiaceae* (see Chapter 1), where pollen presentation is not secondary. In *Loranthaceae* explosive release of a cloud of pollen is characteristic but the pollen is apparently not in contact with organs other than the anthers. Again, growth or movement of stamens can ensure that pollen is presented wherever in the flower it is required without its first being transferred from the anthers to another structure. The explanation of this state of affairs must lie at the door of constraint and canalization.

Burtt (1961), in discussing *Asteraceae*, pointed out (p. 218) that the requirements of the flowering and fruiting stages of the capitulum will certainly interact. This notion was endorsed by Stirton (1983) and an example of such constraint was pointed out by Yeo (1972) as follows: in *Lamiaceae* and *Boraginaceae*, with a maximum of four seeds per flower, adoption of the course of producing large numbers of small seeds per plant (rather than fewer larger seeds) will demand the production of large numbers of small flowers; this will inevitably restrict such plants to certain pollination types (though not necessarily the number of insect species that can pollinate it).

It is postulated in my account of the *Campanulaceae* that the adoption of a bell-shaped flower and the offering of the central style as an alighting place created a problem of pollen presentation. The pollinator's contact with the stigmas is ventral; if the anthers were centralized round the style they would be largely ineffective because they are introrse. On the other hand, if they were spread out to lie against the inside of the bell, they would contact the pollinators dorsally. The introrse anthers thus create structural constraint. Secondary pollen presentation is a way of overcoming the difficulties.

Similarly, if my hypotheses about the importance of compactness of the floral apparatus in *Proteaceae* and *Asteraceae* are valid, one could say that the dense inflorescence had been a constraint leading towards secondary pollen presentation. In the case of *Proteaceae*, however, the inferences that have been made about phylogeny imply the origin of secondary pollen presentation in separate lines, the more primitive of which tend to have looser inflorescences. This seems to indicate that there has been a similar response in different lines to certain selection pressures; if the similarity in response is due to common elements in the construction of the flower one could speak of 'canalization'.

The term 'canalization' is appropriate where flower structure ('Bauplan' in the sense of VOGEL 1954) creates a certain predisposition. Presumably the sympetalous corolla, for example, arose in coevolution with increasing proboscis-length of insect-pollinators. Groups that have acquired this corolla type are not prevented from evolving rotate corollas (more common in the more primitive separate-petalled types) in their subsequent radiations (e.g. *Verbascum* and *Veronica* in *Scrophulariaceae*). However, it would seem likely that the sympetaly itself imposes a canalizing constraint against following this evolutionary path, and in favour of pollination adaptations that make use of the sympetalous construction. Similarly, flowers with separate petals and with stamens free from the petals have repeatedly been able to give rise to the papilionate flower, characterized by division of labour among the petals and independent movement of the androecium.

In the family accounts in Chapter 2 it has been noted that members of *Leguminosae-Papilionoideae*, *Papaveraceae* and *Polygalaceae* possessing the 'papilionate' flower-form are particularly adapted (I earlier said 'committed') to bee-pollination. The initial steps in the evolution of separate-petalled flowers towards pollination by at least the more specialized types of bee are acquisition of zygomorphy and the adoption of a protective role for some of the petals (see previous paragraph). Protection might have been called for either to save the pollen from the depredations of other insects or to control the proportion of the pollen that the bees take for their own needs. Although at least some genera in these bee-pollinated groups presumably trade pollen as well as nectar for pollination, it seems likely that the second of these suggested roles for protection is the more important. Nevertheless, a survey of which bees actually do forage for pollen on which 'papilionate' flower species would be of interest in relation to secondary pollen presentation. In any case, I suggest that once the first steps towards zygomorphy and pollen-protection are taken the plant's evolution will be canalized in this direction, and will readily culminate in the adoption of secondary pollen presentation. This implies either that pollen secondarily presented can be better protected than by analogous forms of primary presentation, or that control of dosing is being selected for (see previous section), and that secondary pollen presentation can do this better. The numerous parallel lines within *Leguminosae* indicate how strong are the selective pressures brought to bear once the initial steps have been taken.

Canna (*Cannaceae*), one of the genera with exposed pollen presentation, is particularly difficult to understand. An easy explanation would be that *Canna* evolved from *Marantaceae*, and the floral differences are the result of change from melittophily to sphingophily, ornithophily and chiropterophily. However, although *Cannaceae* and *Marantaceae* appear to be each other's closest relatives within the

Zingiberales (DAHLGREN, CLIFFORD & YEO 1985; KRESS 1990), the *Cannaceae* seem in some ways to be the less advanced family. If, nevertheless, the suggestion is right, *Cannaceae* would provide a case of canalization. With the adoption of different pollination syndromes the explosive deposition system would have become unnecessary but there would have been no disadvantage in continuing to present the pollen on the style. Other suspected cases of such 'residual' secondary pollen presentation are among the insect-pollinated species of *Bruguiera* (*Rhizophoraceae*) and of *Proteaceae*, certain species of *Verticordia* (*Myrtaceae*), the genera of *Campanulaceae* referred to in the next paragraph and probably some *Rubiaceae*.

While the *Campanulaceae* appear to experience constraint in their possession of introrse anthers, their adoption of a basically bell-shaped flower with the style in the position of the clapper in a bell perhaps contributes an element of canalization. It may be postulated that it was this that made the introrse anthers an embarrassment, to be overcome by the adoption of secondary pollen presentation. Those genera that have departed from this flower-form ('Gestalttyp' – VOGEL 1954) have retained secondary pollen presentation and thus have taken further evolutionary steps now canalized by the pollen presentation system. They present a state analogous to that postulated above for *Canna*.

In the extraordinary case of *Blyxa* (*Hydrocharitaceae*) the inducement to secondary pollen presentation seems to be a need to make the pollen and its presenter look like the styles; this requirement must presumably be rooted in some aspect of the behaviour of the pollinators that the flower exploits. Thus mimicry (intraspecific) is the inducement to the evolution of secondary pollen presentation.

12. Speculations on courses of evolution

Speculations on evolution may be worth making in a few cases. In *Proteaceae* and *Leguminosae* there is evidence not directly based on secondary pollen presentation systems regarding the number of lines of descent in the family. These have indicated, as already stated, that secondary pollen presentation probably arose repeatedly in each family. In the case of *Leguminosae* there is an instance of a strongly papilionate flower in *Cercis* in the primitive subfamily *Caesalpinioideae*, where secondary pollen presentation is not known. This is considered not to be ancestral to subfamily *Papilionoideae*, and it reinforces the evidence for parallel evolution of this flower form in the family. Thus in general we cannot (at least at present) see any evolutionary sequences within these families. However, in *Leguminosae* it is reasonable to assume that the many groups showing secondary pollen presentation arose from various groups having papilionate flowers with primary pollen presentation. Also in this family, explosive systems involving secondary pollen presentation occur only in the tribe *Genisteae* and in *Medicago* (tribe *Trifolieae*). In the *Genisteae* they arose only once; as a secondary non-explosive system is already represented in the tribe it may have arisen by this route. Explosive flowers in *Polygalaceae* and *Papaveraceae* must also have arisen from non-explosive types which already had secondary pollen presentation.

The problem in *Campanulaceae* which I have suggested arose from the combination of introrse anthers with a bell-shaped flower could have been solved in other ways than by exposed pollen presentation on the style: (1), the anthers could have become extrorse and moved to a central position, (2), they could have become poricidal, with dry pollen falling out of the tips, (3), the anthers could have moved to a central position and formed a cylinder round the style, thus establishing the piston-type of secondary pollen presentation which incorporates the additional feature of pollen protection. Presumably this last alternative actually happened and gave us the related family *Lobeliaceae*. It is strongly tempting to believe that this provided the necessary foundation for the origin of the Asteraceous capitulum – its importance lying in its compactness. *Lobeliaceae* have zygomorphic flowers, which suggests that if they do have a close phylogenetic connection with *Asteraceae*, zygomorphy may be primitive in that family, a suggestion already raised in our discussion of *Asteraceae* (Chapter 2).

Goodeniaceae might have suffered from the same problem. Their solution is not one that would have been spontaneously predicted because it has involved the development of a new structure on the style.

Chapter 4

Summary and conclusions

On the adopted definition secondary pollen presentation occurs in 25 families. It takes two main forms, exposed presentation of pollen on a structure other than the thecae, and issue from an enclosing structure, taking place mainly or exclusively in response to pollinator visits. In the second class some flowers have an explosive mechanism which has to be tripped and these in turn fall into two groups, those depositing the pollen directly and those that release a cloud of pollen into the air.

Four functions of secondary pollen presentation are probably of widespread significance. They are harmonization of site of presentation and reception of pollen, protection of pollen, placement of pollen on the vector where it cannot misuse it, and pollen-dosing. Any or all can be involved in a particular species. However, exposed presentation can usually only support the first function, though in *Campanulaceae* firm adhesion to a pollen brush combined with progressive shrinkage of the hairs of the brush results in pollen-dosing. Pollen-dosing takes the form of 'dispensing' in the sense of HARDER & THOMSON (1989). Since all plants with numerous flowers opening over a period also 'package' their pollen, it is not clear that dispensing has been selected for as a means of maximizing the number of pollination events as such; it seems more likely to be primarily a protection arrangement reducing pollen losses, which should itself improve pollination efficiency.

In plants with secondary pollen presentation the style is usually involved in some way with pollen presentation. The following arrangements are important in promoting outcrossing or at least in preventing autogamy from precluding allogamy: (1) slight but sufficient separation between stigmatic surfaces and pollen-presenting areas, (2) dichogamy, and (3) simultaneous receipt and donation of pollen during the first (often the only) pollinator-visit. In the last case stigmatic receptivity is often regulated by a stigmatic membrane that is broken by the first potential pollinator.

The use of the style in secondary pollen presentation is universal in *Asteridae* whereas in other subclasses it is not used in some families and used only in some cases in others. On the other hand the anthers in the other subclasses hardly ever form the cylinder of a piston mechanism, as they frequently do in *Asteridae*. Exposed secondary pollen presentation and non-explosive issue of previously shed pollen are not restricted to particular subclasses. However, pollen issue of the latter kind is frequently caused by growth in the *Asteridae*, whereas in the other subclasses it is usually by a visitor-actuated mechanism. Explosive discharge involving secondary pollen presentation is almost absent in the *Asteridae*.

All the roles that secondary pollen presentation can fill are also capable of being filled without secondary pollen presentation. The existence of secondary pollen presentation must therefore result mainly from constraint and canalization in evolution (though possibly some things could be done better by secondary than by primary presentation). Introrse anthers seem to be such a constraint in *Campanulaceae* and *Asteraceae*, and crowded inflorescences in *Proteaceae* and *Asteraceae*. Lack of space in individual flowers may be a constraint in *Rubiaceae*. The form that secondary pollen presentation takes may be canalized by the basic structure of the flower: in *Asteridae* it is more difficult to arrange for division of labour of the petals and movement of the stamens independently of the petals than in separate-petalled flowers, and this has canalized the evolution of secondary pollen presentation. The papilionate flower (found in *Papaveraceae*, *Leguminosae* and *Polygalaceae*), in which such movements occur, is especially associated with bee-pollination. In some cases, such as *Campanulaceae*, constraint may have led to secondary pollen presentation which may have then canalized further evolution in the group.

There is great scope for further work on secondary pollen presentation, since information on the pollination behaviour of most plants that display it is still far from complete. Appendix 1 pinpoints particular problems in this field and indicates also that general study is required on pollen-dosing and the efficiency of secondary pollen presentation, including evidence from pollen/ovule ratios and the size of pollen residues in the flower.

Epilogue

I now wish again to thank CHRISTIAN WESTERKAMP for continuing to ply me with reprints and references while I have had my head down during the later stages of writing and have therefore not been able to keep a proper watch for new publications. Even so, I have probably missed some things that I ought to have quoted, and for this I apologize to their authors and to my readers.

General references

These references are to works cited in Chapters 1, 3 and 4 and Appendix 1, together with those works cited in Chapter 2 that are of wide scope and/or referred to frequently. Each family in Chapter 2 also has its own bibliography.

BAKER, H. G., HURD, P. D., 1968: Intrafloral ecology. – Annual Rev. Entomol. **13**: 385–414.

BOUTIQUE, R., 1972: *Gentianaceae*. – In BAMPS, P., (ed.): Flore d'Afrique Centrale, Spermatophytes: *Gentianaceae*. – Bruxelles: Jardin botanique national de Belgique.

BRANTJES, N. B. M., 1982: Pollen placement and reproductive isolation between two Brazilian *Polygala* species (*Polygalaceae*). – Pl. Syst. Evol. **141**: 41–52.

BRANTJES, N. B. M., 1983: Regulated pollen issue in *Isotoma*, *Campanulaceae*, and evolution of secondary pollen presentation. – Acta Bot. Neerl. **32**: 213–214.

BRANTJES, N. B. M., DE VOS, O. C., 1981: The explosive release of pollen in flowers of *Hyptis* (*Lamiaceae*). – New Phytol. **87**: 425–430.

BREWBAKER, J. L., 1967: The distribution and phylogenetic significance of binucleate and trinucleate pollen grains in the angiosperms. – Amer. J. Bot. **54**: 1069–1083.

BURTT, B. L., 1961: *Compositae* and the study of functional evolution. – Trans. Bot. Soc. Edinburgh **39**: 216–232.

BURTT, B. L., 1977: Aspects of diversification in the capitulum. – In HEYWOOD & al. 1977.

CAROLIN, R.C., 1960: The structures involved in the presentation of pollen to visiting insetcs in the order *Companales*. – Proc. Linn. Soc. New South Wales **85**: 197–207.

CRUDEN, R. W., 1977: Pollen-ovule ratios: a conservative indicator of breeding systems in flowering plants. – Evolution **31**: 32–46.

DAHLGREN, R. M. T., CLIFORD, H. T., YEO, P. F., 1985: The Families of the Monocotyledons. – Berlin: Springer.

DELPINO, F., 1873 – 4 ['1870']: Ulteriore Osservazione sulla Dicogamia nel Regno Vegetale **2(2)**. Milan (from Atti Soc. Ital. Sci. Nat. **12**).

EAST, E. M., 1940: The distribution of self-sterility in the flowering plants. – Proc. Amer. Phil. Soc. **82**: 449–518.

FAEGRI, K., VAN DER PIJL, L., 1966: The principles of pollination ecology. – Oxford: Pergamon.

FAEGRI, K., VAN DER PIJL, L., 1971: The principles of pollination ecology. 2nd edn. – Oxford: Pergamon.

FAEGRI, K., VAN DER PIJL, L., 1979: The principles of pollination ecology. 3rd edn. – Oxford: Pergamon.

FEEHAN, J., 1985: Explosive flower opening in ornithophily: a study of pollination mechanisms in some Central African *Loranthaceae*. – Bot. J. Linn. Soc. **90**: 129–144.

GANDERS, F. R., 1979: The biology of heterostyly. – New Zealand J. Bot. **17**: 607–635.

GIBBS, P. E., 1986: Do homomorphic and heteromorphic self-incompatibility systems have the same sporophytic mechanism? – Pl. Syst. Evol. **154**: 285–323.

GOTTSBERGER, G., 1988: The reproductive biology of primitive angiosperms. – Taxon **37**: 630–643.

HARDER, L. D., THOMSON, J. D., 1989: Evolutionary options for maximizing pollen dispersal of animal-pollinated plants. – Amer. Naturalist **133**: 323–344.

HESLOP-HARRISON, Y., SHIVANNA, K. R., 1977: The receptive surface of the Angiosperm stigma. – Ann. Bot. (Oxford) **41**: 1233–1258.

HEYN, C. C., SNIR, S., 1986: Selfing and pollen allocation in some Asteraceae. – Proc. Roy. Soc. Edinburgh, B **89**: 181–192.

HEYWOOD, V. H. (ed.), 1978: Flowering plants of the world. – Oxford: University Press.

HEYWOOD, V. H., HARBORNE, J. B., TURNER, B. L. (eds.), 1977: The biology and chemistry of the *Compositae* **1, 2**. – London: Academic Press.

HOEKSTRA, F. A., BRUINSMA, J., 1975: Reduced independence of the male gametophyte in angiosperm evolution. – Ann. Bot. (Oxford) **42**: 759–762.

HOLM, E., 1988: On pollination and pollinators in Western Australia. – Gedved: Eigil Holm.

JUEL, O., 1908: Om pollinationsapparaten hos familjen *Compositae*. – Svensk Bot. Tidskr. **2**: 350–363.

KELLER, S., ARMBRUSTER, S., 1989: Pollination of *Hyptis capitata* by eumenid wasps in Panama. – Biotropica **21**: 190–192.

KENNEDY, H., 1978: Systematics and pollination of the "closed-flowered" species of *Calathea* (*Marantaceae*). – Univ. Calif. Pub. Bot. **71**: 1–90.

KERNER, A., 1873: Die Schutzmittel des Pollens gegen die Nachtheile vorzeitiger Dislocation und gegen die Nachtheile vorzeitiger Befruchtung. – Innsbruck: Wagner'sche Universitäts-Buchdruckerei.

KNUTH, P., 1904, 1905: Handbuch der Blütenbiologie **3(1 & 2)**, (ed. O. APPEL & E. LOEW) [not translated]. – Leipzig: Engelmann.

KNUTH, P., 1908, 1909: Handbook of flower pollination **2, 3**. Trans. [from Band 2 of the German edn.] by J. R. AINSWORTH-DAVIS. – Oxford: Clarendon Press.

KRESS, W. J., 1990: The phylogeny and classification of the *Zingiberales*. – Ann. Missouri Bot. Gard. **77**: 698–721.

KUNZE, H., 1984: Vergleichende Studien an Cannaceen- und Marantaceenblüten. – Flora **175**: 301–318.

LLOYD, D. G., 1979: Parental strategies of angiosperms. – New Zealand J. Bot. **17**: 595–606.

LLOYD, D. G., WEBB, C. J., 1968: The avoidance of interference between the presentation of pollen and stigmas in angiosperms, I. Dichogamy. – New Zealand J. Bot. **24**: 135–162.

LLOYD, D. G., YATES, M. A., 1982: Intrasexual selection and the segregation of pollen and stigmas in hermaphrodite plants, exemplified by *Wahlenbergia albomarginata* (*Campanulaceae*). – Evolution **36**: 903–913.

MORI, S. A., PRANCE, G. T., BOLTEN, A. B., 1978: Additional notes on the floral biology of neotropical *Lecythidaceae*. – Brittonia **30**: 113–130.

MÜLLER, H., 1873: Die Befruchtung der Blumen durch Insekten. – Leipzig.

MÜLLER, H., 1883: The Fertilisation of Flowers. – London: Macmillan.

PAIVA, J., NOGUERA, I., 1990: *Gentianaceae*. – In LAUNERT, E., POPE, G. V., (eds.): Flora Zambesiaca **7(4)**, 3–51. – London: Flora Zambesiaca Managing Committee.

PIPER, J., CHARLESWORTH, B., 1986: The evolution of distyly in *Primula vulgaris*. – Biol. J. Linn. Soc. **29**: 123–137.

PRANCE, G. T., 1976: The pollination and androphore structure of some Amazonian *Lecythidaceae*. – Biotropica **8**: 235–241.

PROCTOR, M., YEO, P., 1973: The Pollination of Flowers. – London: Collins.

RASMUSSEN, F. N., 1986: The gynostemium of *Bulbophyllum ecornutum* (J. J. SMITH) J. J. SMITH (*Orchidaceae*). – Bot. J. Linn. Soc. **91**: 447–456.

SCHICK, B., 1980: Untersuchungen über die Biotechnik der Apocynaceenblüte. I. Morphologie und Funktion des Narbenkopfes. – Flora **170**: 394–432.

SCHICK, B., 1982: Untersuchungen über die Biotechnik der Apocynaceenblüte. II. Bau und Funktion des Bestäubungsapparates. – Flora **172**: 347–371.

SCHMID, R., SCHMID, M. J., 1970: KNUTH's often overlooked "Handbuch der Blütenbiologie, III. Band". – Ecology **51**: 357–358.

SCHRIRE, B. D., 1989: A multidisciplinary approach to pollination biology in the *Leguminosae*. – In STIRTON, C. H., ZARUCCHI, J. L., (eds.): 183–242.

SCOTT ELLIOT, G. F., 1890: Ornithophilous flowers in South Africa. – Ann. Bot. (Oxford) **4**: 265–280.

STIRTON, C. H., 1983: Nocturnal petal movements in the *Asteraceae*. – Bothalia **14**: 1003–1006.

STIRTON, C. H., ZARUCCHI, J. L. (eds.), 1989: Advances in Legume Biology (Monogr. Syst. Bot. Missouri Bot. Gard. **29**).

STUMPF, S., 1987, unpublished: Vergleich der Blütenmorphologie von *Goodeniaceae* und *Lobeliaceae*. – Diplomarbeit anfertigt am Systematisch-Geobotanischen Institut der Georg-August-Universität zu Göttingen 1–149.

TAYLOR, P., 1989: The genus *Utricularia* – a taxonomic monograph. Kew Bulletin additional series XIV. London: HMSO.

THIELE, E.-M., 1988: Bau und Funktion des Antheren-Griffel-Komplexes der Compositen. – Diss. Bot. **117**.

THIEN, L. B., 1974: Floral biology of *Magnolia*. – Amer. J. Bot. **61**: 1037–1045.

VAN DER PIJL, L., 1978: Reproductive integration and sexual disharmony in floral functions. – In RICHARDS, A. J., (ed.): The pollination of flowers by insects 79–88. – London: Academic Press.

VOGEL, S., 1954: Blütenbiologische Typen als Elemente der Sippengliederung. – Bot. Stud. **1**: 1–338. – Jena, Fischer.

VOGEL, S., 1978: Evolutionary shifts from reward to deception in pollen flowers. – In RICHARDS, A. J., (ed.): The pollination of flowers by insects 89–96. – London: Academic Press.

WESTERKAMP, C., 1989: Von Pollenhaufen, Nudelspritzen und Pseudo-Staubblättern: Blütenstaub aus zweiter Hand. – Palmengarten **53**: 146–149.

YEO, P. F., 1972: Miscellaneous notes on pollination and pollinators. – J. Nat. Hist. (London) **6**: 667–686.

Appendix 1: Possibilities for future research

General suggestions

A comprehensive list of questions that should be considered at the outset of a piece of research in floral biology has been presented by SCHRIRE (1989). This is well worth study even though it is concerned specifically with the family *Leguminosae*.

It is noteworthy that SCHRIRE's protocol does not mention secondary pollen presentation, although it does ask when the anthers dehisce, whether the stamen-filaments show expansion, thickening or curvature in relation to the pollination mechanism, and whether hairs on the style assist in the transfer of pollen to the vector. It also asks whether the pollen is available continuously or intermittently, but only in terms of its role as a reward for the pollinator and not in terms of its genetic function.

Quantitative studies of pollen-dosing: how is the pollen-production divided up; how flexible is the dosing in relation to pollinator-visit frequency?

Comparison of pollination efficiency between secondary and primary pollen presenting plants in the same family (i. e. *Leguminosae-Papilionoideae*).

Do plants with secondary pollen presentation have lower pollen/ovule ratios than comparable plants with primary pollen presentation?

How efficient is the mechanism for transferring pollen from anthers to another pollen-presenting structure? Comparison should be made with primary presenters but allowance has to made for further losses at the secondary stage (i.e. from the pollen presenter).

(The above are relevant to determining the function of secondary pollen presentation.)

Importance of pollen longevity. What is the longevity of pollen in pollen magazines and on exposed pollen presenters? Is there any significant relationship with the number of nuclei in the pollen grain?

Is the papilionate flower (in families *Leguminosae*, *Papaveraceae* and *Polygalaceae*) mainly for controlling pollen release? – Find out more about which bees visit which species and to what extent they forage for pollen.

What is the extent of the occurrence of stigmatic membranes?

Special problems

Our knowledge of pollination processes in most plants with secondary pollen presentation is incomplete. Some specific problems are listed here.

Papaveraceae, *Corydalis bulbosa*: study to confirm supposed action of keels on lateral petals (this is not specific to secondary pollen presentation).

Sterculiaceae: too little is known about the floral biology of this family; note should be taken in future studies of the occurrence of transfers of function from one organ to another. Details of the shrinking hairs of *Dombeya* should be obtained to see if there is any resemblance to those of *Campanulaceae*.

Loasaceae: some details are missing from descriptions of secondarily pollen presenting species in the family.

Leguminosae-Papilionoideae: pollen delivery and stigma receptivity in *Lotus* are supposed to be prolonged but initially there is copious stigmatic fluid held in by a stigmatic membrane; how is pollen kept away from the stigmatic fluid after the membrane is ruptured?

How is pollen dispensed from a stylar brush; in flowers with a stylar brush, when is the stigma receptive (known in some cases) and when is it likely to receive pollen?

Do bees scrape pollen directly from stylar brushes (*Xylocopa* is recorded as doing so – on *Lathyrus grandiflorus*).

How and where does the brush touch the pollinator (*Lathyrus* and *Vicia* and species of *Phaseoleae* could be compared)?

It is desirable to have a more precise account than is presently available of the re-loading of the stylar brush in *Pisum*.

More details of pollen presentation in *Harpalyce* are needed.

Myrtaceae: in *Verticordia* the presentation of pollen in a sticky fluid seems appropriate for beetle-pollination but why could Holm (1988) never find pollen on beetles visiting the flowers? The constitution of the oily fluid and how it behaves on drying are unknown.

Vochysiaceae: little is yet known about pollen presentation in this family.

Polygalaceae: more information is needed on the explosive pollen presentation in *Polygala myrtifolia* and it would be useful to know if explosive presentation in *Muraltia* is secondary.

Gentianaceae: some members of *Sebaea*, a genus of herbaceous annuals, have styles with a papillate zone reminiscent of the pollen-carrying brush in *Campanulaceae*, suggesting that secondary pollen presentation may occur here; usually there is also a large gland at the tip of the anther and sometimes two at its base (illustrations can be found in Boutique, 1972, and Paiva & Nogueira, 1990). The family is placed at the beginning of the *Asteridae* in the system used here.

Campanulaceae: is the invagination of the pollen-carrying hairs of the style in any degree a response to pollinator activity in the flower?

What is the significance of the viscid fluid in *Wahlenbergia gracilis*? Possible absence of stylar hairs in some species of *Wahlenbergia* should be investigated.

More information needed on pollen grain structure, pollen nuclear number and occurrence of self-incompatiblity.

The mechanism in *Phyteuma* needs to be clarified.

Cyphiaceae: details of pollen expulsion are unknown.

Lobeliaceae: does the pollen box stretch or are the pollen grains angulated when under pressure?

More information needed on pollen grain structure, pollen nuclear number and occurrence of self-incompatiblity.

Goodeniaceae: is pollen normally applied to the pollinator during withdrawal?

Xyridaceae: pollen presentation needs to be studied.

Appendix 2: Coded summaries of pollen presentation types

Information from Chapter 2, omitting visitors.

Key to Abbreviations

Column 1	**Subfam** = subfamily, **Tr** = Tribe, **gp** = group
Column 2	**a** = anthers, **(a)** = anthers presenting pollen directly, **c** = corolla, **f** = filaments, **k** = calyx, **p** = perianth, **s** = style
Column 3	**rel** = non-explosive release of concealed pollen, **rel(dose)** = the same but in measured doses (these may be suffixed as follows: **-poll** = powered by pollinator, **-trigg** = triggered by pollinator, **irrit** = irritability [movement in response to touch]), **growth** = gradual expulsion by growth, **expl cloud** = release of pollen cloud explosively, **expl depos** = explosive deposition on vector, **exposed** = pollen exposed or accessible in flower
Column 4	**pln** = pollen first, **stg** = stigma first, **sim** = simultaneous
Column 5	initial letters of the words anterior, posterior, dorsal, ventral, lateral, separated by commas (if no comma, place is intermediate, e.g. **ad** = anterodorsal)
Column 6	**n** = nectar, **pln** = pollen
Column 7	see Table 1-3 (page 5)

Column: 1	2	3	4	5	6	7
Plant name	Structure	Method of issue	Sequence	Place of application	Reward	Syndrome
Myristicaceae						
Myristica						
M. fragrans	p, a, (a)?	exposed	na	general	p (in ♂ flowers); nil in (♀ flowers)	cantharo
Papaveraceae,						
Subfamily *Hypecooideae*						
Hypecoum						
H. procumbens	c	rel(dose)-poll	stg	?	?	melitto

Appendix 2 (*Continued*)

Column: 1	2	3	4	5	6	7
Plant name	Structure	Method of issue	Sequence	Place of application	Reward	Syndrome
Subfamily *Fumarioideae*						
Dicentra						
D. spectabilis	s	rel-poll	sim?	v	n	melitto
Adlumia						
A. fungosa	s	rel-poll	sim?	av	n	melitto
Corydalis						
C. bulbosa	s	rel-poll	sim	v	n	melitto
Pseudofumaria						
P. alba & P. lutea	s	expl depos	sim?	v	n	melitto
Fumaria						
F. officinalis	s	rel-poll	sim?	v	n	melitto
Platycapnos						
P. spicatus	s	expl depos	sim?	v	n	melitto
Sterculiaceae,						
Tribe Dombeyeae						
Dombeya						
D. mastersii	c, f, (a)	exposed	pln	?	n, p?	melitto
D. burgessiae	c, a	exposed	stg	v	n, p?	melitto
Tribe Byttnerieae						
Herrania						
H. purpurea, H. nitida & H. albiflora	c, (a?)	exposed	stg?	ad	n	myio, sapromyio
Loasaceae						
Eucnide						
E. urens	f, top of ovary	exposed	sim.	general?	p, n?	melitto
Mentzelia						
M. tricuspis	top of ovary	exposed	sim?	a	p	melitto

Epacridaceae						
Acrotriche						
A. serrulata	c	exposed	pln	a	n	thero
Leguminosae,						
Tr 7, Robinieae						
Robinia						
R. pseudacacia	c, s	rel(dose)-poll	stg	v	n	melitto
Tr 8, Indigofereae						
Indigofera						
I. filiformis	c, f?	expl depos	stg	v	p	melitto
Tr 9, Desmodieae						
Desmodium						
D. canadense &						
D. sessilifolium	c, f	expl depos	stg	a	p	melitto
Tr 10, Phaseoleae						
Phaseolus						
P. coccineus	c, s	rel(dose)-poll	pln?	l	n, p?	melitto
Vigna						
V. caracalla	c, s	rel(dose)-poll	sim?	d	n, p?	melitto
V. vexillata	c, s	rel(dose)-poll	stg	l	n, p?	melitto
Tr 14, Aeschynomeneae						
Stylosanthes						
S. gracilis & S. guianensis	c, a, s	rel(dose)-poll	sim	a, v	n, p	melitto
S. biflora	c, (a), s	rel(dose)-poll	sim	ad, v	n, p?	melitto
Tr 16, Galegeae						
Colutea						
C. arborescens	c, s	rel(dose)-poll	pln	v	n	melitto
Clianthus						
C. puniceus	c, s	rel(dose)-poll	sim?	ad	n	ornitho
Tr 19, Loteae						
Lotus						
L. corniculatus	c, f	rel(dose)-poll	sim or pln	v	n, p	melitto

Appendix 2 (*Continued*)

Column: 1 Plant name	2 Structure	3 Method of issue	4 Sequence	5 Place of application	6 Reward	7 Syndrome
Dorycnium						
D. hirsutum	c, f, (a)	rel(dose)-poll	sim or pln	av	n, p	melitto
Anthyllis						
A. vulneraria	c, f	rel(dose)-poll	pln	v	n, p	melitto
Tr 20, Coronilleae						
Hippocrepis						
H. comosa	c, f	rel(dose)-poll	stg	v	n, p	melitto
Coronilla						
C. vaginalis	c, f	rel(dose)-poll	pln?	v	–	melitto
C. varia	c, f	rel(dose)-poll	stg	v	n?, p?	melitto
Tr 21, Fabeae (Vicieae)						
Vicia						
V. cracca, V. ervilia, V. pisiformis & V. sepium	c, s	rel(dose?)-poll	sim?	v	n, p	melitto
V. faba	s	rel(dose)-poll	pln	v	n, p	melitto
Lathyrus						
L. pratensis	c, f, s	rel(dose)-poll	pln?	v	n, p	melitto
L. grandiflorus	c, s	rel(dose)-poll	stg	l, d	n, p	melitto
Pisum						
P. sativum	c, f, s	rel(dose)-poll	pln?	v	n?, p?	melitto
Tr 23, Trifolieae						
Ononis						
O. spinosa	c, f	rel(dose)-poll	pln	v	p, o	melitto
Medicago						
M. sativa	c?, (a), s	expl depos	sim	v	n, p(?)	melitto
Tr 24, Brongniartieae						

Harpalyce						
H. spp. sects. Harpalyce & Brasilianae	c, a (?), s(?)	expl cloud	sim	d	n	melitto
Tr 27, Podalyrieae						
Podalyria						
P. calyptrata	c, a	rel(dose)-poll	pln	v	n?	melitto
Tr 28, Liparieae						
Liparia						
L. spherica	c, a	rel(dose)-poll	pln	v	?	melitto
Tr 29, Crotalarieae						
Crotalaria						
C. micans	c, a, s	rel(dose)-poll	pln, sim?	v	n, p	melitto
C. capensis	c, a	rel(dose)-poll	pln	v	n	melitto
Tr 32, Genisteae						
Lupinus						
L. luteus	c, a	rel(dose?)-poll	?	v	p	melitto
L. nanus	c, f, a	rel(dose)-poll	pln	v	p	melitto
Genista						
G. tinctoria	c, s	expl depos	stg	v	p	melitto
Cytisus						
C. scoparius	c, a, s	expl depos & expl cloud	stg	general	p	melitto
Lembotropis						
L. nigricans	c, f, a	rel(dose)-poll	pln	v	p	melitto
Rhizophoraceae						
Bruguiera						
B. gymnorhiza	c, a	expl cloud	a	general	n	ornitho
B. parviflora	c, a	expl depos?	a	proboscis?	n	psycho
Ceriops						
tagal	c, a	expl depos?	a	proboscis?	n	phalaeno
Myrtaceae						
Darwinia						
D. vestita	s	exposed	pln	a?	n	entomo

Appendix 2 (*Continued*)

Column: 1	2	3	4	5	6	7
Plant name	Structure	Method of issue	Sequence	Place of application	Reward	Syndrome
D. macrostegia	s	exposed	pln	a?	n	ornitho
Verticordia						
V. huegelii	s	exposed	sim?	d?	n	entomo
V. grandis	s	exposed	sim?	a	n	ornitho
V. habrantha	a	exposed	sim?	a	n	entomo
Actinodium						
A. cunninghamii	s	exposed	sim?	a?, v?	?	entomo
Chamelaucium						
C. uncinatum	s	exposed	pln	a?	n	entomo
Proteaceae,						
Subfam *Proteoideae*, Tr 3, Conospermeae						
Isopogon						
I. anethifolius	s	exposed	pln	l?, v?	?	melitto
Petrophile						
P. longifolia	s	exposed	pln	l?, v?	?	melitto
Subfam *Proteoideae*, Tr 4, Franklandiae						
Adenanthos						
A. obovatus	s	exposed	pln	ad?	n?	ornitho
Subfam *Proteoideae*, Tr 5, Proteeae						
Protea						
P. incompta	s	expl depos?	pln	a?	n?	ornitho
P. longiflora	s	expl cloud?	pln	a?	n?	ornitho
P. amplexicaulis, & P. humiflora	s	exposed	pln	ad	n	thero
P. kilimandscharica	s	expl cloud	pln	a	n	ornitho
Leucospermum						
L. conocarpum	s	expl cloud	pln	a?	n?	ornitho

Subfam *Grevilleoideae*, Tr 10, Embothrieae						
Lomatia						
L. silaifolia	s	exposed?	pln	d?	n?	melitto
Telopea						
T. speciosissima	s	exposed?	pln	av?	n?	ornitho
Subfam *Grevilleoideae*, Tr 11, Grevilleeae						
Grevillea						
G. eriostachya	s	exposed	pln	a	n	ornitho
G. wilsonii	s	exposed	pln	ad?	n	ornitho?
G. fasciculata	s	exposed	pln	ad	n	ornitho
G. leucopteris	s	expl depos	pln	d(?)	n	cantharo
Lambertia						
L. formosa	s	exposed	pln	a	n	ornitho
L. ericifolia	s	exposed	pln	d	n	ornitho
Macadamia						
M. integrifolia	s	exposed				
Subfam *Grevilleoideae*, Tr 14, Banksieae			pln	a?	n(p)	phalaeno
Banksia						
B. marginata	s	expl cloud	pln	a?	n?	ornitho or thero
B. integrifolia	s	exposed	pln	a?	n (n, p for bees)	ornitho
B. coccinea	s	exposed	pln	general	n	ornitho
B. prionotes	s	exposed	pln	av	n	ornitho
B. dryandroides	s	exposed	pln	?	n	thero
Meliaceae						
Turraea						
T. vogelioides	s	exposed	pln	face or proboscis?	n	?
T. obtusifolia	s	exposed	sim	face or proboscis?	n	sphingo
T. floribunda	s	exposed	?	face or proboscis?	n	sphingo

Appendix 2 (*Continued*)

Column: 1	2	3	4	5	6	7
Plant name	Structure	Method of issue	Sequence	Place of application	Reward	Syndrome
Rafflesiaceae						
Cytinus						
C. ruber	p, (a)?	exposed	na	a, general?	n, p	melitto
Vochysiaceae						
Vochysia						
V. spp.	s	exp	sim?	?	n	?
Polygalaceae						
Polygala						
P. comosa & P. vulgaris	s	rel(dose)-poll	stg	av	n	melitto
P. myrtifolia	c, s	expl depos	stg or sim	vl	n	melitto
P. chamaebuxus	c, s	rel(dose)-poll	pln?	av	n, p?	melitto
P. vauthieri & P. monticola	c, s	rel(dose)-poll	stg	avl	n	melitto
Comesperma						
C. virgatum	s	rel(dose)-poll	pln	v	n	melitto
Apocynaceae,						
Subfam *Apocynoideae*						
Apocynum						
A. androsaemifolium	a?, s	exposed	stg	proboscis	n	melitto
Nerium						
N. oleander	a, s	exposed	stg	proboscis	none	sphingo?
Stephanostema						
S. stenocarpum	a, s	exposed.	stg	proboscis?	n?	micro-melitto?
Subfam *Plumerioideae*						
Amsonia						
A. tabernaemontana	s	exposed	stg	proboscis	n	psycho, melitto
Vinca						
V. major, V. minor	(a), s?	exposed	stg	proboscis	n	melitto

Campanulaceae,						
Tr Campanuleae						
Campanula						
C. rotundifolia	s	rel(dose)	pln	v	n, p	melitto
Symphyandra						
S. cretica	s	rel(dose)?	pln	v	n, p	melitto
Adenophora						
A. liliifolia	s	rel(dose)	pln	v	n, p	melitto?
Legousia						
L. speculum-veneris	c?, s	rel(dose)	pln	v	n?, p	micro-melitto
Trachelium						
T. caeruleum	s	rel(dose)	pln	v proboscis?	n	psycho?
Michauxia						
M. campanuloides	s	rel(dose)	pln	a?	n, p?	sphingo?
Petromarula						
P. pinnata	s	exposed	pln	av?	n	melitto
Asyneuma						
A. canescens	s	rel(dose)	pln	probably a, v	n, p?	melitto
Phyteuma						
P. michelii & P. orbiculare	c, s	growth	pln	av	n, p?	melitto, psycho?
Physoplexis						
P. comosa	s	rel(dose), growth?	pln	a, av?	n	psycho
Tr Wahlenbergieae						
Wahlenbergia						
W. gracilis	s	rel(dose)?	pln	a?	n	melitto?
W. albomarginata	s	rel(dose)	pln	a?	n	melitto
Codonopsis						
C. clematidea	s	rel(dose)?	pln	d or v	n	melitto
Canarina						
C. canariensis	s	rel(dose)	pln	a	n	ornitho

Appendix 2 (*Continued*)

Column: 1	2	3	4	5	6	7
Plant name	Structure	Method of issue	Sequence	Place of application	Reward	Syndrome
Jasione						
J. montana	s	rel(dose)	pln	a, v	n, p	entomo
Edraianthus						
spp.	s	rel(dose)	pln	a	n, p	melitto
Cyananthus						
spp.	s	rel(dose)	pln	a, v	n, p	entomo
Tr Platycodoneae						
Platycodon						
P. grandiflorum	s	rel(dose)	pln	v?	n, p?	melitto
Cyphiaceae						
Cyphia						
C. spp.	a, s	growth?	pln	v	n	?
Lobeliaceae,						
Tr Lobelieae						
Lobelia						
L. laxiflora	a, s	growth	pln	ad	n	ornitho
L. cardinalis	a, s	rel-dose	pln	ad	n	ornitho
L. erinus	a, s	growth	pln	ad	n	melitto
L. alata, L. gibbosa, L. gracilis & L. siphilitica	a, s	growth, rel(dose)-poll	pln	ad	n	melitto
Isotoma						
I. petraea	a, s	rel(dose)-poll	pln	ad	n	phalaeno, sphingo
I. axillaris	a, s	rel(dose)-poll	pln	ad	n	psycho
Hippobroma						
H. longiflora	a, s	rel(dose)-poll	pln	ad	n	psycho?

Monopsis						
M. debilis	a, s	growth	pln	a	n	melitto
Goodeniaceae						
Scaevola						
S. hispida, S. hookeri	s	rel, growth	pln	d	?	melitto?
S. suaveolens	c?, s	rel, growth	pln	d, v	?	melitto
S. plumieri	s	growth & rel(dose)-poll	pln	d	n	melitto
S. thesioides	s	rel?, growth	pln	d	?	melitto
Selliera						
S. radicans	s	rel or rel (dose), growth	pln	d	n	melitto?
Dampiera						
D. purpurea	c, s	rel(dose) & growth	pln	ad or v	n	melitto
D. coronata, D. hederacea, D. sericantha & D. teres	s	rel(dose)-trigg	pln	d	n?	melitto?
D. cuneata	s	rel(dose)-trigg	pln	d	n?	melitto
Goodenia						
G. grandiflora	s	rel(dose)-poll & growth	pln	ad	n	melitto
G. cycloptera	s	rel(dose)-trigg & growth?	pln	d	n?	melitto
G. pinnatifida	s, c	growth	pln	d	n	melitto
Lechenaultia						
L. spp.	s	rel-trigg	?	v or d	n?	melitto
L. spp.	s	rel-trigg	?	v or d	n?	ornitho
Brunonia						
B. australis	s	growth	pln	av?	?	?
Rubiaceae,						
Subfam *Rubioideae*, Tr Rubieae						
Phuopsis						
P. stylosa	s	exposed	pln	a	n	psycho

Appendix 2 (*Continued*)

Column: 1	2	3	4	5	6	7
Plant name	Structure	Method of issue	Sequence	Place of application	Reward	Syndrome
Subfam *Coffeoideae*, Tr Acrantherae						
Acranthera						
spp.	s	exposed	pln	proboscis?	?	–
Subfam *Coffeoideae*, Tr Gardenieae						
Burchellia						
B. bubalina	s	exposed	pln?	a?	n	ornitho
Catunaregam						
C. spinosa	s	exposed	pln	a	n	psycho?, melitto?
Gardenia						
G. stanleyana	s	exposed	pln	a	n	sphingo
Griffithia						
spp.	s	exposed	sim	a	n	psycho?
Mitriostigma						
M. axillare	s	exposed	sim	a?	n?	lepidoptero?, melitto?
Oxyanthus						
O. pyriformis	s	exposed	pln	a	n	sphingo
Subfam *Coffeoideae*, Tr Gardenieae, Subtr Gardeniinae						
Tocoyena						
T. brasiliensis & T. formosa	s	exposed	pln	a	n	sphingo
Subfam *Coffeoideae*, Tr Hypobathreae						
Hypobathrum						
H. albicaule	s	exposed	sim	a	n	?
Kraussia						
K. floribunda	s	exposed	pln	?	n?	psycho

Scyphostachys						
S. coffeoides	s	exposed	pln	av?	n	?
Subfam *Coffeoideae*, Tr Pavetteae						
Ixora						
I. platythyrsa	s	exposed	pln	a	n	phalaeno
Pavetta						
P. elliottii	s	exposed	pln	a?	n	?
Subfam *Coffeoideae*, Tr Vanguerieae						
Canthium						
C. laeve	s	exposed	pln	a	n	melitto?
Keetia						
K. gueinzii	s	exposed	pln	a?	n	melitto or micromelitto
Psydrax						
P. odorata & P. obovata	s	exposed	pln	a?	n	melitto or micromelitto
Asteraceae,						
Subfam *Cichorioideae*, Tr 1, Lactuceae						
Hieracium						
H. umbellatum	s	exposed	pln	vl	n, p	entomo
Leontodon						
L. autumnalis	s	exposed	pln	vl	n, p	entomo
Subfam *Cichorioideae*, Tr 2, Mutisieae						
Mutisia						
M. coccinea	s	rel(dose)-trigg	pln	a	n	ornitho
Gerbera						
G. jamesonii	a, s	rel(dose)-irrit	pln	av	n, p	entomo
Perezia						
P. multiflora	a, s	expl depos	pln	av	n, p	entomo

Appendix 2 (*Continued*)

Column: 1	2	3	4	5	6	7
Plant name	Structure	Method of issue	Sequence	Place of application	Reward	Syndrome
Subfam *Cichorioideae*, Tr 4, Arctotideae						
Arctotheca						
A. calendula	s	exposed	pln	vl	n, p	entomo
Subfam *Cichorioideae*, Tr 5, Cardueae, gp 1: connective-appendages short						
Cirsium						
C. arvense	a, s	rel(dose)-irrit? & growth?	pln	vl	n, p	entomo
C. vulgare	a, s	rel(dose)-irrit & growth	pln	avl	n, p	melitto
Carduus						
C. nutans	a, s	rel(dose)-irrit & growth	pln	av	n, p	melitto
Onopordum						
O. acanthium	a, s	rel(dose)-irrit? & growth?	pln	vl	n, p	melitto
Subfam *Cichorioideae*, Cardueae, gp 2: connective-appendages long, united						
Centaurea						
C. jacea, C. cyanus & C. scabiosa	a, s	rel(dose)-irrit & growth?	pln	vl	n, p	melitto
C. collina	a, s	rel(dose)-irrit & growth	pln	vl	n, p	melitto
Subfam *Cichorioideae*, Cardueae, gp 3: deviant genera						
Arctium						
A. minus	a, s	exposed	pln	v	n, p	melitto
Carthamus						
C. tinctorius	s	exposed	pln	av	n, p	melitto

Echinops						
E. sphaerocephalus &						
E. ritro	a, s	growth	pln	av	n, p	melitto
Subfam *Cichorioideae*, Tr 8, Eupatorieae						
Eupatorium						
E. cannabinum	s	exposed	pln	av	n, p	psycho
Subfam *Asteroideae*, Tr 9, Senecioneae						
Senecio						
S. jacobea	a, s	rel(dose)-irrit, growth	pln	av	n, p	entomo
Petasites						
P. albus	s	exposed	na	av	n, p?	entomo
Subfam *Asteroideae*, Tr 11, Heliantheae						
Arnica						
A. montana	c, s	growth	pln	av	n, p	entomo
Zinnia						
Z. haageana	c, a?, s	exposed	pln?	av	n, p?	entomo
Cosmos						
C. atrosanguineus	a, s	growth	pln	av	n, p?	–
Helianthus						
H. annuus	a, s	growth	pln	av, pv	n, p	entomo
Bidens						
B. tripartita	a, s	growth	pln	av	n, p?	entomo
B. cosmoides	s	exposed	pln	a	n	ornitho
Gaillardia						
G. aristata	s	exposed	pln	av	n, p?	entomo
Subfam *Asteroideae*, Tr 12, Inuleae						
Pulicaria						
P. dysenterica	a, s	growth	pln	av	n, p	entomo
Subfam *Asteroideae*, Tr 13, Anthemideae						
Achillea						
A. millefolium	a, s	growth	pln	av	n, p	entomo

Appendix 2 (*Continued*)

Column: 1	2	3	4	5	6	7
Plant name	Structure	Method of issue	Sequence	Place of application	Reward	Syndrome
Leucanthemum						
L. vulgare	a, s	growth	pln	av	n, p	entomo
Tanacetum						
T. vulgare	a, s	rel(dose)-irrit	pln	av	n, p	entomo
Antennaria						
A. dioica	s	rel(dose)-trigg	pln	av	n, p	entomo
Leontopodium						
L. alpinum	s	exposed	pln	av	n, p	entomo
Subfam *Asteroideae*, Tr 14, Ursinieae						
Ursinia						
U. anthemoides subsp. versicolor	a, s	growth	pln	av	n, p	melitto
Subfam *Asteroideae*, Tr 17, Astereae						
Aster						
A. linosyris	a, s	growth	pln	av	n, p	entomo
A. simplex	a, s	growth	pln	av	n, p	entomo
Bellis						
B. perennis	a, s	growth	pln	av	n, p	entomo
Hydrocharitaceae						
Blyxa						
B. octandra	c	exposed	na	feet	water (for Dipt.), perch (for Odonata)	special
Cannaceae						
Canna						
C. indica	s	exposed	sim	?	n	ornitho

Marantaceae						
Calathea						
Various species	s	expl depos	stg	v	n	melitto
Thalia						
T. geniculata	s	expl depos	stg	v	n	melitto
Liliaceae						
Albuca						
A. canadensis	s	exposed	sim?	v?, l?	n, p?	melitto

Index of plant names

The main reference for a taxon at the rank of family or below is printed in bold type; references to illustrations are in italic. Names of taxa above the rank of order in Chapter 2 are not indexed; they will be found in the list of contents.

M. Hesse, F. Ehrendorfer (eds.)

Morphology, Development, and Systematic Relevance of Pollen and Spores

(Plant Systematics and Evolution, Supplementum 5)

1990. 122 figures. VII, 124 pages.
Cloth öS 980,–, DM 138,–
Reduced price for subscribers to "Plant Systematics and Evolution":
Cloth öS 882,–, DM 124,20
ISBN 3-211-82182-1

T.J. Mabry, G. Wagenitz (eds.)

Research Advances in the Compositae

(Plant Systematics and Evolution, Supplementum 4)

1990. 20 figures. V, 124 pages.
Cloth öS 980,–, DM 138,–
Reduced price for subscribers to "Plant Systematics and Evolution":
Cloth öS 882,–, DM 124,20
ISBN 3-211-82174-0

C. Puff

A Biosystematic Study of the African and Madagascan Rubiaceae-Anthospermeae

(Plant Systematics and Evolution, Supplementum 3)

1986. 126 figures. IX, 535 pages.
Cloth öS 1680,–, DM 240,–
Reduced price for subscribers to "Plant Systematics and Evolution":
Cloth öS 1512,–, DM 216,–
ISBN 3-211-81919-3
Prices are subject to change withouth notice

Springer-Verlag Wien New York

MIX
Papier aus verantwortungsvollen Quellen
Paper from responsible sources
FSC® C105338
www.fsc.org

If you have any concerns about our products,
you can contact us on
ProductSafety@springernature.com

In case Publisher is established outside the EU,
the EU authorized representative is:
Springer Nature Customer Service Center GmbH
Europaplatz 3, 69115 Heidelberg, Germany

Printed by Libri Plureos GmbH
in Hamburg, Germany